The Politics of Chemistry

Agustí Nieto-Galan argues that chemistry in the twentieth century was deeply and profoundly political. Far from existing in a distinct public sphere, chemical knowledge was applied in ways that created strong links with industrial and military projects, national rivalries and international endeavours that materially shaped the living conditions of millions of citizens. It is within this framework that Nieto-Galan analyses how Spanish chemists became powerful ideological agents in various political contexts, from liberal to dictatorial regimes, throughout the century. He unveils chemists' position of power in Spain, their place in international scientific networks and their engagement in fierce ideological battles in an age of extremes. Shared discourses between chemistry and liberalism, war, totalitarianism, religion and diplomacy, he argues, led to developments in all of these fields.

Agustí Nieto-Galan is Professor of History of Science at the *Universitat Autònoma de Barcelona*. He has written widely on the history of chemistry and natural dyestuffs, the history of science popularisation and the urban history of science (eighteenth to twentieth centuries). In 2009 and 2018, he was awarded the 'ICREA-Acadèmia' Research Prize by the Catalan Government.

SCIENCE IN HISTORY

Series Editors
Simon J. Schaffer, University of Cambridge
James A. Secord, University of Cambridge

Science in History is a major series of ambitious books on the history of the sciences from the mid-eighteenth century through the mid-twentieth century, highlighting work that interprets the sciences from perspectives drawn from across the discipline of history. The focus on the major epoch of global economic, industrial and social transformations is intended to encourage the use of sophisticated historical models to make sense of the ways in which the sciences have developed and changed. The series encourages the exploration of a wide range of scientific traditions and the interrelations between them. It particularly welcomes work that takes seriously the material practices of the sciences and is broad in geographical scope.

The Politics of Chemistry
Science and Power in Twentieth-Century Spain

Agustí Nieto-Galan
Universitat Autònoma de Barcelona

CAMBRIDGE
UNIVERSITY PRESS

University Printing House, Cambridge CB2 8BS, United Kingdom

One Liberty Plaza, 20th Floor, New York, NY 10006, USA

477 Williamstown Road, Port Melbourne, VIC 3207, Australia

314–321, 3rd Floor, Plot 3, Splendor Forum, Jasola District Centre, New Delhi – 110025, India

79 Anson Road, #06–04/06, Singapore 079906

Cambridge University Press is part of the University of Cambridge.

It furthers the University's mission by disseminating knowledge in the pursuit of education, learning, and research at the highest international levels of excellence.

www.cambridge.org
Information on this title: www.cambridge.org/9781108482431
DOI: 10.1017/9781108687614

© Agustí Nieto-Galan 2019

This publication is in copyright. Subject to statutory exception and to the provisions of relevant collective licensing agreements, no reproduction of any part may take place without the written permission of Cambridge University Press.

First published 2019

Printed in the United Kingdom by TJ International Ltd. Padstow Cornwall

A catalogue record for this publication is available from the British Library.

Library of Congress Cataloging-in-Publication Data
Names: Nieto-Galan, Agustí, author.
Title: The politics of chemistry : science and power in twentieth-century Spain / Agusti Nieto-Galan (Universitat Autonoma de Barcelona).
Description: Cambridge ; New York, NY : Cambridge University Press, 2019. | Series: Science in history | Includes bibliographical references and index.
Identifiers: LCCN 2019008743 | ISBN 9781108482431 (alk. paper)
Subjects: LCSH: Chemistry – Social aspects – Spain – History – 20th century. | Chemistry – Political aspects – Spain – History – 20th century. | Science – Social aspects – Spain – History – 20th century. | Science – Political aspects – Spain – History – 20th century. | Science and state.
Classification: LCC QD39.7 .N54 2019 | DDC 338.946/06–dc23
LC record available at https://lccn.loc.gov/2019008743

ISBN 978-1-108-48243-1 Hardback

Cambridge University Press has no responsibility for the persistence or accuracy of URLs for external or third-party internet websites referred to in this publication and does not guarantee that any content on such websites is, or will remain, accurate or appropriate.

To the memory of Professor Juan Julio Bonet Sugrañes (1940–2006), a liberal chemist.

Contents

List of Figures		*page* ix
Preface		xi
Acknowledgements		xvi
Chronology		xx
List of Abbreviations		xxiv
	Introduction	1
	Biographies of Power	3
	A Political Chemistry	10
1	Dreams of Modernity	20
	Cosmopolitanism	26
	Laboratories and Schools	33
	Useful Chemistry	40
2	A Republican Science	52
	A New Enlightenment	55
	Nobel Visitors	65
	The Silver Age of Industry	73
3	War Weapons	83
	A Chemical Civil War	84
	A Damaged Community	93
	Tortured Skills	104
4	Totalitarian Ambitions	112
	Fascist Chemistry	116
	Chemistry and Religion	126
5	Autarchic Ambiguities	136
	'Our' Chemicals	138
	'Technical' Chemistry	149
	Chemical Diplomacy	155
6	Technocratic Progress	163
	'Neutral' Expertise	167

	Cold War Allies	175
	Corporate Chemistry	181
7	**Liberal Dissent**	**189**
	Chemists in Exile	192
	Internal Refugees	203
	Conclusion: The Moral Ambiguity of Chemistry	**216**
	Pure–Applied Chemistry	218
	Modernisation Paradoxes	221
	A Troubled Identity	223
	Chemists as Intellectuals	226
	History and Memory	228
	Addendum: Juan Julio Bonet Sugrañes (1940–2006)	232
	Bibliography	236
	Index	268

Figures

1.1	Francisco Giral in a public meeting in support of the Second Spanish Republic, c. 1978.	*page* 21
1.2	The new chemistry laboratories of the Faculty of Science in Madrid in 1929.	37
1.3	Enrique Moles with the delegates of the 1930 Meeting of the *Real Sociedad Española de Física y Química* in Seville.	39
1.4	Chemistry professors and students at the *Escola Industrial*, Barcelona, in 1917.	42
1.5	The building of the Chemistry Pavilion (*Palacio de la Química*) at the 1929 Barcelona International Exhibition.	47
2.1	Chemistry laboratories at the *Instituto Nacional de Física y Química* (INFQ), Madrid 1932.	59
2.2	The 1933 Summer School in Santander.	67
2.3	(a) The opening ceremony of the 9th International Conference of the International Union of Pure and Applied Chemistry (IUPAC), Madrid, April 1934, and (b) the emblem of the conference.	69
2.4	Gilbert Newton Lewis during his plenary lecture at the IUPAC Conference, Madrid, 1934.	70
2.5	The geography of chemistry in Spain in 1934.	74
3.1	A poster produced by the Catalan *Consell de Sanitat de Guerra* (Wartime Health Council) for the protection of the civil population from chemical threats, c. 1938.	88
3.2	Mussolini offering poison gases to Franco in 1937.	91
3.3	Jesús Yoldi in his laboratory at the Faculty of Science of the University of Granada in the 1920s.	94
3.4	Fernando Calvet at the award ceremony of his DPhil in Oxford in 1928.	104
4.1	A portrait of José María Albareda in the journal *Arbor* in 1966.	114
4.2	A portrait of Antonio de Gregorio Rocasolano in an obituary following his death in 1941.	119
4.3	Emilio Jimeno's chart for a totalitarian chemistry in 1940.	122

List of Figures

4.4 Gumersindo García, the president of the *Sindicato Nacional de Industrias Químicas*, in 1941.   124
4.5 Eduardo Vitoria in the laboratory together with Manuel Sanz at the *Instituto Químico de Sarriá* in 1956.   132
5.1 The gasogen 'solution', as described in the early 1940s as a fuel for cars and tractors.   141
5.2 Front cover of the *Catálogo del Libro Español de Química (1920–1955)*, a special issue of the journal *Afinidad*, May 1955.   148
5.3 Barnett F. Dodge, Professor of Chemical Engineering at Yale, in a cartoon in 1954.   152
5.4 'Foreign *Savants* as Guests of the Spanish Science', under the auspices of the CSIC in Madrid in 1950.   159
6.1 Manuel Lora-Tamayo as new Minister of Education and Science at the *Palacio del Pardo* in Madrid in 1962.   169
6.2 The US ambassador, Robert F. Woodward, visits the *Instituto Químico de Sarriá* in Barcelona in 1963.   177
6.3 Manuel Lora-Tamayo, Helmut Allardt and Adolf Butenandt in an interview with Franco in the *Palacio del Pardo*, Madrid, in 1963.   180
6.4 *Sociedad petrolífera española Shell* in 1966.   184
7.1 Antonio García Banús in Caracas in 1955.   201
7.2 Ibys advertisement for the antacid Alugel in 1943.   205
7.3 Zeltia advertisement in the journal *Ión* in 1942.   208
7.4 Miguel Masriera in 1956 at the BBC.   211
8.1 Addendum: Juan Julio Bonet Sugrañes in his office at the Steroids Laboratory of the IQS in the 1980s.   234

Preface

In the autumn of 2002, I received a phone call from Ms Nuria Alomar. She introduced herself as the granddaughter of the chemist Antonio García Banús (1888–1955), whose name at that time sounded totally alien to me. She told me the fascinating story of a brilliant chemist who had left the country in 1936, at the beginning of the Spanish Civil War (1936–9), and never returned, ending his days in Venezuela in 1955. Although García Banús was an organic chemist, I could not understand why, having taken a chemistry degree myself in the 1980s with a specialisation in that field, I had never heard of him. Ms Alomar's personal papers and publications by her grandfather helped me to write an article on García Banús in 2004, which appeared in *The British Journal for the History of Science*.[1] After that experience, I soon became aware of the lack of historical knowledge about the role of chemistry in twentieth-century Spain – a century full of extremes, in which chemistry and politics intermingled profoundly. García Banús's case was to me a sort of ur-phenomenon, a seminal driving force to pursue other case studies that could one day represent a big picture of at least up until the 1980s. I am pleased to see that, many years later, that bigger picture, with all its virtues and limitations, will be unveiled through the chapters of this book.

Years after my work on García Banús, I came across other prominent Spanish chemists who seemed to have had very different profiles. José María Albareda (1902–66) had a distinguished international career in soil chemistry and later became a key figure in the research policies of General Francisco Franco's dictatorship (1939–75),[2] as well as being one of the early members of the Opus Dei congregation. Manuel Lora-Tamayo (1904–2002) made valuable contributions to organic chemistry during the dictatorship, built a prestigious research school and even became

---

[1] Agustí Nieto-Galan, 'Free Radicals in the European Periphery. "Translating" Organic Chemistry from Zurich to Barcelona in the Early Twentieth Century', *The British Journal for the History of Science*, 37 (2004), 167–91.

[2] Antoni Malet, 'José María Albareda (1902–1966) and the formation of the Spanish Consejo Superior de Investigaciones Científicas', *Annals of Science*, 66 (2009), 307–32.

Minister of Education and Science in the 1960s.[3] More recently, I came across the archive of the chemist Miguel Masriera (1901–81), a pupil of García Banús, who had begun a brilliant international career in the 1920s. Masriera went into exile as a result of the Civil War, and although he returned to Spain during the dictatorship, he suffered academic marginalisation and so survived professionally as a scientific journalist.[4]

García Banús went into exile in Latin America and Masriera was internally marginalised by the dictatorship, whereas Albareda and Lora-Tamayo became key agents in building Franco's regime from the inside. Although all of these different worlds seemed to me to be incommensurable a priori, the intersections of the four stories inspired not only my curiosity as a citizen, but also my deeper interest as a historian of science.

With all of these cases in mind, my colleague and friend José Ramón Bertomeu-Sánchez recently mentioned to me the case of Max Aub (1903–72), a Spanish writer and intellectual who also left the country in 1939, at the end of the Civil War, but returned to Spain for a short visit in 1969 with Franco's dictatorship still in place. On 23 August 1969, Aub landed at Barcelona Airport and stayed for three months in Spain. His experience was much more shocking and painful than he could have expected. After his bitter disappointment with the wilful forgetting of the liberal values of the 1930s and the conformism to the regime among the people he met and interviewed, Aub decided to return to Mexico. In 1971, he published *La gallina ciega* (in English, *Blind Man's Buff*), an acclaimed book that described his daily experience in those three months in Spain and the 'blindness' of the Spanish population in the face of the totalitarian web of the dictatorship.[5]

Max Aub's tragic experience reminded me of that of the chemist Francisco Giral (1911–2002), who departed into exile, having already won a chair in organic chemistry in Spain in 1936, and went on to have a very distinguished scientific career in Mexico focusing on the chemistry

---

[3] Agustí Nieto-Galan, 'Reform and repression. Manuel Lora-Tamayo and the Spanish University in the 1960s', in Ana Simões, Kostas Gavroglu and Maria Paula Diogo (eds.), *Sciences in the Universities of Europe, Nineteenth and Twentieth Centuries* (Dordrecht: Springer, 2015), pp. 159–74.

[4] Agustí Nieto-Galan, 'From Papers to Newspapers: Miguel Masriera (1901–1981) and the Role of Science Popularisation under the Franco Regime', *Science in Context*, 26(3) (2013), 527–49.

[5] Max Aub, *La gallina ciega. Diario español. Edición, estudio introductorio y notas de Manuel Aznar Soler* (Barcelona: Alba Editorial, 1995). I am deeply grateful to José Ramón Bertomeu for his suggestion to use *La gallina ciega* as a metaphor for the adventurous journey of the Spanish chemical community before and after the Civil War.

Preface

of natural products with strong links with industry.[6] After Franco's death in 1975, Giral reclaimed his old chair and, following a long and painful struggle with bureaucracy, he finally obtained an organic chemistry position at the University of Salamanca in 1977, where he stayed until his retirement in 1981. We can compare Aub's three months of disappointment in the Francoist atmosphere with the everyday life of Giral's four years of academic unease following the recovery of his chair. Like Aub, Giral also went back to Mexico, where he died in 2002.

With all of these biographies in mind, I found it hard to distinguish García Banús's free radical molecules and his irreversible exile, Masriera's cosmo-chemistry and his articles in censored newspapers, Albareda's study of the chemical composition of soil and his enthusiastic collaboration with the dictatorship, Lora-Tamayo's diene synthesis and his leadership in the science policies of the Franco regime, and Giral's natural products and his disappointing return from exile. Similar atoms and molecules seemed to weigh down opposing research programmes and political ideologies, and the integration of these chemicals into the dramatic political events of twentieth-century Spain seemed to be an almost impossible endeavour. This was, however, the challenge that drove me to write this book, with the hope that the pages that follow will reveal unexpected links, contradictions, polarities and essential intersections between chemistry and politics, as well as between chemistry and culture in a broad sense. García Banús, Albareda, Lora-Tamayo, Masriera, Giral and many other names that appear in the pages of this volume will trace the history of a scientific community in its struggle for professional identity throughout the twentieth century. It is through the window of these eminent names that this book reviews the history of twentieth-century Spain – and the history of twentieth-century chemistry – from a political perspective.

Preliminary scans of the primary sources for this research project soon highlighted the considerable amount of power that chemists had throughout the century in different political regimes. This is therefore a reflection on the nature of the co-production of chemistry and power in twentieth-century Spain. The book invites the reader to reflect on the political dimension of chemistry through chemists' public addresses and opinions on their own contemporaneity. It also aims to steer debates on the role of experts in different political regimes and ideological contexts, with a particular focus on the extremes of the twentieth-century totalitarian regimes.

---

[6] Francisco Giral, *Ciencia española en el exilio (1939–1989). El exilio de los científicos españoles* (Barcelona: CIERE/Anthropos, 1994).

This is not an exhaustive sociological study of the Spanish community of chemists in the twentieth century, but a new look at the political dimension of chemistry. Although in the following chapters the reader will find general comments and analyses of specific institutions, labs, research schools, chemical industries and university departments, the book focuses mainly on the most important agents: research school leaders, university professors, internationally distinguished scholars, industrial chemists, policymakers and chemistry teachers and popularisers in a broad sense. Equally, it pays particular attention to the role of Spanish chemists as 'co-producers of power', but also to their position as 'chemists in power': from their academic service as deans and rectors to their presence on city councils and in government cabinets. Departing from the analysis of a specific scientific community, at the same time it offers a more general approach to the culture of chemistry in the twentieth century through the international networks it created. It can therefore contribute to the enrichment of the role of chemistry in the big picture of science in the twentieth century, but also to the way in which chemistry and its practitioners shaped science and politics in twentieth-century Spain.

The book relies on an expansive range of sources: books, articles and leaflets, personal and institutional correspondence, archive documents, obituaries and public addresses, academic peer-reviewed papers and national and international secondary literature, but also on local publications, which provided very useful data. I have also benefitted from a huge amount of local secondary literature on Spanish chemistry throughout the twentieth century. In spite of the varying quality of these sources, all of the obituaries, commemorative publications, biographies, historical introductions to scientific papers and textbooks, as well as a considerable number of well-researched academic publications, have helped me to weave my own narrative. I take responsibility for the use of all of these ideas and concepts from other publications.

Unless explicitly stated, I have used the Spanish versions of first names of Catalan chemists, since this is the way in which, for most of century, they were referred to as authors of papers, books and public presentations. In personal and family relations, they used their Catalan names – for example, Enrique, 'Enric'; Fernando, 'Ferran'; and José, 'Josep'. In the Republican years (1931–9), including during the Civil War, the Catalan language was used more publically, but it was later forbidden during Franco's dictatorship.

The specific nature of chemistry as a science of matter – closely linked to laboratory culture and with its strong links to industry and its flexible, dynamic boundaries between the physical and the biomedical sciences – has probably not helped historians, and historians of science in particular, to integrate it into the broader scenario of the twentieth century. At an academic level, there is no doubt that names such as Linus Pauling and Gilbert Newton Lewis do not belong so much within the popular scientific culture of the century as Marie Skłodowska-Curie and Albert Einstein do for the physical sciences and James Watson and Francis Crick do for molecular biology. At the industrial level, perhaps only plastics and fuel derivatives seem to count for historians from a generalist point of view when attempting to describe the twentieth century through chemicals. Nevertheless, despite these evident historiographic limitations, the use of chemical weapons in the First World War, and the involvement of the chemical industry in the Nazi regime are two striking examples to argue against those defending a non-political, neutral, technocratic status of chemistry. It is precisely the shock of the chemical war and the cruelty of the chemical holocaust that have driven me to pursue further reflection on the links between chemistry and politics – in this case, in Spain.

Acknowledgements

This book is the result of a long period of research and I feel deeply grateful to many friends and colleagues who have accompanied me over all these years and who I would like to mention here. I am indebted to my colleague Xavier Roqué at the *Centre d'Història de la Ciència* (CEHIC), *Universitat Autònoma de Barcelona* (UAB), for his support in the organisation of a series of research seminars on 'Science and Francoism' (2012–14). The seminars created a fruitful, dynamic, 'work-in-progress' atmosphere for debating new historiographical trends when tackling the role of science in Francoism, which covers a good part of the history of science in Spain in the twentieth century. Other colleagues at the CEHIC have been extremely supportive in formal and informal discussions, which stimulated further reflections for the writing of this book. Among them, I would like to mention Pedro Ruiz-Castell, Matteo Realdi, Lino Camprubí, Carlos Tabernero and Annette Mulberger. At the *Universitat Pompeu Fabra* (UPF), Antoni Malet, Albert Presas and Daniele Cozzoli's excellent work on science in totalitarian regimes has also been very useful. They have invited me to many workshops and seminars, from which my research has benefitted enormously. At the *Institució Milà i Fontanals* (CSIC), my friend and colleague José Pardo-Tomás deserves further recognition for making me aware of the links between architecture and dictatorship in his own research, but also for his illuminating discussions on the way in which the Spanish transition from dictatorship to democracy took place after 1975. Josep Lluís Barona and Alfons Zarzoso have been extremely helpful in disseminating the historiography of Spanish exile, while Borja de Riquer and Carme Molinero have offered their expertise in the historiography of Francoism. Ángel Toca's seminal papers on the history of chemical engineering and industrial chemistry in Spain have been of great value for starting to sketch out some of the main ideas of this book. Some years ago, I had the pleasure of supervising Ángel's PhD on the establishment of Solvay in Spain, and I have since then benefitted from his remarkable research skills.

Acknowledgements

I have also enjoyed the privilege of sharing research questions and projects with Joaquim Sales, who, from his professional chemistry background and his passion for history, has been working with me over the last few years on the biographies of other important Spanish chemists, such as Enrique Moles (1883–1953), Emilio Jimeno (1886–1976), Fernando Calvet (1903–88) and José Pascual (1895–1979). His recent work as a chemist–historian has been crucial for the recovery of many sources and data from the community of twentieth-century chemists at the University of Barcelona. I am particularly indebted to him, but also to my friend and colleague José Ramón Bertomeu-Sánchez, for their generous, detailed, critical comments on earlier versions of the manuscript. Equally, Xavier Roqué, Fernando Vidal, Oliver Hochadel, Clara Florensa and Santiago Gorostiza have provided useful thoughts for the preparation of the final version of this book.

Other chemists (and friends), such as Santiago Álvarez, Gemma Arsequell, Gregori Valencia, Miquel Seco and José Antonio Chamizo, have been extremely supportive during these years, in which the local history of chemistry was always present in our informal conversations. Nuria Alomar, Beatriz Moles, Francisco Barnés, José Nogareda, Manuel Tremoleda, Artur Bladé and José, Ángela, Adela and Carmen Giral have all generously contributed through interviews, providing new data on some of their relatives or even valuable autobiographical accounts. Jesús Movellán was also very supportive through personal interviews with Francisco Giral's relatives in Mexico. I would also like to thank Emilio Lora-Tamayo for allowing me access to some of his father's professional papers.

I am truly indebted to the anonymous referees who critically read my manuscript. They made very useful suggestions from which the book has benefitted. I am also grateful to the editors of the Cambridge University Press (CUP) series 'Science in History', Simon J. Schaffer and James A. Secord, for their early support for my project. Equally, Lucy Rhymer, CUP Senior Commissioning Editor for History of Science and Medicine, has been extremely helpful in all the process of assessment of my manuscript. Fiona Kelso's patient stylistic corrections have turned my draft manuscript into a readable English text, and I am truly indebted to her for her meticulous work.

I have presented papers at conferences and workshops on many of the chemists who appear in the book, particularly on García Banús, Moles, Lora-Tamayo, Masriera, Jimeno and Pascual. I have also contributed to several commemorative events, such as the centenary of chemistry studies at the Science Faculty of Granada and the 160th anniversary of Enrique Moles's birth. In Granada, I am particularly indebted to Pedro Luis

Mateo for his kind invitation and for the information he generously provided me with on the scientific repression of the city during and after the Spanish Civil War.

In the autumn of 2016, and thanks to a semester's sabbatical, my stay in Oxford as Visiting Senior Member of Linacre College was vital to outlining and writing the first draft of this manuscript. There, I had the support of my old mentor and friend, Robert Fox, whose company and fruitful, informal discussions have been crucial to my inspiration. In Oxford, the Bodleian Library staff have been extremely supportive in helping me find printed sources and manuscript material, particularly the papers of the Society for the Protection of Science and Learning (SPSL).

In terms of archive facilities, I am also very grateful to Laia Miret from the *Institut d'Estudis Catalans*, who has generously helped me to review Masriera's papers. I am also indebted to Mireia Bachs' support in the *Biblioteca de Ciència i Tecnologia* at the *Universitat Autònoma de Barcelona*. The *Universitat de Barcelona* archive was crucial for the reconstruction of García Banús's research school, the *Universidad de Navarra* for access to Lora-Tamayo's and Albareda's correspondence and the *Reial Acadèmia de Ciències i Arts* in Barcelona for Pascual's papers. The *Archivo General de la Administración* (AGA) and other Spanish public archives have also been crucial for obtaining the necessary biographical data of many of the chemists appearing in this book. The online archive of the *Junta para Ampliación de Estudios e Investigaciones Científicas* (JAE) at the *Residencia de Estudiantes* in Madrid has become an excellent resource for the reconstruction of the academic lives of Spanish chemists in the early decades of the twentieth century. Printed sources at the *Biblioteca de Catalunya* (Barcelona) were also decisive in the writing of this book.

This research had been funded through my ICREA-Acadèmia award (2009 and 2018) and thanks to other research projects such as HAR2012-36204-C02-02: 'Scientific Authority in the Public Sphere in Twentieth-Century Spain' (*Ministerio de Economía y Competitividad*); 2014 SGR 1414 and 2017 SGR 1138: 'Science, Technology and Medicine in Modern Catalonia (18th–20th centuries)' (*AGAUR-Generalitat de Catalunya*); HAR2009-12918-C03-02 (subprograma HIST): 'Science and Expertise in the Public Sphere: Barcelona (1888–1992)' (*Ministerio de Economía y Competitividad*); and HAR2015-66364-C2-1-P: 'Natural vs. Artificial: Industrial Waste, Expertise and Social Responses in 20th-Century Spain' (*Ministerio de Economía y Competitividad*).

I have dedicated this book to my mentor in chemistry, Juan Julio Bonet Sugrañes (1940–2006), who in my youth helped me to understand the international dimension of the profession and the liberal ethos of his research school in steroid compounds. As a brilliant chemist–historian, he made a profound mark on me and on many chemists, friends and colleagues.

Finally, and once again, my deepest gratitude goes to my family, to Montserrat and Martí, who have patiently witnessed the slow and sometimes painful elaboration of these pages, often during weekends and vacation periods, and even during hours of the day when they probably expected a bit more company.

Chronology

1898	Spanish colonial crisis. War against the USA and loss of Cuba, Puerto Rico and the Philippines
1899	Eugenio Mascareñas on the backwardness of chemistry in Spain
1903	Foundation of the *Real Sociedad Española de Física y Química* (RSEFQ) and the journal *Anales de la RSEFQ*
1905	Foundation of the *Laboratorio Químico del Ebro* (LQE)
1906	Santiago Ramón y Cajal wins Nobel Prize in Physiology or Medicine
1907	Foundation of the *Junta para Ampliación de Estudios e Investigaciones Científicas* (JAE)
	Foundation of the *Institut d'Estudis Catalans* (IEC)
1908	Foundation of the *Asociación Española para el Progreso de las Ciencias* (AEPC)
1909	Barcelona Tragic Week
1911	New JAE labs and *pensionados*
1914–18	First World War
1915	Antonio García Banús gains chair of organic chemistry in Barcelona
1917	Ernest Fourneau visits Spain
1921	Defeat of the Spanish Army in Annual (Morocco)
	First issue of the journal *Afinidad: Revista de química teórica y aplicada*
1923–30	Primo de Rivera's dictatorship
1924	First issue of the journal *Química e Industria*
1926	Enrique Moles and Miguel Catalán travel to foreign labs
	First issue of the journal *Ciència: Revista catalana de ciència i tecnologia*
1927	Heinrich Wieland in Spain
	Moles gains chair of inorganic chemistry in Madrid
1929	International Conference on Industrial Chemistry at the Barcelona International Exhibition

1930	Crisis of the monarchy and Primo de Rivera's dictatorship
1931	Victory of the Republican parties in local elections Second Spanish Republic Foundation of the right-wing movement *Acción Española*, with the chemist Antonio de Gregorio Rocasolano among its members
1932	Opening of the *Instituto Nacional de Física y Química* (INFQ, the 'Rockefeller'): Blas Cabrera (magnetism), Julio Palacios (X-rays), Miguel Catalán (spectroscopy), Julio Guzmán (electrochemistry), Antonio Madinaveitia (organic chemistry), Enrique Moles (physical chemistry)
1933	Foundation of the 'Autonomous' University of Barcelona (UAB) International Summer School of Chemistry in Santander Conservative hegemony (CEDA)
1934	9th International Conference of Pure and Applied Chemistry, Madrid (April) Political crisis: revolt in Asturias and Catalonia (October)
1935	Luis Bermejo, Antonio Madinaveitia and José Giral as nominators of the Nobel Prize of Chemistry First issue of the journal *Metalurgia y Construcciones Mecánicas*, edited by Emilio Jimeno.
1936	Electoral victory of the Popular Front (leftist Republicans and socialists) Franco's coup d'état and leftist revolution Spanish Civil War The chemists' involvement in the War The Republican government moves to Valencia
1937	Guernica bombing of civilian population by Nazi air force
1938	Battle of the Ebro Bombing in Barcelona of civilian population Franco's provisional government in Burgos
1939	Franco's victory in the Civil War Civil servant purges Foundation of the *Consejo Superior de Investigaciones Científicas* (CSIC) with chemist José María Albareda as General Secretary University purges, repression and exile of chemists J. Giral, F. Giral and A. Madinaveitia escape to Mexico as refugees
1940	First issue of the journal *Ciencia: Revista hispano-americana de ciencias puras y aplicadas*

1941	Foundation of *Instituto Nacional de Industria* (INI)
	Manuel Lora-Tamayo gains chair in organic chemistry in Madrid
	Enrique Moles jailed
	Foundation of the *Instituto de Química* in México, Madinaveitia becomes its scientific director in exile
	'Applied chemistry' at the *Patronato Juan de la Cierva* (PJC)
	First issue of the journal *Ión: Revista española de química aplicada*
1942	Iberian pact of Franco–Salazar
1943	A new university act: *Ley de Ordenación Universitaria* (LOU)
1944	Liberation of Paris
1945	J. Giral becomes president of the Spanish Republic in exile
	Miguel Masriera translates Arthur Eddington
1946	UN vetoes Francoist Spain
	Spanish chemists travel abroad (Manuel Lora-Tamayo and José Pascual)
1947	First issue of the journal *Revista de Ciencia Aplicada*
1950	First US Ambassador to Spain
	Spain joins the Food and Agriculture Organization and the World Health Organization
1952	The *Congreso Eucarístico* is held in Barcelona
	Spain joins UNESCO
1953	Vatican Concordat
	US–Spain cooperation agreement
	Spain joins the Organisation for Economic Co-operation and Development (OECD)
	Death of Enrique Moles in Madrid
1955	Spain joins the UN
	International Conference on Science Popularisation (UNESCO, Madrid)
	Death of García Banús in exile in Venezuela
	Chemical Industry Exhibition and Spanish Chemistry Book Exhibition in Barcelona
1956	First TV broadcast in Spain
	Spain joins the International Labour Organization (ILO)
1957	Treaty of Rome
	Launching of Sputnik 1
	The VIII International Conference on Astronautics is held in Barcelona
	US agreement to nuclear cooperation
1958	University student protests
	De Gaulle becomes president of the French Republic

1959	President Eisenhower visits Spain
	Official opening of the Valley of the Fallen (*Valle de los Caídos*)
	Stabilisation Plan, the liberalisation of the economy in a dictatorial context
1960	Franco visits the *Instituto Químico de Sarriá* (IQS), Barcelona
	US funding of new chemical instrumentation
	International Conference of Industrial Chemistry, Barcelona
	International chemical corporations established in Spain
1962	*Plan de Desarrollo* (Development Plan), Lora-Tamayo becomes Minister of Education and Science and gains Honorary Doctorate from the Sorbonne (Paris)
1963	Execution of the communist leader Julián Grimau
	Chemist Adolf Butenandt in Spain
1964	Franco opens a new petrochemical complex in Huelva
1965	Liberal professors expelled from universities
1966	Palomares nuclear crisis
	Death of José María Albareda
	University student revolts
1967	New university crisis
1968	Lora-Tamayo's resignation as Minister of Education and Science
	Student revolt in Paris
1969	Apollo XI, first man on the moon
	Max Aub visits Spain from exile
1970	Nixon visits Madrid
1971	Max Aub publishes *La gallina ciega*
1973	Franco's Prime Minister, Luis Carrero Blanco, killed by the Basque pro-independence organisation, *Euskadi Ta Askatasuna* (ETA)
1975	Franco's death and the transition towards a monarchic liberal democracy
1976	F. Giral and Augusto Pérez Vitoria return to Spain to recover their Republican university chairs
1978	Referendum for the approval of the new Spanish Constitution
	Amnesty Act
1981	F. Giral returns to Mexico
1994	F. Giral publishes *Ciencia española en el exilio (1939–1989)*

Abbreviations

AC	*Acción Católica*
ACNP	*Asociación Católica Nacional de Propagandistas*
ACIV	*Anales del Centro de Investigaciones Vinícolas*
AEPC	*Asociación Española para el Progreso de las Ciencias*
ANQUE	*Asociación Nacional de Químicos de España*
ARSEFQ	*Anales de la Real Sociedad Española de Física y Química (Anales)*
BIQ	*Boletín del Instituto de Química (Mexico)*
BUPUEE	*Boletín Informativo de la Unión de Profesores Universitarios Españoles en el Extrnajero*
CAMPSA	*Compañía Arrendataria del Monopolio de Petróleos Sociedad Anónima*
CEDA	*Confederación Española de Derechas Autónomas*
CID	*Centro de Investigación y Desarrollo*
CIG	*Comissió d'Indústries de Guerra*
CIV	*Centro de Investigaciones Vinícolas*
CNIQ	*Cámara Nacional de Industrias Químicas*
CNR	*Consiglio Nazionale delle Ricerche*
CNRS	*Centre nationale de la recherche scientifique*
CNT	*Confederación Nacional del Trabajo*
CSIC	*Consejo Superior de Investigaciones Científicas*
DAAD	*Deutscher Akademischer Austauschdienst* (German Academic Exchange Service)
ENCS	*Empresa Nacional Calvo Sotelo*
ENPETROL	*Empresa Nacional de Petróleo*
ETH	*Eidgenössische Technische Hochshule* (Federal Polytechnic School)
FNCE	*Fábrica Nacional de Colorantes y Explosivos*
FNICER	*Fundación Nacional para Investigaciones Científicas y Ensayos de Reformas*
IBYS	*Instituto de Biología y Sueroterapia*
IIT	*Instituto de Investigaciones Técnicas*

List of Abbreviations

IUPAC	International Union of Pure and Applied Chemistry
ILE	*Institución Libre de Enseñanza*
INFQ	*Instituto Nacional de Física y Química*
INI	*Instituto Nacional de Industria*
IQ	*Institut de Química, Universitat Autònoma de Barcelona*
IQA	*Institut de Química Aplicada*
IQM	*Instituto de Química, México*
IQS	*Instituto Químico de Sarriá*
IRI	*Istituto per la Reconversione Industriale*
JAE	*Junta para Ampliación de Estudios e Investigaciones Científicas*
KWG	*Kaiser Wilhelm Gesellschaft*
KWI	*Kaiser Wilhelm Institut*
LIB	*Laboratorio de Investigaciones Bioquímicas*
LIF	*Laboratorio de Investigaciones Físicas*
LQE	*Laboratorio Químico del Ebro*
MCM	*Metalurgia y Construcción Mecánica*
MPG	*Max-Planck Gesellschaft*
MPI	*Max-Planck Institut*
PJC	*Patronato Juan de la Cierva*
QI	*Química e Industria*
RACAB	*Reial Acadèmia de Ciències i Arts de Barcelona*
RACEFN	*Real Academia de Ciencias Exactas, Físicas y Naturales, Madrid*
RSEFQ	*Real Sociedad Española de Física y Química*
SEU	*Sindicato Español Universitario*
SPSL	Society for the Protection of Science and Learning
UAB	*Universitat Autònoma de Barcelona (1933–39)*
UPUEE	*Unión de Profesores Universitarios Españoles en el Extranjero*

Introduction

In his novel *Cat's Cradle* (1963), American writer Kurt Vonnegut (1922–2007) drew inspiration from his own personal experience working at the public relations department of General Electric (GE). He noticed that many GE expert scientists often felt indifferent to the applications of their own discoveries.[1] Nobel Prize winner for chemistry, Irving Langmuir (1881–1957), who spent a good part of his career at GE, inspired the novel's main character, Dr Felix Hoenikker. In an interview in 1980, Vonnegut emphasised that 'Langmuir was absolutely indifferent to the uses that might be made of the truths he dug out the rock and handed out to whoever was around'.[2] Vonnegut's criticism was very much in tune with the Cold War context in which no moral values were attributed to 'pure' chemistry, and even its application to industry and war was often justified through the need for national economic prosperity and patriotism, particularly in totalitarian regimes.[3] Some years earlier, in 1958, the philosopher Hannah Arendt had published *The Human Condition*,[4] a book that continued her earlier intellectual struggle against the consequences of the totalitarianism of the Second World War. Arendt denounced the risks of holding back from civic responsibilities, leaving control of the public world to technocratic experts, and therefore promoted the need to preserve a 'civilized life' with a tangible expression of human freedom.[5]

Arendt's reflection on civic responsibility could even be extrapolated to any kind of political regime and extended half a century to the present. In fact, historians of science and science and technology studies (STS)

---

[1] I am indebted to Jaume Sastre for making me aware of this novel.
[2] Robert Musil, 'There Must Be More to Love than Death: A Conversation with Kurt Vonnegut', *The Nation*, 231(4) (1980), 128–32, p. 129. See also: Tom McCartan (ed.), *Kurt Vonnegut: The Last Interview and Other Conversations* (New York: Melville House, 2012).
[3] Robert Bud, 'Introduction', *Isis, Focus: Applied Science*, 103 (2012), 515–17.
[4] Hannah Arendt, *The Human Condition* (Chicago: University of Chicago Press, 1958).
[5] Dana R. Villa (ed.), *The Cambridge Companion to Hannah Arendt* (Cambridge: Cambridge University Press, 2000), p. 5.

scholars largely agree that the ways in which we know and represent the world (nature and society) are inseparable from the ways in which we choose to live.[6] So in our case, chemistry, just like any other human activity, must be deeply embedded in the social practices that shape its identity (being a chemist), norms (nomenclature, academic disciplines), conventions (research funds, grants), discourses (public addresses, publications), instruments (labs, experimental culture) and institutions (universities, research centres, academies). Therefore, 'doing' chemistry merges into 'doing' politics. Since chemists create a political context for their own practices, the chemical product becomes a political element in a specific context.[7] Similarly, factories, laboratories and other places of chemistry act as mediators between experts, political ideology and propaganda in a technology-based consumer society.[8]

This book analyses the ways in which, from the early years of the formation of a modern scientific community, twentieth-century Spanish chemists talked about the natural world – chemicals, reactions, industrial processes – and then immediately faced issues of authority, credibility and power in society.[9] It approaches the conflicting views about professional identity and also provides conflicting views of the political identity of Spain throughout the twentieth century: from the Bourbon Restoration, which provided low levels of liberal democracy, and Primo de Rivera's dictatorship (1923–30), to the liberal values of the Second Republic (1931–39), the Civil War (1936–39), the totalitarian scientific culture of Franco's dictatorship (1939–75) and the resistance to it.[10]

Beyond discussion about chemistry and chemists on one side and political regimes on the other, the book attempts to build up a narrative in which politicians and chemists, as historical actors, often merge into a single character, making both dimensions indistinguishable, in a crucible that fired the real power of chemistry.

---

[6] Sheila Jasanoff, *States of Knowledge. The Co-Production of Science and Social Order* (London: Routledge, 2004).
[7] Ibid., p. 29.
[8] Carsten Reinhardt, 'Sites of Chemistry in the Twentieth Century', *Ambix*, 62(2) (2015), 109–13. See also Carsten Reinhardt, Harm Schröter, 'Academia and Industry in Chemistry: The Impact of State Intervention and the Effects of Cultural Values', *Ambix*, 51(2) (2004), 99–106.
[9] Jasanoff, *States of Knowledge*, p. 29. In this discussion, Jasanoff draws inspiration from Simon Schaffer and Steven Shapin's *Leviathan and the Air Pump. Hobbes, Boyle, and the Experimental Life* (Princeton: Princeton University Press, 1985). The book has been re-edited with a new introduction by the authors (a paperback edition appeared in 2017).
[10] 'Historical work may gain profundity and relevance through more explicit attention to questions of power, culture and normativity'. Jasanoff, *States of Knowledge*, p. 5. See, in particular, chapter 2: 'Ordering Knowledge, Ordering Society'.

Biographies of Power

Why did Italian chemists such as Giovanni Battista Bonino (1899–1985) so enthusiastically embrace Mussolini's fascism? In the First World War, patriotic (nationalist) values placed chemistry and its uses in war and industry at the service of the nation, so many scientists had a prominent position in the *Consiglio Nazionale delle Ricerche* (CNR)[11] and comfortably supported Mussolini's regime when he came to power in the 1920s.

Some years ago, historian Mark Walker defined scientists working under Nazi Germany as 'fellow travellers' in the following terms:

> If we want to understand how National Socialism affected German science, we cannot restrict ourselves to the few scientists who enthusiastically embraced the Third Reich, and those even fewer scientists who actively and consistently resisted it. Instead we must also include those many scientists who neither resisted nor joined Hitler's movement, rather who *went along for the ride*.[12]

Walker added that 'Expert advice [...] can hardly be divorced from all civil responsibility'.[13] Similarly, as Robert Proctor has rightly pointed out for Nazi Germany, political initiatives arose very often from within the scientific community, so scientists (and chemists in our case) designed and enforced the regime. Thus, 'science is, among other things, a social activity and the politics of those who practice it is part of that science'.[14]

Ute Deichmann has presented a prosopographic approach to biologists during the Nazi era that, without going into intricate detail, allows her to provide a very useful picture of levels of collaboration with the regime, freedom of research and continuities and discontinuities in scientific practices.[15] Nazi Germany also provides similar big pictures for the case

---

[11] Andreas Karachalios, *I chimici di fronte al fascismo. Il caso di Giovanni Battista Bonino (1899–1985)* (Palermo: Istituto Gramsci Siciliano, 2001). Roberto Maiocchi's work on the role of chemistry during the Italian autarchy will be particularly valuable. Roberto Maiocchi, *Scienza e fascismo* (Rome: Carocci, 2004).

[12] Mark Walker, *Nazi Science. Myth, Truth and the German Atomic Bomb* (Cambridge, MA: Perseus, 1995), p. 4 (emphasis added).

[13] Marc Walker, *German National Socialism and the Quest for Nuclear Power* (Cambridge: Cambridge University Press, 1989), p. 5.

[14] Robert Proctor, *Racial Hygiene. Medicine under the Nazis* (Cambridge, MA: Harvard University Press, 1988), p. 9; Kristie Macrakis, *Surviving the Swastika. Scientific Research in Nazi Germany* (Oxford: Oxford University Press, 1993). Other studies, such as Jeffrey Allan Johnson's analysis of chemists in Imperial Germany, have reflected different 'adaptations' to specific political regimes. Jeffrey Allan Johnson, *The Kaiser's Chemists. Science and Modernization in Imperial Germany* (Chapel Hill: University of North Carolina Press, 1990). Johnson discusses the chemists' 'conservative modernization' in the early twentieth century and the continuity of many chemical practices pre- and post-First World War.

[15] Ute Deichmann, *Biologists under Hitler* (Cambridge, MA: Harvard University Press, 1996).

of physicists[16] and medical doctors,[17] but again chemists seem to be thin on the ground when looking for an overview of the behaviour of the chemical community under a particular political regime.

What was the impact of the expulsion of Jewish scientists? What sort of influence did National Socialist ideology and politics come to exert on scientific research in Germany? In what way did scientists during this time participate in the crimes of National Socialism? What impact did Nazi rule have on the development of science after 1945? Using Deichmann's framework, we could easily apply these questions to the community of chemists in Spain: what was the impact of the exiled chemists after the Civil War? What sort of role did dictatorial regimes (Primo de Rivera, Franco and also liberal periods) play in the scientific achievements of Spanish chemists? How responsible were Spanish chemists during Franco's repressive dictatorship after the Civil War?[18]

As this book will make clear, the political positions of many of our chemists had a lot to do with their biographical profiles,[19] family origins, religious backgrounds, social classes and ideologies, early training, the professional networks in which they were integrated in permanent tension between their individual aspirations and dreams and their final contributions to specific collective endeavours.[20] It is therefore perhaps not too late to refer once again to Steven Shapin and Arnold Thackray's seminal paper on the use of prosopography in the history of science, the spirit of which is very much present here when trying to describe the complex and as yet unknown landscape of twentieth-century Spanish chemistry.[21]

There has not been a great deal of research on the history of specific chemical communities and chemical societies in the twentieth century. Anita K. Nielsen and Sona Strabonova have recently analysed European chemical societies in terms of their chronology, membership figures, journals, financial support, professionalisation profiles, debates about

---

[16] Walker, Mark, *Nazi Science*; Phillip Ball, *Serving the Reich. The Struggle for the Soul of Physics under Hitler* (London: Vintage Books, 2013); John Cornwell, *Hitler's Scientists. Science, War, and the Devil's Pact* (London: Penguin Books, 2003).
[17] Proctor, *Racial Hygiene*. [18] Deichmann, *Biologists under Hitler*, pp. 319–20.
[19] On the historiography of scientific biographies, see: Michael Shortland, Richard Yeo, *Telling Lives in Science: Essays on Scientific Biography* (Cambridge: Cambridge University Press, 1996); Theodore Porter, 'Is the Life of the Scientist a Scientific Unit?', *Isis*, 97(2) (2006), 314–21; Mary Jo Nye, 'Scientific Biography: History of Science by Another Means', *Isis*, 97(2) (2006), 322–9; Mott Green, 'Writing Scientific Biography', *Journal of the History of Biology*, 40(4) (2007), 727–59.
[20] Steven Shapin, *The Scientific Life: A Moral History of a Late Modern Vocation* (Chicago: University of Chicago Press, 2008), p. xvii
[21] Steven Shapin, Arnold Thackray, 'Prosopography as a Research Tool in History of Science: The British Scientific Community, 1700–1900', *History of Science*, 12 (1974), 1–28.

'pure' and 'applied' chemistry, discipline-making and foreign relations. Their coverage extends to 14 European countries, from the Big Three (France, Germany and the United Kingdom) to Austria, Belgium, Denmark, Hungary, the Netherlands, Norway, Poland, Portugal, Russia, Sweden and the Czech Republic, but the case of Spain (together with Italy) is regrettably absent.[22] In the 1990s, in a collective volume devoted to the professionalisation of chemistry in the nineteenth century,[23] historian Helge Kragh highlighted a series of issues to be taken into account when approaching the history of a European chemical community in the twentieth century: the growing influence of the United States from the 1920s onwards, and even more evident after 1945; and the progressive emancipation of chemistry from pharmacy, medicine and industry, but at the same time its continuous changing status and boundaries with physics, the life sciences, the new petrochemical industry and many other intersections occurring in the last decades of the century, which are beyond the scope of this volume. To that complex picture, Kragh added the never-ending tension between the science faculties and engineering and polytechnic schools, which, in a way, reflected the interests of different groups regarding their professional hegemony with specific academic or industrial interests.[24]

In fact, professionalisation is a dynamic process in which concepts such as education, organisation, status, autonomy, power and influence have to be taken into account in different historical contexts.[25] In the early 1990s, Mary Jo Nye used several parameters to analyse the *identity* construction of a chemical community: (1) the genealogy of the group, and

---

[22] Anita K. Nielsen, Sona Strabonova (eds.), *Creating Networks in Chemistry. The Founding and Early History of Chemical Societies in Europe* (Cambridge: RSC Publishing, 2008).

[23] David Knight, Helge Kragh (eds.), *The Making of the Chemist* (Cambridge: Cambridge University Press, 1998). In the 1970s, a pioneering study on the history of the Royal College of Chemistry in Britain perceived professional chemists as fulfilling certain requirements, such as acquiring academic qualifications, social responsibility, stable remuneration, corporate identity and scientific authority in society. Colin A. Russell, Noel G. Coley, Gerrylynn K. Roberts, *Chemists by Profession. The Origins and Rise of the Royal Institute of Chemistry* (Milton Keynes: Oxford University Press and the Royal Institute of Chemistry, 1977), p. 3. For a more recent approach to the British community of chemists, see: Gerrylynn K. Roberts, Anna E. Simmons, 'British Chemists Abroad 1887–1971: The Dynamics of Chemists' Careers', *Ambix*, 60(1) (2009), 103–28.

[24] Helge Kragh, 'Afterword: The European Commonwealth of Chemistry', in Knight, Kragh (eds.), *The Making of the Chemist*, pp. 329–341, 329.

[25] In the 1990s, Jack Morrell defined science professionalisation according to the following elements: (1) full-time positions linked to particular knowledge; (2) specialist qualifications and examinations; (3) standard training procedures; (4) boundaries and specialisation; (5) solidarity and self-awareness; and (6) reward systems for best practices. Jack Morrell, 'Professionalization', in Robert Olby et al. (eds.), *Companion to the History of Modern Science* (London: Routledge, 1990), pp. 980–9.

therefore its own history – and the way in which the members of the community themselves perceived it; (2) a core literature of publications – academic papers, textbooks, popular articles and all sorts of publishing strategies; (3) specific rituals of commemoration, anniversaries, obituaries, festivals, public lectures and exhibitions that provide excellent data for the understanding of a particular worldview of a group and of some of its most significant members; (4) a home base of teaching and research institutions (laboratories and lecture halls, but also industrial plants); (5) external recognition by members of other competing scientific communities at home and abroad; and (6) shared values in terms of the genuine scientific culture of the group, but also in terms of political ideology and research priorities.[26]

Therefore, it is within this framework that this book approaches the professional identity of Spanish chemists in the twentieth century. They became active, constant creators of historical accounts of their own professional status under different political regimes and audiences – from the chemistry 'heroes' of the Latin American colonial period to the supposedly glorious experiments on the composition of atmospheric air during the Enlightenment.[27] Inaugural lectures for new academic years, commemorative practices, autobiographies, public addresses, popular lectures, textbooks and popular articles were extensively used to discuss the core identity of their profession.[28] Many of our Spanish chemists behaved as nineteenth-century 'chemist–historians', as Colin A. Russell depicted them some decades ago,[29] often for the legitimation of their discipline and its professional boundaries in order to please or even to endorse ideological guidelines of contemporary politics, or in other cases even as a way to justify their research lines and scientific production. As discussed some years ago by Pnina G. Abir-Am, commemorative practices in science – in chemistry, in our case – often use 'the past in the service of political agendas of the present while appealing to a captive audience: academic and administrative, officers, alumni, students, and

---

[26] Mary Jo Nye, *From Chemical Philosophy to Theoretical Chemistry: Dynamics of Matter and Dynamics of Disciplines, 1800–1950* (Berkeley: University of California Press, 1993). For a study into the identity of a chemistry discipline, see the case of polymer science. Yasu Furukawa, *Inventing Polymer Science. Staudinger, Carothers and the Emergence of Macromolecular Chemistry* (Philadelphia: University of Pennsylvania Press, 1998), p. 9.

[27] Jan Golinski, *Science as Public Culture: Chemistry and the Enlightenment in Britain, 1760–1820* (Cambridge: Cambridge University Press, 1992).

[28] Mary Jo Nye, *From Chemical Philosophy to Theoretical Chemistry*.

[29] Colin A. Russell, '"Rude and Disgraceful Beginnings": A View of History of Chemistry from the Nineteenth Century', *The British Journal for the History of Science*, 21(3) (1988), 273–94.

the science-education conscious public'.[30] In the search for its own identity, groups and nations look for the sense of themselves through subjective narcissistic narratives, just as regimes use historical myths to legitimise their established power.[31] It is precisely in the actors' accounts that we find reflections on the continuity or discontinuity of historical actors, scientific institutions, intellectual traditions and material resources. They also inform us about the progressive evolution or sometimes sudden breakdowns in a particular chemical culture, in our case in the context of the colonial crisis of 1898, Primo de Rivera's dictatorship of 1923, the Republican regime in 1931, the Civil War and the different periods of Franco's dictatorship from 1939 to 1975.

Historians have not sufficiently explored the role of chemistry in twentieth-century Spain. There is still a lot to do in terms of prosopography, institutions and teaching and research practices.[32] A brief look at the final chapter of Manuel Lora-Tamayo's *La investigación química española* (1981) clearly shows the crucial importance of that prosopography and the enormous amount of historical research that has still to be done.[33] There is considerable research on the dynamism of the early decades of the century up to the Republican 1930s,[34] as well as more recent work on the role of science in the Francoist regime,[35] but chemistry has been notably absent from all of these narratives, with only some minor references to its intersection with physics in science faculties at the *Junta para Ampliación de Estudios e Investigaciones Científicos* (JAE) and the *Instituto Nacional de Física y Química* (INFQ), as well as some quantitative studies.[36] In the recent

---

[30] Pnina G. Abir-Am, Clark A. Elliot (eds.), *Commemorative Practices in Science, Osiris (2nd Series)*, 14 (1999), p. 15.

[31] Adam Budd (ed.), *The Modern Historiography Reader. Western Sources* (London: Routledge, 2009), pp. 43–6; Michael Ignatieff, 'The Nightmare from Which We Are Trying to Awake', in *The Warrior's Honor: Ethnic War and the Modern Conscience* (London: Chatto and Windus, 1998), pp. 166–99.

[32] As José Manuel Sánchez Ron stated some years ago, there is still not enough historical research on chemistry in the first half of the twentieth century in Spain. José Manuel Sánchez Ron, *Cincel, martillo y piedra. Historia de la ciencia en España (siglos XIX y XX)* (Madrid: Taurus, 1999), p. 244. See also: José Manuel Sánchez Ron, *Un siglo de ciencia en España (Diciembre de 1998–Marzo de 1999)* (Madrid: Residencia de Estudiantes, 1998).

[33] Manuel Lora-Tamayo, *La investigación química española* (Madrid: Alhambra, 1981); Joaquim Sales, *La química a la Universitat de Barcelona* (Barcelona: Publicacions i Edicions de la Universitat de Barcelona, 2011).

[34] Luis Enrique Otero Carvajal, José María López Sánchez, *La lucha por la modernidad. Las ciencias naturales y la Junta para Ampliación de Estudios* (Madrid: Residencia de Estudiantes, 2012).

[35] Lino Camprubí, *Engineers and the Making of the Francoist Regime* (Cambridge, MA: MIT Press, 2014).

[36] See, for instance, Gerardo Palao, 'Influencias extranjeras en la investigación química española (1904–1965)', *Llull*, 13 (1990), 131–52.

commemorative book *Tiempos de investigación*, which aimed to celebrate – not without controversy – the 'continuity' of 100 years (1907–2007) of scientific research in Spain, there is no single chapter devoted to chemistry, nor is there a chapter on the chemistry of the post-war Francoist period.[37] We have details about university academic chairs and positions in all fields of knowledge in the early decades of the century, but there is no focused analysis of the chemical community and its identity.[38]

In the history of physics, seminal work on the JAE period by José Manuel Sánchez Ron and Antoni Roca-Rosell has been recently complemented by the fresher approach of Xavier Roqué and Néstor Herrán on the ways in which physics and physicists co-produced Franco's regime, moulded that scientific community and controlled scientific institutions under the values of the dictatorship.[39] In the history of the biological sciences, María Jesús Santesmases has published prolifically on the ways in which new disciplines such as biochemistry emerged and were 'accommodated' in the context of the dictatorship, as well as the role played, for instance, by the 1959 Nobel Prize winner, Severo Ochoa.[40] Amparo Gómez, Antonio Canales and Brian Balmer have discussed extensively the science policies that were in force in twentieth-century dictatorships, including Franco's Spain.[41] Antoni Malet has described in detail the complex process that led to the creation of the *Consejo Superior de Investigaciones Científicas* (CSIC) in 1939 and the role of José María Albareda as a science policy-maker of the regime. Equally, Albert Presas has focused his research on the German scientific relations with Franco's Spain and the ways in which nuclear power and 'atoms for peace' campaigns co-produced part of the values of the regime.[42] The nuclear question has also drawn the attention of other scholars such as Ana Romero, Sánchez Ron and, more recently, Roqué and Herrán.[43] They have all described in detail the ways in which expertise and power intersected in the local context during Franco's dictatorship, but also at

---

[37] José Manuel Sánchez Ron provided some data on chemists. J. M. Sánchez Ron, 'Las ciencias físicas y químicas en la JAE', in Miguel Ángel Puig-Samper (ed.), *Tiempos de investigación JAE–CSIC. Cien años de ciencia en España* (Madrid: CSIC, 2007), 103–44.
[38] Otero Carvajal, López Sánchez, *La lucha por la modernidad*.
[39] Xavier Roqué, Nestor Herrán (eds.), *La física en la dictadura. Físicos, cultura y poder en España 1939–1975* (Bellaterra: Universitat Autònoma de Barcelona, 2012).
[40] María Jesús Santesmases, 'Severo Ochoa and the Biomedical Sciences in Spain under Franco, 1959–1975', *Isis*, 91 (2000), 706–34.
[41] Amparo Gómez, Antonio Canales, Brian Balmer (eds.), *Science Policies and Twentieth-Century Dictatorships: Spain, Italy, and Argentina* (Farnham: Ashgate, 2015).
[42] Albert Presas, 'Science on the Periphery. The Spanish Reception of Nuclear Energy: An Attempt at Modernity?', *Minerva*, 43 (2005), 197–218.
[43] Roqué, Herrán (eds.), *La física en la dictadura*.

the international level during the Cold War as an excellent way to place Spanish examples in the mainstream historiography.[44]

From history of technology, Lino Camprubí has recently reconstructed the role of engineers in the making of Francoist Spain after the Civil War.[45] In his view, far from the old analysis of the role of scientists in dictatorships as 'autonomous' actors who 'resisted' the regime's hostility, engineers in particular co-constructed and co-produced the regime's power: 'It is through concrete material products [dams, churches, coal silos, agronomy labs] that the state takes over its territory and that becomes part of the landscape'.[46] In a similar vein, Tiago Saraiva's recent book on the technoscience of fascism focuses on the ways in which these regimes (in his case, Italy, Portugal and Germany) created a new material world through science. Food, plants and animals – fascist pigs – are described as a material, alternative modernity that opposed that of liberal democracies and communism.[47]

From Saraiva's reification of fascism to Camprubí's structural assemblage of early Francoism, and borrowing from other ideas of co-production (also from Roqué and Herrán), this book focuses more on specific historical actors, their biographical profiles and their political positions in everyday scientific practice, as well as on their statements in the public sphere. It covers a longer historical period, with different political regimes – liberal and dictatorial – and times of deep turbulence (such as the Civil War), and contrasts the ethos of the chemical community with a particular emphasis on conflicting and even fiercely opposing views on the ways in which chemistry (but also society) had to evolve. It also attempts to enrich the present historiography of Francoism through a systematic analysis of the chemists' political positions.[48]

In the framework of that complex intersection between science and ideology, it is time to revisit historian Thomas Glick's old thesis of the 'civil discourse' of Spanish science in the early twentieth century. In Glick's view, which many other historians of Spanish science have

---

[44] Lino Camprubí, Xavier Roqué, Francisco Sáez de Adana (eds.), *De la Guerra Fría al calentamiento global. Estados Unidos, España y el nuevo orden científico mundial* (Madrid: La Catarata, 2018).
[45] Camprubí, *Engineers and the Making of the Francoist Regime*. [46] Ibid., p. 164.
[47] Thiago Saraiva, *Fascist Pigs. Technoscientific Organisms and the History of Fascism* (Cambridge, MA: MIT Press, 2016), pp. 3–6.
[48] Paul Preston, *Franco* (London: Fontana Press, 1995); Enrique Moradiellos, *La España de Franco (1939–1975). Política y sociedad* (Madrid: Síntesis, 2000); Borja de Riquer, *Historia de España (IX): La dictadura de Franco* (Barcelona/Madrid: Crítica/Marcial Pons, 2010), p. xxiii; Ismael Saz, *Las caras del franquismo* (Granada: Comares, 2013); Enrique Moradiellos, *Franco: Anatomy of a Dictator* (London: I. B. Tauris, 2018). Although in recent decades this historical period has become a main target of research for historians, the role of chemistry in the shaping of the regime needs further study.

endorsed for decades, at the beginning of the twentieth century, Spanish society decided to place science apart from political struggle. Memories of the devastating consequences of late nineteenth-century debates on Darwinism and their political appropriations were still fresh, together with the appropriation of science as being the main cause of the crisis of 1898 and the further-reaching image of the backwardness of the country. Glick defended the case that scientific policies of the early twentieth century had achieved considerable 'autonomy' from political upheaval, in a sense preserving scientific knowledge in a cage or ivory tower to avoid contamination with the miseries of everyday life, which led to notable developments in 'pure', 'academic' research.[49] It is, however, hard to subscribe to Glick's thesis of political neutrality, even in the supposedly liberal, cosmopolitan, 'modern' ethos of the JAE chemists. Moreover, we can easily trace the temptations of the rhetoric of an apolitical chemistry among our chemists in dictatorial periods, such as that of Primo de Rivera's dictatorship (1923–30) and during several periods of Franco's dictatorship, particularly in the 'technocratic' trends of the 1960s. Even in the totalitarian dreams of the early autarky of the 1940s, chemistry – and the chemical industry in particular – served and shaped the regime through public statements of 'technical' or 'objective' competence. Therefore, the supposed neutrality of the 'civil discourse' should now be replaced by an approach focusing on the scientists' (chemists' in our case) 'civic responsibility' in the building of any particular political regime. As Robert Proctor discussed some decades ago in his *Value-Free Science?*, chemistry is political because it is closely linked to industrial and military power, because there are always alternative research fields, production processes and consumer choices and, finally, because chemistry is mainly devoted to the making of materials for our everyday lives.[50]

A Political Chemistry

Although historians have paid considerable attention to the role of chemistry in the second Industrial Revolution,[51] there is still a lot of work to be

---

[49] Thomas Glick, *Einstein in Spain: Relativity and the Recovery of Science* (Princeton: Princeton University Press, 1988), ch. 1. See also: Thomas Glick, 'Dictating to the Dictator: Augustus Trowbridge, the Rockefeller Foundation, and the Support of Physics in Spain, 1923–1927', *Minerva*, 43(2) (2005), 121–45.
[50] Robert Proctor, *Value-Free Science? Purity and Power in Modern Knowledge* (Cambridge, MA: Harvard University Press, 1991), p. 266.
[51] For the German science-based industry, see: Georg Meyer-Thurow, 'The Industrialization of Invention: A Case Study from the German Chemical Industry', *Isis*, 73(3) (1982), 363–81; John Lesch (ed.), *The German Chemical Industry in the Twentieth Century* (Dordrecht: Kluwer, 2000).

done on its contribution in the twentieth century,[52] a period in which chemistry evolved from its traditional nineteenth-century 'applications' to agriculture, industry, medicine, life sciences and pharmacy, to new subjects such as quantum chemistry, molecular biology, environmental science and the industries of oil, plastics and pharmaceuticals.[53] It was also a century of profound instrumental revolution in which techniques imported from physics such as X-ray diffraction, infrared, ultraviolet and mass spectroscopy, nuclear magnetic resonance and, later, computing radically changed the material culture of chemistry.[54] Its boundaries with physics, biology, mathematics and technology evolved towards a new 'interconnected path of scientific specialties'.[55]

In spite of all these deep, significant transformations, chemistry has been notably absent in the big picture of twentieth-century science. A few years ago, historian of technology, David Edgerton, pointed out how the big picture of the last century mainly focused on scientific issues such as particle physics, the atomic bomb, eugenics and molecular biology, or was taken from the perspective of the history of technology through automobiles, aviation, bombs, space rockets and computers.[56] Equally, among the titles of the 46 chapters of John Krige and Dominique Pestre's *Science in the Twentieth Century*, the word 'chemistry' appears only once, where it is linked to polymers, with two additional chapters on atomic, molecular science and on the pharmaceutical industry.[57] Nor does much chemistry appear in Jon Agar's big picture of science in the twentieth century and beyond.[58] In a similar vein, historian of chemistry, Seymour Mauskopf, pointed out some years ago that:

---

[52] David Edgerton, 'Not Counting Chemistry: How We Misread the History of Twentieth-Century Science and Technology', *Distillations* (2008), Science History Institute. www.sciencehistory.org/distillations/magazine/not-counting-chemistry-how-we-misread-the-history-of-20th-century-science-and (last accessed 29 September 2018).

[53] José Ramón Bertomeu-Sánchez, Thorburn Burns, Brigitte Van Tiggelen (eds.), *Neighbours and Territories. The Evolving Identity of Chemistry. The 6th International Conference on the History of Chemistry* (Louvain-la-Neuve: Mémosciences, 2008), p. 11.

[54] Carsten Reinhardt, *Shifting and Rearranging: Physical Methods and the Transformation of Modern Chemistry* (Sagamore Beach: Science History Publications, 2006). See also: Peter Morris (ed.), *From Classical to Modern Chemistry: The Instrumental Revolution* (London: RSC/Science Museum, 2002).

[55] Theoretical chemistry, nuclear chemistry, cosmo-chemistry and, more recently, solid-state chemistry, biotechnology and nanotechnology. Carsten Reinhardt (ed.), *Chemical Science in the Twentieth Century: Bridging Boundaries* (Weinheim: Wiley-VCH, 2001). See also: Ernst Homburg, 'Shifting Centres and Emerging Peripheries: Global Patterns in Twentieth Century Chemistry', *Ambix*, 52(1) (2005), 3–6.

[56] Edgerton, 'Not Counting Chemistry'.

[57] John Krige, Dominque Pestre (eds.), *Science in the Twentieth Century* (Amsterdam: Harwood Publishers, 1997).

[58] Jon Agar, *Science in the Twentieth Century and Beyond* (Cambridge: Polity Press, 2012).

12   Introduction

Chemistry did not participate directly in the great and dramatic transformative scientific revolutions of the beginning of the twentieth century associated with such celebrated names as Einstein, Bohr and Freud. And in contrast to physics, biology and the behavioural sciences, chemistry has not seemed to have much import from the great philosophical or existential issues of our time. Hence it has not been integrated in cultural history the way these sciences were.[59]

When approaching the history of twentieth-century chemistry, we are therefore facing a relatively unexplored field that requires a better-equipped historiographic framework for the analysis of the complex relationships between chemistry, industry and the economy, its material culture (from labs to instruments and places of chemical practice), its flexible disciplinary boundaries (which have accompanied chemistry throughout its history, as well as in the twentieth century) and finally its political dimension.[60] This book therefore attempts to invigorate the historiography of chemistry from all of these perspectives, and all of the cases appearing in the following chapters support this research agenda, with the final aim being to contribute to a history of twentieth-century chemistry that can offer a better understanding of the period in all its complexity.

To complicate matters, not so much attention has been paid to twentieth-century chemistry beyond certain cases in the Big Three European nations (France, the United Kingdom and Germany) and the United States and its intimate relationship with the innovative research centres and universities that had promoted the discipline since the nineteenth century.[61] The Justus von Liebig research school in the early decades of the nineteenth century, prestigious laboratories such as those of Wilhelm Ostwald and Walther Nernst and, later, dynamic centres such as the Federal Polytechnic Institute (ETH) in Zurich, the Sorbonne and the Pasteur Institute in Paris, as well as outstanding groups in Berlin, Leipzig and Munich – all had a profoundly European, cosmopolitan character that merits further examination.[62]

---

[59] Seymour Mauskopf (ed.), *Chemical Sciences in the Modern World* (Philadelphia: University of Pennsylvania Press, 1993), p. xii.
[60] 'Still, historians of modern chemistry are but a tiny minority among the professional historians of modern science'. Carsten Reinhardt (ed.), *Chemical Sciences in the Twentieth Century*, p. ix.
[61] Nye, *From Chemical Philosophy to Theoretical Chemistry*. John E. Lesch (ed.), *The German Chemical Industry*; William Brock, *The Fontana History of Chemistry* (London: Fontana Press, 1992).
[62] On networks, see, for instance: Michel Callon (ed.), *La science et ses réseaux. Genèse et circulation des faits scientifiques* (Paris: La Découverte, 1988). See also: Bruno Latour, *Science in Action. How to Follow Scientists and Engineers through Society* (Cambridge, MA: Harvard University Press, 1987).

Chemistry research schools in the twentieth century grew through complex networks in which Spanish chemists also had their say.[63] In that sense, this book focuses on Spain, but also provides further reflection on the role of chemistry as an international profession in the twentieth century.[64] This is the study of what is still a relatively unknown scientific community, its ethos, rituals and collective endeavours and the political weight associated with them. It shows how specific academic research fields, industrial projects and claims of identity contributed to the construction of specific political regimes and how these regimes shaped research priorities and experimental practices.

Spanish chemists became a significant scientific community in the twentieth century. In spite of dramatic fragmentation after the Civil War, they retained their identity, but not without severe tensions, disagreements and significant changes to their research priorities and workforce. They wrote, talked and lectured about their own history and the ways in which they became 'citizen scientists',[65] public intellectuals[66] and ideological agents in liberal and dictatorial periods. On a journey from the foundation of the *Real Sociedad Española de Física y Química* (RSEFQ) in 1903 to the end of General Franco's dictatorship in 1975, our historical analysis of chemicals and the research, teaching and industrial production of the actors involved provides a new perspective on the role of chemistry as a profession in the twentieth century.

Evidently, our chemists were neither pure nor neutral. José Giral (1879–1962), the father of Francisco Giral, was a fierce opponent of Primo de Rivera's dictatorship. He was a Minister of the Navy in the 1930s in two of the governments of the Second Spanish Republic, and Prime Minister in the first months of the Civil War. But J. Giral also became Prime Minister of the Republican Government in exile from 1945 to 1954. Lora-Tamayo – one of the most prominent Spanish chemists of the twentieth century – became Minister for Education and Science in Franco's cabinet during the turbulent 1960s; José María Fernández-Ladreda (1885–1954), a reputable chemist and army officer,

---

[63] On a revision of centres and peripheries at the European level, see: Kostas Gavroglu et al., 'Science and Technology in the European Periphery. Some Historiographic Reflections', *History of Science*, 46 (2008), 1–23.

[64] Inside Spain, dynamic centres and peripheries also contributed over the different periods to the making of the chemical community. See: Mary Jo Nye, *Science in the Provinces. Scientific Communities and Provincial Leadership in France, 1860–1930* (Berkeley: University of California Press, 1986).

[65] Mark Walker (ed.), *Science and Ideology: A Comparative History* (London: Routledge, 2003).

[66] Edward Said, *Representations of the Intellectual. The 1993 Reith Lectures* (New York: Vintage Books, 1996).

was the Minister for Public Works – also in Franco's cabinet – in the 1950s. Many chemists became university rectors (J. Giral, Emilio Jimeno, Francisco Buscarons [1906–89], Albareda, Antonio de Gregorio Rocasolano [1873–1941] and Ángel Vián [1914–99]), but also city mayors (Jesús Yoldi [1894–1936] and Luis Bermejo [1880–1941]). Others, such as Albareda, controlled the scientific policies of Franco's regime with an iron grip for decades.

The history of twentieth-century Spanish chemists becomes a useful historiographical tool to discuss twentieth-century political issues such as modernity, republicanism, war, totalitarianism, autarchy, technocracy and liberalism, which I have used as organising concepts throughout the whole volume and as mediating words for the analysis of the intersection between chemistry and power and the role of chemistry in the co-production of politics in the twentieth century. Following this Introduction, Chapter 1, 'Dreams of Modernity', describes how, during the initial decades of the twentieth century, the Spanish chemical community grew, fighting backwardness to develop an ambitious plan of modernisation. Under the banner of the JAE, cosmopolitanism and the connection to a European network of chemical expertise became crucial targets for the academic and political authorities during the final decades of the Bourbon Restoration and Primo de Rivera's dictatorship. This dream of modernity materialised on several fronts: reforms in chemical training (teaching and laboratories); new internationally orientated research schools, fellowships and exchange programmes; and major projects to link academic chemistry to industry.

Chapter 2, 'A Republican Science', describes the role that chemists played in the construction of the Second Spanish Republic (1931–39). The chapter explores how many became allies of the leftist policies of the new regime, with its dreams of secularism, universalism and peace, as opposed to the dark images of chemical weapons of the First World War. It covers the opening in 1932 of the INFQ – nicknamed 'The Rockefeller' after the funding used to create it – and its leading role in chemical research in the 1930s; the university reforms in teaching and experimental chemistry; the International Conference on Pure and Applied Chemistry held in Madrid in 1934; and further attempts to link academic chemistry to industrial growth.

Chapter 3, 'War Weapons', discusses how the Civil War (1936–39) dramatically interrupted plans for modernisation, forcing chemists to get involved in war. The chapter also discusses the ways in which violence, repression and exile divided the chemical community. In the three years of the Civil War, the Republican order progressively faded away and a new dictatorship gained power and social support. War dramatically changed chemists' daily practices and brought division and suffering to a politically polarised community. Some chemists in the new regime became active repressors, whereas others suffered exile and marginalisation. Moreover, in spite of certain tacit trends of continuity, dynamic fields such as physical chemistry, spectroscopy and early biochemistry, which had flourished in the 1930s, had now lost momentum and were reshaped and reorganised.

Chapter 4, 'Totalitarian Ambitions', describes how chemists such as Albareda, Lora-Tamayo, Jimeno, Antonio Rius Miró (1890–1973) and Pascual co-constructed the new regime, to the extent that totalitarian political ambitions often converged with totalitarian chemistry projects. In public, aggressive rhetoric against liberal values helped them to reinforce their own professional positions and enthusiastically ascribe to the moral, political and religious values of the dictatorship. However, some pre-war research lines continued. This chapter particularly addresses the ways in which religion, especially theological arguments from Opus Dei and the Society of Jesus, shaped how chemists contributed to the legitimisation of the regime.

Chapter 5, 'Autarchic Ambiguities', analyses how, in the context of the international isolation of Francoist Spain in the early 1940s, autarchic discourses and projects for economic and industrial self-sufficiency gained prominence and guided new policies. Raw materials, how they were geographically distributed and the economics of their exploitation became state decisions to which chemists contributed through reframing academic disciplines and industrial endeavours. This chapter also stresses the distinction between *autarky* (referring to economic self-sufficiency) and *autarchy* (referring to autocratic political rule)[67] and emphasises the paradoxical role of autarchic chemists as ambassadors for the international legitimisation of the regime.

Chapter 6, 'Technocratic Progress', focuses on the 1960s internationalisation of the economy and the ways in which, in public, and in the context of the Cold War, chemistry became a 'neutral' subject that contributed to the non-democratic modernisation of the country, again with major religious alliances. The chapter explores the apolitical, technocratic role of chemists and discusses how they undertook academic reforms and searched for international alliances for their own professional interests. It also situates a good part of that apolitical discourse and practise in the context of the new international, corporate chemical industry, which was established throughout the country from the early 1960s on.

---

[67] Tiago Saraiva, Norton Wise, 'Autarky/Autarchy: Genetics, Food Production, and the Building of Fascism', *Historical Studies in the Natural Sciences*, 40(4) (2010), 419–28.

Chapter 7, 'Liberal Dissent', describes some of chemists' strategies of resistance to totalitarianism and the co-construction of new, intermediate spaces. Going back chronologically, it explores the liberal ethos of the Spanish chemical community in exile after the Civil War and the ways in which it evolved, particularly in the Latin American context. It also points out some chemists' attempts to protect liberal values in the chemical industry, in the universities and in the public sphere in hostile, anti-liberal contexts such as Franco's dictatorship, as well as the ways in which many of them survived as internal refugees.

Finally, the concluding chapter discusses how chemists became intellectuals of a complex, polyhedral process of modernisation that embraced the early years of modernity up to the Republican dream of the 1930s, but also the ethos of the exiled chemical community. The chapter highlights other reactionary forms of modernity following the Civil War, which provided expertise and an ambitious agenda of applied chemistry in a radical break with the pre-war Republican culture. The concluding chapter also highlights how the case of the chemists has influenced the present tension between scholarly, academic history and the popular memory of the past. From the lessons of twentieth-century Spain, this final chapter also tackles the issue of the political weight of chemistry and its moral ambiguity as a profession at a more general level.

In twentieth-century Spain, chemistry as a profession and chemists as key historical actors have strengthened the projects of modernisation – in terms of educational reforms, international links, research policies and industrial endeavours – in tune with the liberal values of cosmopolitanism and pacifism, but also in a reactionary form during dictatorial periods. The material, pragmatic nature of the chemical sciences contributed to the frequent rhetoric of political neutrality, which facilitated the chemists' role in the construction of political regimes. Natural products – alkaloids, terpenes and carbohydrates – became paradigms of God's creation for Catholic chemists, objects of international academic research for some of the most prominent *pensionados* (recipients of grants for foreign study) of the JAE and tools of industrial and national progress for the exiled community after the Civil War. At the same time, luminaries such as Enrique Moles used his atomic weights experimental culture from his university chair at the end of Primo de Rivera's dictatorship, at the elitist, republican INFQ, and for the making of chemical weapons during the war. He even attempted (unsuccessfully) to place his 'objective' chemistry in the framework of the new dictatorial state in the early 1940s.[68]

---

[68] Agustí Nieto-Galan, Joaquim Sales, 'Exilio y represión científica en el primer franquismo. El caso de Enrique Moles', *Ayer. Revista de Historia Contemporánea*, 114(2) (2019), 279–311.

Lora-Tamayo combined his diene synthesis with his duties as a minister of Franco's cabinet and, in public, detached his chemistry from any political affiliation. In a way, chemistry acted as a joker in hands of the different actors appearing in the following chapters. 'Neutral', 'objective' molecules often became flexible mediators linking personal and professional ambitions with a particular political ideology. This is the kind of paradox that this book aims to explore in depth.

A political analysis of chemistry in the past is inevitably linked to the political concerns of our present. This book brings to the fore the values of the new presentism, in which detailed case studies approaching complex mechanisms of co-production between science and power help us to assess the present crisis of liberal democracies and to revisit the inherent political baggage of scientific expertise. 'Militant history', 'history matters' and 'new presentism' are some of the appealing labels that have recently encouraged historians of science to find new tools of dialogue and influence in our present.[69] However, chemistry has probably been one of the fields to suffer from the lack of a more refined analysis of historical case studies and present professional concerns such as pollution, waste management and food and drug adulteration.[70] As Naomi Oreskes has recently emphasised, we move towards a 'motivational presentism' in which historians should openly admit why they choose a specific topic of research and the methodological tools they use as a result of their political position in the present[71] as citizen historians – if we may extrapolate here Mark Walker's term 'citizen scientists' – for the analysis of the intersection between science and ideology in different political regimes.[72]

Some years ago, in an interview, the British historian, Paul Preston, discussed the personal reasons that brought him to work on twentieth-century Spain, and in particular on the Civil War, the subsequent repression and Franco's dictatorship. Preston openly admitted that his work had been mainly devoted to making his readers more aware of the justice of the Republican cause and the evil of Franco and Francoism.[73] There

---

[69] See, for instance: *The Activist History Review*. 'Everyone involved in *TAHR* is dedicated to a simple principle: that the past is relevant to the present. We hail from a wide variety of personal, political, intellectual, and disciplinary backgrounds both in and out of academia. The issues of today were formed historically, and the only appropriate solutions to those issues are ones informed by a comprehensive understanding of how they came to be'. https://activisthistory.com/304-2 (last accessed 25 November 2017).
[70] Joachim Schummer, Bernadette Bensaude-Vincent, Brigitte Van Tiggelen (eds.), *The Public Image of Chemistry* (Singapore: World Scientific Publishing, 2007).
[71] Naomi Oreskes, 'Why I Am a Presentist,' *Science in Context*, 26 (2013), 595–609.
[72] Walker, *Science and Ideology*.
[73] Josep M. Muñoz, 'Paul Preston: El compromís emocional de l'historiador', *L'Avenç*, 338 (2008), 17–24.

is no doubt about Preston's professional prestige as a historian, a fact that bears no contradiction with the assumption that there is no full objectivity or impartiality in history. Similarly, in a recent interview, the Catalan historian of Francoism, Borja de Riquer, described his personal memories of Franco's dictatorship in the 1950s as a driving force for his future historical research. Ice sales, old tramways, gasogen taxis, grey and dirty cities, war mutilations and the contradictions between silence, poverty and moral misery in everyday life, plus the bourgeois pro-Franco values of his own family, profoundly influenced his calling towards history as a profession.[74]

In a country that today still has more than 100,000 unidentified bodies of victims of the Civil War and the fierce repression of the dictatorship, writing a respectable history of the twentieth century is a civic duty.[75] A country that has been unable to face a process of revision and reconciliation with its dictatorial past needs further historical research to give a voice to so many silenced actors. Through the accounts of the most representative Spanish chemists, this book offers new data and arguments relating to debates on the cultural and political history of twentieth-century Spain, on the weakness of its democratic institutions, on its long-term authoritarian tradition and on the uncomfortable history of exile, repression, iniquity and forgetfulness that still casts a shadow on the present.

After the end of Franco's dictatorship in 1975, and during the new democratic period, several commemorative practices have presented the Second Republic and the Civil War as 'unfortunate' events that did not hinder the progress of science beyond personal biographies and institutional transformations.[76] This kind of commemorative approach risks presenting the history of twentieth-century science in Spain (chemistry in our case) as a history of cumulative, apolitical progress. It is a history in which a kind of scientist spirit would supposedly have survived the 'miseries' of several political regimes.[77] As the philosopher, Josep Ramoneda, set out some years ago in the general press:

---

[74] Borja de Riquer, 'Viure i historiar el franquisme', *L'Avenç*, 417 (2015), 24–33.

[75] Paul Preston, *The Spanish Holocaust: Inquisition and Extermination in Twentieth-Century Spain* (London: Harper Press, 2012).

[76] *50 años de investigación en física y química en el edificio Rockefeller de Madrid 1932–1982. Noviembre de 1982. Recuerdos históricos* (Madrid: CSIC, 1982); Puig-Samper, *Tiempos de investigación*; Carlos González, Antonio Santamaría (eds.), *Física y química en la Colina de los Chopos. 75 años de investigación en el edificio Rockefeller del CSIC (1932–2007)* (Madrid: CSIC, 2009).

[77] Agustí Nieto-Galan, 'La memoria histórica de la ciencia en España', *Sin permiso*, 7 December 2008. www.sinpermiso.info/textos/la-memoria-histrica-de-la-ciencia-en-espaa (last accessed 22 September 2018).

It is true that the history of Spain is as it is, and democratic tradition is scarce. However, this fact does not legitimise exercises in confusion, which aim to let people think that in the end a general election is the same as a coup d'état, and that Francoism is as legitimate as the Second Republic.[78]

In fact, too often ideas of continuity and homogeneity serve the institutional, corporate and political interests of the present. This book takes a critical position in the face of these approaches throughout the history of the twentieth century. The lessons of the past that appear in the following chapters aim to help contemporary professional chemists to reflect on their own identity and deontological status, but also to bring to the fore the forgotten names of past generations, which have been sadly dismissed through the filters of different political regimes.

---

[78] Josep Ramoneda, 'Los avatares de la memoria', *El País* (22 July 2007, *Domingo*).

1 Dreams of Modernity

In 1980, at the Great Hall of the University of Salamanca, the chemist Francisco Giral gave the inaugural address for the new academic year. Using historical examples of the 'glorious' Spanish past – a recurrent rhetorical strategy of the chemical community in its public rituals – F. Giral presented the main American substances used in colonial times: alkaloids, tobacco, dyestuffs, metals and minerals,[1] and the ways in which those products had shaped the Spanish economy for centuries. At the end of the talk, he turned his eye to his own experience in Latin America as a political refugee at the end of the Civil War.[2] F. Giral grew up in a republican culture (Figure 1.1),[3] in an intellectual world that included, among many others, humanists such as Miguel de Unamuno and Claudio Sánchez Albornoz, prestigious scientists such as Santiago Ramón y Cajal (winner of the Nobel Prize in Physiology or Medicine in 1906), naturalist Ignacio Bolívar, physicist Blas Cabrera and chemists such as Miguel Catalán (1894–1957), Enrique Moles and Antonio Madinaveitia (1890–1974),[4] the latter three being his main mentors. After his PhD in organic chemistry, supervised by Madinaveitia, F. Giral went to Heidelberg to work with Richard Kuhn (1900–67), winner of the Nobel Prize for Chemistry in 1938, with whom he published several papers.[5]

---

[1] Francisco Giral, *Comentarios químico-farmacéuticos a la historia española en América* (Salamanca: Universidad de Salamanca, 1980).
[2] Ibid., p. 99.
[3] Carolyn P. Boyd, *Historia Patria. Politics, History and National Identity in Spain, 1875–1975* (Princeton: Princeton University Press, 1997). Francisco Giral, Currículum, 1939. *Colegio de México*. Fondo antiguo. Sección personal.
[4] Francisco Giral, *Vida y obra de José Giral Pereira* (México: UNAM, 2004), p. 122. María Fernanda Mancebo, 'Tres vivencias del exilio en México: Max Aub, Adolfo Sánchez Vázquez y Francisco Giral', *Migraciones y Exilios*, 5 (2004), 85–102, p. 86.
[5] Francisco Giral, *La ciencia española en el exilio (1939–1989). El exilio de los científicos españoles* (Barcelona: CIERE/Anthropos, 1994). Richard Kuhn, Francisco Giral, 'Einfluß der Kettenlänge auf den Geschmack aliphatischer ω-Betaine', *Zeitschrift für physiologische Chemie*, 231 (1935), 208–9.

Figure 1.1 Francisco Giral, seated on the left, in a public meeting in support of the Second Spanish Republic, c. 1978. Notice the picture of President Manuel Azaña at the feet of the speakers.
Archivo Gráfico Carta de España. Dirección General de Migraciones. Ministerio de Empleo y Seguridad Social. Public domain archive.

After decades of exile, F. Giral's public address in Salamanca expressed deep nostalgia for the so-called 'Silver Age of Spanish Science'[6] and all of the mentors who had marked his youth. Historians of Spanish science largely agree on the use of that label to describe the dynamism of the early decades of the twentieth century. The concept was extrapolated decades ago from the 'Silver Age of Spanish Culture', in which, under the banner of the *Institución Libre de Enseñanza* (ILE), a liberal educational project, intellectuals, writers and artists made their mark on the country in terms of artistic creativity, which also had its own resonance in science.[7]

---

[6] José Manuel Sánchez Ron, *1907–1987. La Junta para Ampliación de Estudios e Investigaciones Científicas 80 años después.* 2 vols. (Madrid: CSIC, 1988), II, p. 259.
[7] For the Silver Age of Spanish Culture, see: José Carlos Mainer, *La edad de plata (1902–1939). Ensayo de interpretación de un proceso cultural* (Madrid: Cátedra, 1983). See also Ramón Menéndez Pidal, José María Jover Zamora (dirs.), *Historia de España. XXXIX. La Edad de Plata de la Cultura Española (1898–1936)* (Madrid: Espasa Calpe, 1993).

F. Giral's address in Salamanca proved that, in the 1930s, chemistry had made considerable progress from its backward state in around 1900. In fact, at the beginning of the twentieth century, it was a young and fragile academic discipline. Chemistry courses at the science faculties occupied a minor position with no clear identity, and chemistry served as a complementary field in faculties of pharmacy and medicine in subjects such as physiological chemistry, pharmacology and biological chemistry. Although earlier university reforms had already created science faculties, only a limited number of chemistry subjects appeared in the curricula, with no clear demarcation between chemistry and physics.[8] General chemistry subjects in the faculties of medicine and analytical, inorganic and organic chemistry courses in the faculties of pharmacy completed a general picture in which few students became true chemists and even fewer continued to gain PhDs. In spite of several reforms of the official new curricula, there was a gap between rhetoric and reality.[9]

Late nineteenth-century chemists' held their academic chairs with a serious lack of experimental equipment and few international collaborations.[10] Standard training for the chemical industry was elusive and poorly defined in universities and technical schools, and traditional apprenticeships in the factory usually prevailed.[11] Weak demand from the chemical industry did not help to encourage a more robust, professional chemistry education.[12] The attribution of the 1898 Spanish colonial crisis to the

---

[8] Antonio Moreno, *Una ciencia en cuarentena. Sobre la física en la universidad y otras instituciones académicas desde la Ilustración hasta la crisis finisecular del XIX* (Madrid: CSIC, 1989).

[9] Siro Arribas, *Introducción a la historia de la química analítica en España* (Oviedo: Universidad de Oviedo, 1985).

[10] Eugenio Mascareñas (1853–1934), José Rodríguez Carracido (1856–1928), José Ramón Luanco (1825–1905), José Rodríguez Mourelo (1857–1932) and Antonio Casares (1812–88). For biographical profiles of these chemists, see: Jordi Mora, *Unidad de la materia y diversidad ideológica. Discursos ontológicos en la España de la segunda mitad del siglo XIX*. PhD thesis (Bellaterra: Universitat Autònoma de Barcelona, 2014). For the Casares saga, see: Ignacio Suay-Matallana, *Análisis químico y expertos en la España contemporánea Antonio Casares Rodríguez (1812–1888) y José Casares Gil (1866–1961)*. PhD thesis (Valencia: Universitat de València, 2014). See also: Ignacio Suay-Matallana, 'Between Chemistry, Medicine and Leisure: Antonio Casares and the Study of Mineral Waters and Spanish Spas in the Nineteenth Century', *Annals of Science*, 73(3) (2016), 289–302.

[11] James F. Donnelly, 'Representations of Applied Science: Academics and Chemical Industry in Late-Nineteenth-Century England', *Social Studies of Science*, 16 (1986), 195–234. James F. Donnelly, 'Defining the Industrial Chemist in the United Kingdom, 1850–1921', *Journal of Social History*, 29 (1996), 779–96.

[12] Jordi Nadal, 'La debilidad de la industria química española en el siglo XIX. Un problema de demanda', *Moneda y Crédito*, 176 (1986), 33–70.

scientific backwardness of the country also did not help, leading to a regenerationist movement that combined nationalist pride after the 1898 humiliation with a dream of modernity through ambitious educational reforms and the internationalisation of Spanish culture (and science). This was a project in which the material progress of chemistry in terms of new laboratories, research schools and foreign collaborations had its say in the new political ambitions of the early twentieth century. Chemistry therefore became a clear materialisation of the country's dream of modernity.[13]

It was in that climate of reaction against the pessimism of 1898 that, in 1903, a group of chemists (together with physicists) founded the *Real Sociedad Española de Física y Química* (RSEFQ) and soon launched a new journal, the *Anales de Real Sociedad Española de Física y Química* (*ARSEFQ, Anales*), which can be considered landmarks in the construction of the identity of Spanish chemists in the early twentieth century. RSEFQ membership soon rose to close to 300 in the early 1910s, approached 500 fellows in the 1920s and reached more than 1,000 in the early 1930s.[14] The RSEFQ paved the way for the new generation of chemists of the Silver Age, who achieved considerable international prestige. In the period 1903–37, the journal *Anales* published around 560 different academic authors, 26 of whom were women.[15]

In addition to the RSEFQ, those early years of the twentieth century witnessed the creation of other scientific societies in which chemistry had a role. The *Junta para Ampliación de Estudios e Investigaciones Científicas* (JAE), as a new government body to promote high-quality research and education, was founded in 1907, while the *Institut d'Estudis Catalans* (IEC) was originally conceived in the same year as an ambitious research institution by the political Catalanism in Barcelona.[16] One year later, in

---

[13] On the controversy around the '*polémica de la ciencia española*', see: Agustí Nieto-Galan, 'The Images of Science in Modern Spain. Rethinking the "Polémica"', in Kostas Gavroglu (ed.), *The Sciences in the European Periphery during the Enlightenment* (Dordrecht: Kluwer, 1998), pp. 65–86.

[14] For accurate figures of membership at the RSEFQ up to 1930, see: Camen Magallón, *Pioneras españolas en las ciencias. Las mujeres del Instituto Nacional de Física y Química* (Madrid: CSIC, 2004), p. 132.

[15] Antonio Moreno, 'A grandes males, grandes remedios. Una sociedad española para el adelanto de la física y la química', *Anales*, 2ª etapa, abril–junio, 2003, 244–65, p. 255. See, for example, R. D. Coghill, Dorotea Barnés, 'Estudio del ácido nucleico del bacilo de la difteria', *Anales*, 30 (1930), 208–21. The research was carried out at the Sterling Chemical Laboratory of the University of Yale.

[16] Antoni Roca-Rosell, 'Ciencia y sociedad en la época de la Mancomunitat de Catalunya (1914–1923)', in José Manuel Sánchez Ron (ed.), *Ciencia y sociedad en España: De la Ilustración a la Guerra Civil* (Madrid: El Arquero/CSIC, 1988), 223–52, p. 229.

1908, the *Asociación Española para el Progreso de las Ciencias* (AEPC) was founded and began a successful series of itinerant conferences.[17]

Early significant chemistry papers in the *Anales* during the first decades of the century were those by José Muñoz del Castillo (1850–1926) on radioactivity, Juan Fagés (1862–1911) on the theory of acidity, Eduardo Vitoria (1864–1958) on magnesians and organic synthesis and José Prats (1873–?) on synthetic dyestuffs.[18] The RSEFQ also welcomed honorary members such as Svante Arrhenius (1859–1927), awarded the Nobel Prize for Chemistry in 1903 and one of the key founders of physical chemistry through his theory of ionic dissociation; Henry-Louis Le Chatelier (1850–1936), a leading figure in the study of chemical equilibrium; Georges Urbain (1872–1938), a prestigious professor of chemistry at the Sorbonne in Paris; Heike Kamerlingh-Onnes (1853–1926), also a Nobel laureate for his studies on low temperatures; and Richard Willstätter (1872–1942), a prestigious organic chemist and Nobel Prize winner in 1915 for his studies on plant pigments. Physicist and Nobel Prize winner in 1902, Pieter Zeeman (1865–1943), and Santiago Ramón y Cajal (1852–1934), the recent Spanish Nobel Prize winner in Physiology or Medicine in 1906, completed the list of prestigious names.[19] The positive impact of Cajal's Nobel Prize and the foundation in 1907 of the JAE led to a new dynamism in the chemical community.

As described in detail in the JAE archive, from 1907 to 1936, around 100 chemists enjoyed fellowships for their academic education abroad, working in first-class research schools and labs under the leadership of several Nobel laureates in chemistry.[20] A new generation of young chemists, who soon came together through the RSEFQ and published regularly in *Anales*, reported on their international experiences in prestigious research schools abroad and prompted ambitious educational reforms for chemical training. In the liberal spirit of the JAE, the ideal that a new, more modern chemistry could contribute to the modernisation of a wounded post-1898 country, in a time of social agitation and political

---

[17] Elena Ausejo, *Por la ciencia y por la patria. La institucionalización científica en España en el primer tercio del siglo XX: la Asociación Española para el Progreso de las Ciencias* (Madrid: Siglo XXI, 1993). See also: Elena Ausejo, 'La Asociación Española para el Progreso de las Ciencias en el Centenario de su creación', *Revista Complutense de Educación*, 19(2) (2008), 295–310.

[18] On the life and work of Muñoz del Castillo, see: Néstor Herrán, *Aguas, semillas y radiaciones: el laboratorio de radiactividad de la Universidad de Madrid, 1904–1929* (Madrid: CSIC, 2008). On Vitoria and Prats, see the 'Useful Chemistry' section of this chapter.

[19] Antonio Moreno, 'A grandes males'.

[20] JAE archive: http://archivojae.edaddeplata.org/jae_app (last accessed 15 March 2017). See also: Luis Enrique Otero Carvajal, José María López Sánchez, *La lucha por la modernidad. Las ciencias naturales y la Junta para Ampliación de Estudios* (Madrid: Residencia de Estudiantes, 2012).

decadence, progressively gained support among the chemical community. In the 1920s, Primo de Rivera's dictatorship and its industrially based, state-driven, authoritarian conception of modernity and the nation-state seemed also to be in tune with these innovations in chemistry, especially if they were efficiently applied to industry.[21]

In the last few decades of the nineteenth century, other European countries faced similar debates on chemical backwardness in the face of overwhelming German hegemony in academia and in industry.[22] But the Spanish case – at least in its public rhetoric – embodied a specific dramatism. The JAE's ambitious plan, from which a young generation of chemists benefitted notably, struggled to counterbalance public pessimism. Moreover, the construction of a new European chemical network became a crucial goal of the academic and political authorities in the last decades of the Bourbon Restoration and Primo de Rivera's dictatorship, reaching a peak in the early 1930s with the Second Spanish Republic. At the same time, that dream of modernity had an industrial dimension, which also merits further examination.

Although the idea of the rise of modern chemistry can be easily associated with the nineteenth-century post-Lavoisierian times of professionalisation in universities and industries in many European countries,[23] modernity was perceived by contemporary Spanish authors as arriving late. Opposing that trend, chemists tried to increase their influence in the science, pharmacy and medicine faculties; they were also keen to improve theoretical and experimental training and to find new ways to efficiently appropriate foreign chemistry-based industry patterns in order to create further industrialisation. Above all, and perhaps embracing all of the rest, early twentieth-century Spanish chemists desperately searched for internationalisation, for the opening up of new European networks in order to exchange formulae, experiments, laboratory equipment, chemistry students and professors, for a new cosmopolitan conception of the profession that questioned the

---

[21] Alejandro Quiroga, *Making Spaniards. National Catholicism and the Nationalisation of the Masses during the Dictatorship of Primo de Rivera (1923–1930)*. PhD thesis (London: London School of Economics and Political Science, 2004): https://core.ac.uk/download/pdf/46518522.pdf (last accessed 15 March 2017).

[22] For the Franco-German scientific rivalry, see: Chris Manias, 'The *Race prussienne* Controversy: Scientific Internationalism and the Nation', *Isis*, 100(4) (2009), 733–57. See also: Georg Meyer-Thurow, 'The Industrialization of Invention: A Case Study from the German Chemical Industry', *Isis*, 73(3) (1982), 363–81. Robert Fox, *Science Without Frontiers: Cosmopolitanism and National Interests in the World of Learning, 1870–1940* (Corvallis: Oregon State University Press, 2016).

[23] David Knight, Helge Kragh (eds.), *The Making of the Chemist: The Social History of Chemistry in Europe, 1789–1914* (Cambridge: Cambridge University Press, 1998).

backward, isolated image of a scientific culture, which seemed to have lost its place in the making of modern nations.

Cosmopolitanism

In the inaugural address for the 1899–1900 academic year, Eugenio Mascareñas (1853–1934), at that time Professor of Inorganic Chemistry at the University of Barcelona, called for urgent improvements in the training of chemistry undergraduates in terms of experimental laboratory skills.[24] Mascareñas denounced the 'regrettable state of experimental sciences in … [the] country and their longstanding, punishable and shameful neglect'.[25] He also stressed the uselessness of the creation of new chemistry chairs under the label of 'practice', since they lacked well-equipped labs and simply became supplementary exercises for the professors' extolment of theories.[26] José Casares Gil (1866–1961), one of the most prominent figures in Spanish chemistry at the turn of the century, seconded these pessimistic overtones. In 1911, Casares denounced once again the lack of experimental training and highlighted the need for better-equipped labs.[27] From his period of training abroad, Casares relied on several German textbooks to guide the experimental work,[28] and also pointed out the close link between state investment, chemical research and political power at the international level: 'The state invested capital for the promotion of chemistry, and in general, for the physical sciences, produces the highest revenues. In practice, today *rivalry among nations is a rivalry about science, in particular, a competition on the applications of chemistry*'.[29]

In fact, national rivalry among European nations and the German chemical hegemony surely contributed to the outbreak of the First World War, but Spain remained neutral and took advantage of its status to increase supplies that provided interesting revenues for the chemical

---

[24] Eugenio Mascareñas, *Consideraciones generales acerca de la enseñanza y estudio particular del estado en que se halla la de las Ciencias Experimentales en España* (Barcelona: Hijos de Jaime Jepús, 1899), p. 65.

[25] Ibid., p. 58. [26] Ibid., p. 65.

[27] José Casares, *El problema de la investigación científica en España. Discurso inaugural: Sección 3ª, Ciencias Fisicoquímicas* (Madrid: Asociación Española para el Progreso de las Ciencias, 1911), p. 531.

[28] Hans Pechmann, *Volhards Anleitung zur qualitativen chemischen Analyse* (Munich: Chemisches Laboratorium des Staates, 1901). A Spanish version was translated from German by García Banús: *Introducción al análisis químico cualitativo según Volhard* (Madrid: Calpe, 1921); Casares also referred to other titles such as: *Anleitung zur Quantitiven chemischen Analyse, Anleitung zur darstellung organiche preparate* and *Die Praxis des organisehen Chemikers*.

[29] José Casares, *El problema de la investigación científica*, p. 562 (emphasis added).

industry.[30] In spite of this, at the end of the war, laments of backwardness continued, as the prestigious pharmacist and organic chemist Obdulio Fernández (1883–1982), one of the founding members of the RSEFQ, publicly demonstrated in his 1918 speech at the *Real Academia de Ciencias Exactas, Físicas y Naturales* (RACEFN) in Madrid.[31] Fernández described the failure to produce aniline black dye in Spain during the war and the lack of academic and industrial resources to fill the gap in the dyestuffs market, which had been controlled until then by German firms:

> We lacked aniline black, among other reagents, and suddenly wanted to produce it by purchasing aniline; but since this was not found, the idea arose of purchasing nitrobenzene and reducing it to obtain aniline. But since nitrobenzene was not available either, not even for perfuming bleaches it became essential to make it using nitric acid and benzene, which could not be found either, and so disappointment spread as future manufacturers realised that aniline black, and other important chemical products, were relegated to sixth or seventh place in a chain which nobody had intended to construct.[32]

In 1924, chemist José Uhthoff, Professor of Organic Chemistry at the *Institut de Química Aplicada* (IQA), a Barcelona technical school for the training of industrial chemists, published a devastating paper.[33] Uhthoff wondered if 'our' chemists were different from the others, and his answer was: 'Yes, mostly very different, and except for honourable exceptions, *very bad*'.[34] After that provocative statement, he provided an exhaustive list of Spanish chemists' weaknesses: lack of theory, delay in the arrival of the latest innovations, resistance to novelty, weak experimental training, failure of industrial projects, lack of original papers and lack of teamwork in the lab and at the institutional level. Uhthoff went even further, describing the main causes of that backwardness as being intrinsic to the Spanish, 'southern' character. In his view, chemistry was a strongly rational science that required patience, constancy and teamwork, all being opposite virtues to the selfish, over-passionate, superficial Spanish behaviour. The argument was in tune with contemporary debates on the

---

[30] Francisco Romero, 'Spain and the First World War', in Sebastian Balfour, Paul Preston (eds.), *Spain and the Great Powers in the Twentieth Century* (London: Routledge, 1999), pp. 32–52.

[31] Obdulio Fernández, *Influencia de la industria en el desarrollo de la ciencia pura. Los Laboratorios de ensayos* (Madrid: RACEFN, 1918). Quoted by Raúl Rodríguez Nozal, 'En torno al establecimiento de una industria químico-farmacéutica en España con anterioridad a la Guerra Civil', in *Actas VIII Congreso de la Sociedad Española de Historia de las Ciencias y de las Técnicas* (Logroño: Univeridad de la Rioja, 2004), 739–55, pp. 743–4.

[32] O. Fernández, *Influencia de la industria*, pp. 14–15.

[33] José Uhthoff, 'El químico en nuestro país. Lo que es y lo que debe ser', *Química e Industria*, I(2) (1924), 39–41.

[34] Ibid., p. 39 (emphasis added).

causes of the so-called '*polémica de la ciencia española*', or controversy of Spanish science, which had, since the late eighteenth century, described Spaniards as being unskilled for the practice of experimental science.[35] Uhthoff believed that the country suffered from very weak educational scientific training that provided students with only superficial explanations of chemical theories, which was combined with badly equipped, dirty, poor-quality labs. The problem was said to lie in a lack of efficient educational policies: badly paid lecturers and ineffective reforms. In Uhthoff's view, the modernisation of chemistry had to be an example for the modernisation of the country – a way to produce new experts and professionals beyond the old habits and routines.[36]

Mascareñas, Casares, O. Fernández, Uhthoff and many others used the rhetoric of backwardness for the sake of the improvement of their professional status, academic authority and social recognition, which were weak both inside and outside the science faculties. However, they did contribute in practice to the establishment of chemistry as a more 'modern' profession in terms of international standards, and they steered government policies and industrial efforts to place them at the core of the Silver Age. The idea of better training (for teachers and students) in better labs with international standards was at the heart of the founding spirit of the JAE, which provided predoctoral and postdoctoral fellowships for research stays abroad (*pensionados*) and invitations to foreign scientific luminaries (with frequent visits from Nobel Prize winners) to give research seminars and full courses in Spain. The JAE also envisaged ambitious educational reform and the introduction of modern foreign-style research and teaching laboratories.

As far as chemistry is concerned, files of 95 *pensionados* and more than 200 fellowships during the period 1907–39 – in other words, from the foundation of the JAE to the end of the Spanish Civil War – are well conserved in the archives.[37] The geography of the chemistry *pensionados* unveils the international networks in which Spanish chemists were inserted in the early decades of the twentieth century. Germany (mainly

---

[35] Nieto-Galan, 'The Images of Science in Modern Spain'.
[36] Uhthoff, 'El químico en nuestro país', p. 41. Along a similar line, geographical, cultural and even racial determinism had been used to explain this chemical backwardness some years earlier, in 1918, by pharmacist and later prominent political figure José Giral, who, in a public address at the University of Salamanca, had stressed the Latin lack of patience, organisation and dedication to experimental work in the lab. José Giral, *Industrias químicas. Discurso leído en la solemne apertura del curso académico de 1918 a 1919* (Salamanca: Universidad de Salamanca, 1918), p. 62.
[37] Residencia de Estudiantes. *Archivo de la Junta para Ampliación de Estudios e Investigaciones Científicas (1907–1939)* (JAE Archive): http://archivojae.edaddeplata.org/jae_app/JaeMain.html (last accessed 15 December 2018).

Berlin, Munich and Leipzig), Switzerland (Zurich and Geneva), France (Paris), the UK (London, Oxford and Edinburgh) and the USA (Washington, DC and New York) led the ranking of the countries in which these Spanish chemists were trained. Particular cities, university departments and research institutes acquired specific importance. This was the case, for example, of Willstätter's laboratory at the *Eidgenössische Technische Hochschule* (ETH), the Federal Polytechnic Institute in Zurich,[38] Urbain at the Sorbonne in Paris, Ernest Fourneau (1872–1949) at the Pasteur Institute also in Paris and Heinrich Wieland (1877–1957) at the University of Munich, and these became key nodes in the network. But other major names such as Wilhelm Ostwald (1853–1932) in Leipzig, Phillippe Guye (1862–1922) in Geneva, Christopher Ingold (1893–1970) in London, Robert Robinson (1886–1975) and Frederick Daniel Chattaway (1860–1944) in Oxford, Hermann Staudinger (1881–1965) in Zurich and Paul Sabatier (1854–1941) in Toulouse also contributed to that new cosmopolitan 'republic of chemistry', as it was perceived from the Spanish perspective.

Willstätter, the Nobel Prize winner for chemistry in 1915 for his work on plant pigments, especially chlorophyll, had been a student of Adolf von Bayer (1835–1917) and a contemporary of Emil Fischer (1852–1919) and Fritz Haber (1868–1934), and he played a fundamental role in the chemistry of natural products and the early development of synthetic organic chemistry. Willstätter soon became one of the central figures and a moral and political reference for many Spanish liberal chemists. He welcomed numerous *pensionados* to his laboratory and visited Spain several times until the 1930s (see Chapter 2). In 1917, during the First World War, Georges Urbain managed to travel to Spain,[39] to give a course in Madrid on coordination complexes over 15 lectures, which chemist Ángel del Campo (1881–1944) translated and later published.[40] Urbain's stay represented a great opportunity for del Campo to introduce himself to Alfred Werner's (1866–1919) coordination chemistry.[41] A decade earlier, del Campo had been a *pensionado* with Urbain at the Laboratory of Mineral Chemistry at the Sorbonne in Paris, and in 1912, he had translated William Ramsay's (1852–1916) textbook of general chemistry into Spanish, with Ramsay at that time being a great international luminary due to his earlier

---

[38] On the early origins of the ETH, see: Peter Ramberg, 'Chemical Research and Instruction in Zurich, 1831–1871', *Annals of Science*, 72(2) (2015), 170–86.
[39] *La Vanguardia*, 5 January 1917, p. 10.
[40] Among others, León Le Boucher and Emilio Jimeno attended the course.
[41] In 1913, Alfred Werner was awarded the Nobel Prize for Chemistry 'in recognition of his work on the linkage of atoms in molecules by which he has thrown new light on earlier investigations and opened up new fields of research especially in inorganic chemistry': www.nobelprize.org/prizes/chemistry/1913/summary (last accessed 19 March 2019).

discovery of noble gases.[42] Urbain offered his course to what he regarded as a bright, promising audience of university undergraduate and graduate students. He praised the scientific collaboration between Madrid and Paris as a clear sign of the scientific 'revival' of the country.[43] After gaining a chair in analytical chemistry at the Faculty of Pharmacy in Madrid, in 1915, del Campo began his collaboration with Miguel Catalán (1894–1957). Both men strengthened their joint venture after Urbain's visit. It preceded Catalán's stay at Imperial College London with Alfred Fowler (1868–1940), Professor of Astrophysics, which led him to the internationally acknowledged discovery of multiplets – a crucial step towards the link between spectral lines and the complex electronic structure of atoms – and some remarkable work on spectroscopy.[44]

Fourneau visited Spain in 1917. He was at that time the director of the Laboratory of Therapeutic Chemistry at the Institute Pasteur in Paris. His mastery also encouraged the younger generation of Spanish organic chemists who aimed to link organic synthesis with pharmacological applications. Invited by the JAE and the Faculty of Pharmacy, he gave a theoretical–practical course that Madinaveitia, one of the leaders in organic chemistry, later translated and published as *Síntesis de medicamentos orgánicos* – a short, widely read book in scientific circles.[45] Similarly, Wieland, Nobel Prize winner for chemistry in 1927 and at that time director of the Chemistry Laboratory of the University of Munich, visited Spain that same year to lecture at the universities of Barcelona, Madrid and Seville. In Madrid, he talked about the chemistry of bile acids – his main research topic, for which he received the Nobel Prize – and the alkaloid 'lobelia'. At the RSEFQ, Wieland lectured on several redox processes.[46] Wieland was another key player in the international network of silver-age chemists. José Pascual and Fernando Calvet, two of García Banús's most brilliant pupils, were Wieland's *pensionados* in the 1920s, and later became leading figures who influenced the way in which chemistry research evolved in Spain in the following decades.[47]

---

[42] William Ramsay, *Química moderna teórica y sistemática . . . Traducción del inglés por Ángel del Campo Cerdán* (Madrid: Romo, 1912).
[43] Georges Urbain, *Conferencias acerca de la química física de los complejos minerales dados en la Facultad de Ciencias de la Universidad de Madrid* (Madrid: Eduardo Arias, 1917), p. xi.
[44] José Manuel Sánchez Ron, *Miguel Catalán. Su obra y su mundo* (Madrid: CSIC/Fundación Ramón Menéndez Pidal, 1994).
[45] Ernest Fourneau, Antonio Madinaveita, *Síntesis de medicamentos orgánicos* (Madrid: Calpe, 1921). See also: Manuel Lora-Tamayo, *La investigación química española* (Madrid: Alhambra, 1981), p. 139. José Ranedo and Cándido Torres also worked with Fourneau in Paris.
[46] *Anales*, 27 (1929), 140–1.
[47] 'Fernando Calvet Prats', *Anthropos*, 35 (1984), 1–27. Calvet submitted his DPhil at Queen's College, Oxford, in 1927 'on the condensation products of phenolic substances

Wieland held a chemistry chair in Freiburg from 1921 to 1935 and worked on addition reactions, dehydrogenation processes and the catalytic uses of metals, azo-compounds and alkaloids.[48]

José Sureda (1890–1984) epitomised the profile of the best JAE *pensionados*.[49] In the 1913–14 academic year, Sureda joined Wieland in Munich to work on benzene derivatives and hydracines. He also carried out some practical exercises on dyestuffs, and later visited several dyestuff factories. In 1916, Sureda had a new *pension* in Geneva with Professor Amé Pictet (1857–1937), and later went back to Zurich to work with Staudinger – the luminary of the chemistry of macromolecules, and winner of the Nobel Prize for chemistry in 1953 – on some diazo compounds. During the 1916–17 academic year, Sureda took several courses in Zurich: organic chemistry, carbohydrates, war chemistry, alkaloids and atomic structure.[50] Graduate courses, research training and industrial visits often constituted a standard programme for the internationalisation of the young *pensionados* abroad, which also included publications in foreign journals on new research achievements, later translated into Spanish to appear in the *Anales*.

Aiming to strengthen links with industry, José Agell (1882–1973) was also keen to welcome foreign luminaries to give lectures at his technical school in Barcelona. By 1914, he had invited Dieter Delarue, director of *Aktiengesellschaft für Anilinfabrikation* (AGFA) in Berlin, who had escaped from France during the war. From November 1914 to March 1915, Delarue taught a full course on chemical synthesis, with daily lectures and flexible time in the lab to make dyestuffs, drugs and perfumes. Up until the mid-1920s, Agell invited other prestigious foreign lecturers such as Jean Perrin (1870–1942), the 1926 Nobel Prize winner for physics, Giuseppe Bruni (1873–1946), the director of the prestigious *Politecnico di Milano*, Nobel Prize winner Paul Sabatier, Antonio de Gregorio Rocasolano (who had also orientated his research towards industrial

---

with chloral. Many of these decompose under the influence of reagents in a variety of ways and Mr. Calvet has shown great skill in not only elucidating the structure of the original condensation products, but also in following out the complex changes they undergo ... although the fundamental idea of the work originates with Dr. Chattaway, the actual investigation was initiated by the candidate'. Report of the examiners from the viva voce examination, 2 December 1927. Queen's College. University Archives, Oxford, UR2/8, p. 9.

[48] Bernhard Witkop, 'Remembering Heinrich Wieland (1877–1957), Portrait of an Organic Chemist and Founder of Modern Biochemistry', *Medicinal Research Reviews*, 12(3) (1992), 195–274. James K. Laylin (ed.), *Nobel Laureates in Chemistry, 1901–1992* (Washington, DC: ACS/CHF, 1993), pp. 164–8.

[49] Josep Sureda, *Antologia científica* (Palma de Mallorca: Conselleria d'Innovació i Energia, 2003).

[50] JAE Archive. JAE/140-670.

applications at the University of Zaragoza), Wilhelm Ostwald's son, Karl Wilhelm Wolfgang (known as Wo. Ostwald), and Enrique Moles. Ostwald gave five lectures on colloid chemistry as a new branch of physical chemistry.[51]

Fighting against backwardness also included the active participation of several chemists in international chemical societies. For instance, this was the case for Moles's active role at the International Union of Pure and Applied Chemistry (IUPAC), particularly in international discussions for reaching a consensus on atomic weights, measures and constants.[52] In the early years of his career, Moles was already aware of the first international attempts to create a committee on atomic weights. In 1921, at the Brussels Conference, he was already active in the Atomic Weights Commission of the recently founded IUPAC. A report by the Spanish Delegation – made up by physicist Blas Cabrera and chemists del Campo, José Rodríguez Mourelo (1857–1932) and Moles himself – was enthusiastically welcomed by the other delegates.[53] In the following years, Moles played an important role in further IUPAC conferences. Similarly, O. Fernández regularly attended IUPAC meetings. Together with Moles, he travelled to the 7th IUPAC Conference in Washington, DC, and the 72nd Conference of the American Chemical Society (ACS) in Philadelphia, both in 1926. Both men then took the chance to visit the Sesquicentennial International Exhibition of Philadelphia and several chemical labs and industries across the country. They also visited Theodore Richards – Gilbert Newton Lewis's (1875–1946) mentor – in Harvard.[54] In the same cosmopolitan spirit, *pensionados* abroad often became members of national chemistry societies for the sake of their integration into the new scientific community. For instance, this was the case for García Banús, who, during his stay at the ETH in Zurich, was admitted to the *Deutsche Chemische Gesellschaft* as a foreign member.[55]

---

[51] 'El curso del professor Ostwald en el Instituto de Química aplicada de Barcelona (Mayo 1924)', *Química e Industria*, 1(9) (1924), 221–5.
[52] On the history of the IUPAC, see: Roger Fennell. *History of IUPAC, 1919–1987* (Oxford: Blackwell Science, 1994).
[53] Blas Cabrera, Ángel del Campo, José Rodríguez Mourelo, Enrique Moles, 'Informe de la Comisión española de pesos atómicos', *Anales*, 20 (1922), 25–33; 21 (1923), 57–66; 22 (1924), 367–81.
[54] Obdulio Fernández, Enrique Moles, 'La reunión de la Sociedad química americana y la 7ª Conferencia de la Unión internacional de química: Un viaje a Norteamérica', *Anales*, 24 (1926), 684–706.
[55] Agustí Nieto-Galan, 'Free Radicals in the European Periphery: "Translating" Organic Chemistry from Zurich to Barcelona in the Early Twentieth Century', *The British Journal for the History of Science*, 37 (2004), 167–91, p. 172.

The JAE project for the modernisation of chemistry enjoyed continuity from the last decade of the Bourbon Restoration to Primo de Rivera's dictatorship. The latter's industrialist rhetoric of modernisation coexisted with the cosmopolitan JAE policies of the internationalisation of chemistry training. The apparent 'neutrality' of chemistry paved the way, with the Chemistry Palace (*Palacio de la Química*) at the 1929 Barcelona International Exhibition being probably the clearest example of the contribution of the profession to the military regime. Moreover, Primo de Rivera continued earlier negotiations with the Rockefeller Foundation for the creation of a new research institute for physics and chemistry in Madrid. At the same time, however, liberal chemists such as José Giral and Antonio Madinaveitia were politically active against what they perceived as the backward and corrupt structures of the monarchy.

Laboratories and Schools

As has been recently described by historian Peter Morris, the laboratory plays a central role in the way in which a specific culture of chemistry grows in a particular historical context. In its broad conception, the chemistry laboratory includes the buildings but also the fittings, so the material culture of the lab tells us a lot about the kind of chemistry practiced there.[56] In the Spanish case, in spite of reforms of the curricula, the miserable situation in the science faculties, but also the poor lab equipment in engineering and technical schools, acted as a driving force for the numerous attempts to modernise.

In 1911, Casares considered the creation of new laboratories outside the university to offer flexible experimental training at different levels. This needed to be a fully practical endeavour in which students would begin with qualitative essays, later moving on to quantitative experiments, to be followed by inorganic and organic synthesis. With 20–30 benches, the labs would not serve particular industrial needs, nor would they provide degrees, but they would simply serve for training students. The labs would help future candidates for JAE *pensionados*, who needed more solid training before entering a foreign research group. They would also welcome any chemistry trainee who could cover the fees for all experiments.[57] Casares' idea was not far from the aims of the new labs of the JAE, which opened throughout the 1910s. They included new chemistry premises that were also used for the complementary training

---

[56] Peter J. T. Morris, *The Matter Factory: A History of the Chemistry Laboratory* (London: Reaktion Books, 2015).
[57] Casares, *El problema de la investigación científica*.

of young chemists. The new laboratories at the *Residencia de Estudiantes* in Madrid covered general chemistry, microscopic anatomy, physiological chemistry, general physiology, histology, physiology of the nervous centres, serology and radiology. The general chemistry lab offered two general training courses for 22 students covering experiments on glasswork and apparatus, inorganic synthesis, analytical problems, elementary organic analysis, separation processes, experimental measurement of constants, organic synthesis and volumetric and gravimetric analysis.[58]

In addition, the *Laboratorio de Investigaciones Físicas* (LIF) began its work in 1910. Although physics (metrology, electricity and spectroscopy) originally monopolised its research lines and experimental areas, the physical chemistry section opened the door to fruitful collaborations with chemists. Names such as del Campo, Julio Guzmán (1883–1956) and Mariano Marquina, among others, found their place in that new experimental setting. In the period of 1918–20, Marquina enjoyed a spell as a *pensionado* at Columbia University, New York, working on the synthesis of camphor and attending courses in industrial chemistry.[59] Later, organic chemistry labs joined the LIF, in which Madinaveitia began to make his mark.[60] In the 1920s, the LIF focused on magnetochemistry, atomic weights – a field that gave Moles and his pupils great international recognition – and spectroscopy, which allowed the development of several interdisciplinary alliances such as the successful joint venture between del Campo and Catalán.[61]

In Barcelona, at the *Escola Industrial*, chemistry professors Antonio Ferran and Agell had managed to convince local officials to support the building of new labs. In fact, the poor lab equipment in the capital of Catalonia also caused serious concerns, since its powerful textile firms acted as a driving force for the chemical industry, and demand for experimental training in industrial chemistry was high. Once settled in the

---

[58] In 1912, José Sureda and Julio Blanco (later, in 1913, José Ranedo and Juan B. Sánchez) were in charge of the general chemistry lab. The microscopic anatomy lab was led by Luis Calandre. In 1915, a new physiological chemistry lab was put in the hands of Antonio Madinaveitia and J. R. Sacristán. In 1916, the general physiology lab was directed by Juan Negrín and the lab for the physiology of nervous centres was put in hands of D. G. R. Lafora. In 1919, the histology lab was led by Pío del Río Ortega, and in 1920, the serology and radiology labs were led by Paulino Suarez. 'Los Laboratorios de la Residencia', *Residencia. Revista de la Residencia de Estudiantes*, 1 (1934), 26–30. See also: Santos Casado, 'Ciencia y conciencia bajo los tilos. Los laboratorios de la Residencia de Estudiantes y el exilio de 1939', in *Los científicos del exilio español en México* (Morelia: Universidad Michoacana de San Nicolás de Hidalgo: SMHCT, SEHCT, 2001), pp. 125–46.
[59] JAE Archive. JAE/92-172.
[60] Biological chemistry was still in the hands of the old master, José Rodríguez Carracido.
[61] On Catalán, see Sánchez Ron, *Miguel Catalán*.

basement of the school, the new lab officially became the *Laboratori d'estudis superiors de química*. With 1500 m² of floor space, three sections were devoted to qualitative, gravimetric and volumetric analysis, and others to organic chemistry, synthesis, electricity, microscopic metallography and microscopy. Students had to work in the lab for a semester carrying out a set of fairly flexible experiments to complete their theoretical training. That seminal lab soon became the starting point for Agell's more ambitious project, the *Escola de Directors d'Indústries Químiques*, aiming to train future managers of the chemical industry.[62]

At the university level, there were also major moves towards the modernisation of the chemistry labs. García Banús's outstanding training in organic chemistry as a *pensionado* at the ETH Zurich, particularly his work on free radical chemistry,[63] helped him to obtain a chair in organic chemistry at the University of Barcelona in 1915. He used the pattern he had become familiar with at the ETH as his local strategy to establish experimental organic chemistry at his own faculty.[64] He created a new laboratory that, under the name *Laboratorio de Química Orgánica de la Facultad de Ciencias de Barcelona*, became the natural place for his research school.[65] New vacuum reaction flasks, filters, inert gases, cryoscopy apparatus and glassware for photochemical reductions revolutionised the lab and enabled him to train his students in vacuum distillation, microflask usage and recrystallisation techniques.[66] Since free radicals required an inert atmosphere and very strict experimental conditions, Banús imported new chemical instruments from Zurich.[67] Many of his pupils abroad published prestigious papers in international journals, and as *pensionados* gained access to some of the main nodes of the international network, particularly Wieland's group at the University of

---

[62] Alexandre Galí, *Història de les institucions i del moviment cultural a Catalunya, 1900–1936* (Barcelona: Fundació Alexandre Galí, 1981), I, p. 213.
[63] With Julius Schmidlin.
[64] John H. Brooke, 'Methods and Methodology in the Development of Organic Chemistry', *Ambix*, 34 (1987), 147–55.
[65] For a broader perspective of organic chemistry in Spain in the early decades of the twentieth century, see: Juan Vergara, *La química orgánica en España en el primer tercio del siglo XX*. PhD thesis (Valencia: Universittat de València, 2004).
[66] Antonio García Banús, 'Estudio sobre los derivados organomágnesicos IV. Contribución al estudio de los difenilisocromanos', *Anales*, 26 (1928), 372–98. In 1925, Banús publíshed a historical paper in *Anales*, in which he described the history of free radicals. Starting from Lavoisier, and mentioning names such as Liebig, Wöhler, Frankland and Kolbe, he ended with the tribiphenylmethyl molecule of Gomberg and Schlenk and placed himself at the end of the timeline. Antonio García Banús, 'Los radicales libres en química orgánica', *Anales*, 23 (1925), 147–63. Nieto-Galan, 'Free Radicals'.
[67] García Banús, 'Los radicales libres en química orgánica', p. 153.

Freiburg.[68] Masriera, another key name in García Banús's school, had also worked at the ETH with Hermann Staudinger.[69]

In 1929, Moles was also able to celebrate the construction of new laboratories for physics and chemistry at the Faculty of Science of the Central University in Madrid (Figure 1.2). Just two years after his appointment as full professor of inorganic chemistry, he had managed to modernise the experimental setting for its teaching and research needs.[70] Some 120 students could be accommodated in the new labs of analytical, organic, inorganic, physical and theoretical chemistry, whereas the new physics labs covered electricity and magnetism, optics and acoustics, under the supervision of physicist Blas Cabrera. The new library held 25 chemistry journals and a wide range of books and journals from Japan, the United States, the USSR, France and Britain, and it also contained the former library of the RSEFQ. Moles was particularly grateful to another chemist, Luis Bermejo, at that time the Rector of the Central University, which had invested 1 million pesetas in the new premises. In fact, those new labs represented a first attempt to find new patterns of construction for a new university building and for the new *Instituto Nacional de Física y Química* (INFQ), already under design (see Chapter 2). The new labs also benefitted from further economic support from private donors, public administrations and private firms.[71]

The modernisation of laboratory culture went hand in hand with the establishment of other internationally orientated research schools as embryos of the emerging scientific culture in the 1910s and 1920s. It was at the LIF that Moles had already begun to build up his own research school of physical chemistry, adapting the knowledge he had gained

---

[68] These were the cases of Juan Ferrer, Juan Guiteras, José Pascual and Fernando Calvet. See, for instance: H. Wieland, E. Honold, J. Pascual, 'Untersuchungen über die Gallensäuren XVII. Über Iso-desoxychlosäure', *Zeitschrift für physiologische Chemie*, 130 (1923), 326–37; H. Wieland, F. Calvet, W. Moyer, 'Contribución al esclarecimiento de las reacciones coloreadas típicas de los alcaloides del "strychnos"', *Anales*, 30 (1928), 50–8; H. Wieland, F. Calvet, 'La metilación del ácido vomicínico y descripción de algunos derivados de la vomicina', *Anales*, 30 (1928), 59–70. In the 1930s, the group attracted other brilliant students such as Francisco Boqué, Manuel Tremoleda, José Ribalta, José Monche and Eduardo De Salas, the latter integrating some of Pascual and Calvet's earlier results in the synthesis of diphenylidenes.

[69] Agustí Nieto-Galan, 'From Papers to Newspapers: Miguel Masriera (1901–1981) and the Role of Science Popularisation under the Franco Regime', *Science in Context*, 26(3) (2013), 527–49.

[70] Enrique Moles, 'Los nuevos laboratorios de química de la Facultad de Ciencias de Madrid', *Anales*, 31 (1929), 33–49.

[71] For example, Doña María Pelayo, *Diputación Provincial de Madrid*, *Sociedad Industrial Asturiana*, *Laboratorios Gamir*, *Sociedad Ibérica del Nitrógeno* and *Sociedad Fábrica de Mieres*.

9. = Facultad de Ciencias. = Laboratorio de Análisis químico.

11. = Facultad de Ciencias. = Laboratorio de Química orgánica.

Figure 1.2 The new chemistry laboratories of the Faculty of Science in Madrid, which were built in 1929. (a) Analytical chemistry; (b) Organic chemistry.
De l'Espagne chimique (bref aperçu). Notes sur l'enseignement et la pratique de la chimie en Espagne. Avril 1934 (Madrid: IX Congreso Internacional de Química, 1934), pp. 16 (9, 11). Reproduced with permission of the *Biblioteca de Catalunya*, Barcelona.

abroad, particularly in Leipzig and Geneva.[72] These were pioneering years for the introduction of the new discipline, and labs were required for new kinds of experiments, such as the determination of atomic weights, which Moles had tried to introduce into the Spanish academic system. At the LIF, Moles was active in teaching graduate courses in physical chemistry in collaboration with Guzmán, as well as in magneto-chemistry with Cabrera.[73] After gaining his chair in inorganic chemistry in Madrid in 1927, he developed a training programme that demanded of his students strict experimental accuracy and encouraged the use of lab notebooks, practical reports, exhaustive bibliographical searches and research seminars.[74] This was just the first stage of Moles's prestigious research school for physical chemistry, which flourished in the 1930s (see Chapter 2).

Concerning other emerging research schools, Rocasolano's case is of particular interest.[75] After his international training in food and agricultural chemistry in Paris, Rocasolano obtained a chair in general chemistry in Zaragoza, and later became Rector in the period 1929–31. He focused on colloid chemistry and the assimilation of nitrogen via bacteria. In 1918, Rocasolano founded the *Laboratorio de Investigaciones Bioquímicas* (LIB) and soon produced important papers in his *Trabajos del LIB*, as well as textbooks on chemistry applied to medicine and the natural sciences. A treatise on biochemistry appeared in 1928, and a year later he published *Química para médicos y naturalistas* with Luis Bermejo.[76] Rocasolano progressively built up a solid research school, with his own lab, a focused research subject and academic publications (its own journal and textbooks). He trained a group of young chemists who, after the Civil War, had a great impact on Franco's new chemical culture (see Chapter 4). Rocasolano also led the foundation of the *Academia de Ciencias de Zaragoza* in 1916 and encouraged frequent contact with corresponding members. His contribution to the collective volume *Colloid Chemistry*,

---

[72] After his stay with Wilhelm Ostwald (1910) at the Physico-chemical Instiute in Leipzig, and with Charles-Eugène Guye (1926) in Geneva. Enrique Moles, *'Determinació de pesos moleculars de gasos pel mètode de les densitats límits'*, Enric Moles i Ormella. Traducció, introducció i comentaris de Joaquim Sales i Agustí Nieto-Galan (Barcelona: Societat Catalana de Química, 2013).

[73] Magallón, *Pioneras españolas*, p. 208.

[74] For an updated version of Enrique Moles' life and work, see: Enrique Moles, *'Determinació de pesos moleculars'* ... Traducció, introducció i comentaris de Joaquim Sales i Agustí Nieto-Galan.

[75] José Luis Cebollada, 'Antonio de Gregorio Rocasolano y la Escuela Química de Zaragoza', *Llull*, 11 (1980), 189–216. Tomeo Mariano, *Rocasolano. Notas bio-bibliográficas* (Zaragoza: La Académica, 1941).

[76] Antonio de Gregorio Rocasolano, Luis Bermejo, *Química para médicos y naturalistas* (Madrid: Ramona Velasco, 1929).

Figure 1.3 Enrique Moles (first row, fifth from the left), with the delegates of the 1930 Meeting of the *Real Sociedad Española de Física y Química* in Seville. *Anales*, 28 (1930), 532.
Reproduced with permission of the *Real Sociedad Española de Química*.

Theoretical and Applied (1926)[77] was a sign of the importance of his work for the international chemical community. After the official opening of the new lab in 1920, between the Faculties of Science and Medicine, 12 benches were equipped for work on biochemistry and colloid chemistry.[78] Together, they formed a true research school.

The cases of García Banús, Moles and Rocasolano show how, in the early decades of the twentieth century, a new chemical culture entered university departments in science faculties. Appropriating foreign patterns and benefitting from the *pensionados* policies of the JAE, many of these new group leaders created a genuine culture of chemistry far removed from the backwardness of the beginning of the century, with obvious limitations in terms of international standards, but nevertheless important in terms of the modernisation of the profession (Figure 1.3). If we submit some of these groups to the old prerequisites for a research

---

[77] Jerome Alexander (ed.), *Colloid Chemistry, Theoretical and Applied, by Selected International Contributors* (New York: Chemical Catalog, 1926–50).

[78] Ángel Toca, 'Química en provincias: Antonio Rius Miró en Zaragoza (1922–1930)', *Actes d'Història de la Ciència i de la Tècnica*, 3(1) (2010), 79–91, p. 83.

school defined by Gerald Geison, we soon find that the early groups at the LIF and García Banús's, Moles's and Rocasolano's schools met many of them. They had a charismatic leader, an informal setting with institutional power and social cohesion. In appropriating foreign research cultures, all of these groups defined a focused research programme with rapidly exploitable experimental techniques, and they managed to build up a pool of potential recruits, with their students publishing under their own names and going on to successful professional careers.[79] Although research programmes often had to be adapted to the constraints of the local context to publish, mainly in the *Anales*, and teaching and popularising chemistry often came before research, those new research schools realised a good part of the dream of modernity, and interestingly combined national interests with cosmopolitan values as a way to invigorate the modern chemical community, which often went with the tide of monarchic reforms and Primo de Rivera's dictatorial projects of industrialisation.

Useful Chemistry

In the early decades of the century, 'pure' academic chemistry was not necessarily opposed to industrial projects, which aimed at realising a good part of the dream of modernity. In 1918, in his address at the Madrid Academy of Science, O. Fernández showed great concern for the way in which academic chemistry could effectively contribute to the reorganisation and progress of industry.[80] He asked for a more 'scientific' approach to industrial areas such as metallurgy, food and tanning, and again stressed the need to establish German-like industrial research labs across the country, with the Bureau of Standards in Washington, DC, being another useful pattern to imitate.[81] With a nationalist overtone, in which chemical hegemony was again closely related to political hegemony among nations, O. Fernández referred to the British publication, *Science and the Nation* (1917),[82] which, under the pressure of the First World War, considered the distinction between 'pure' and 'applied' science as

---

[79] Gerald L. Geison, Frederick L. Holmes (eds.), *Research Schools. Historical Reappraisals. Osiris. Second Series*, 8 (1993).

[80] Obdulio Fernández, *Discurso leído ante la Real Academia de Ciencias Exactas, Físicas y Naturales* (Madrid: Imprenta clásica española, 1918).

[81] Jimeno and other *pensionados* had been trained in that institution in the early twentieth century.

[82] Albert Charles Seward (ed.), *Science and the Nation: Essays by Cambridge Graduates with an Introduction by the Right Honourable Lord Moulton, K.C.B, F.R.S.* (Cambridge: Cambridge University Press, 1917), p. 17.

'vague and artificial'.[83] *Science and the Nation* contained a detailed, critical analysis of the causes of the loss of British hegemony to Germany in the dyestuffs industry, a field in which, at that time, Spain was significantly dependent on from abroad, but argued to overcome any pure–applied dichotomy and to forge new fluid, flexible relations between academia and industry.[84]

José Prats was probably one of the best representatives of the debate at that time on the alliance between academic chemistry and industrial research.[85] In the early twentieth century, Prats trained as a chemist at the *Escola d'Enginyers Tèxtils de Terrassa* (Catalonia), a prestigious textile engineering school, and in 1894, he became Professor of Industrial Chemistry and held a chair in dyeing and finishing. In 1908, Prats obtained a JAE *pension* to travel to Mulhouse and Zurich in order to learn the latest methods of organic synthesis, mainly focused on new artificial dyestuffs. During his trip, Prats combined industrial and academic experience. In the winter of 1909, he stayed at the *Gesellschaft für Chemische Industrie* in Basel, where he experimented with new dyes and published papers in the *Anales* and *Chemiker Zeitung*.[86] Prats later entered ETH technical school in Zurich, where he took several chemistry courses, particularly organic chemistry with Willstätter and dyestuffs theoretical chemistry with Julius Schmidlin (García Banús's supervisor for his research on free radicals).[87] Prats naturally combined theoretical lessons with several dyestuff syntheses in the lab. In his summer holidays, he worked at the *Farbenfabriken* in Elberfeld and prepared a book on dye samples of cotton, wool and silk. He later visited dyeing factories in Mulhouse, Krefeld and Lyons. In his report to the JAE, *Síntesis Química*, he highlighted that, in order to overcome Spanish backwardness in chemistry and the chemical industry, the country had to appropriate the German, Austrian and Swiss model of science-based industry.[88] He

---

[83] 'and, [...] it does not exist as a guiding principle in the minds of those classes to whom we must look for the force which will place science in its right position in England'. Seward (ed.), *Science and the Nation*, p. ix.

[84] Graeme Gooday, 'Vague and Artificial. The Historically Elusive Distinction between Pure and Applied Science', *Isis*, 103(3) (2012), 546–54.

[85] Ricard Duran, 'Colorants Artificials, 1856–1936: Josep Prats i Aymerich', in Josep Batlló et al. (eds.), *Actes de la VIII Trobada d'Història de la Ciència i de la Tècnica* (Barcelona: SCHCT, 2006), pp. 487–94.

[86] José Prats, 'Neue formeln um eine Mischung von Olem von gegebenen Gehalt zu bereiten', *Chemiker Zeitung*, 31 (1910), 264–72.

[87] José Prats, *Síntesis Química. Memoria que en cumplimiento de las disposiciones vigentes eleva a la superioridad José Prats Aymerich, Profesor numerario de la Escuela superior de Industrias y de Ingenieros Textiles de Tarrasa. Pensionado en el Extranjero para el Estudio de dicha ciencia desde diciembre de 1908 hasta agosto de 1909.* JAE Archive. JAE/117-553.

[88] JAE/117-553.

Figure 1.4 Chemistry professors and students at the *Escola Industrial*, Barcelona, in 1917 (Agell is the fifth on the left). Frederic Ballell. S0 024 089. Reproduced with permission of the *Arxiu Fotogràfic de Barcelona*.

made an argument for the teaching of new subjects such as chemical synthesis and dyestuffs chemistry, for a dramatic increase in the time devoted to experimental laboratory training and for integrating the curricula of engineering, chemistry and pharmacy. Prats' textbooks on industrial organic chemistry enjoyed wide circulation.[89] He also engaged in several international reforms of the organic chemical nomenclature.[90]

The pure–applied chemistry debate also took other forms in the Silver Age. In the 1910s, Agell had founded the *Escola de Directors d'Indústries Químiques* in Barcelona, taking advantage of the recent laboratory reforms at the *Escola Industrial* (Figure 1.4). Agell enjoyed notable autonomy in defining his project within the framework of that school, and was keen to attract public but also private funds. Moles' stay in Barcelona in 1924 and his personal relationship with Agell had strengthened the links between Moles's international 'pure' chemistry academic profile – at that time still

---

[89] José Prats, *Apuntes de química general* (Barcelona: J. Nova, 1911). See also: José Prats, *Las materias colorantes orgánicas en España desde el punto de vista analítico y fiscal* (Madrid: Eduardo Arias, 1911); *Química de las materias colorantes: apuntes redactados según las explicaciones del Dr. José Prats Aymerich* (1910) (manuscript); *Programas de química de las materias colorantes y tintorería, estampados y aprestos de tecnología de las materias colorantes. Escuela Superior de Industrias y de Ingenieros de Industrias Textiles de Tarrasa* (Barcelona: Imprenta Pedro Ortega, 1910). JAE/117-553.

[90] JAE 117–553.

Useful Chemistry 43

struggling after his years training in physical chemistry abroad to gain a chair in the Spanish university system – and Agell's industrially orientated school.[91] Agell designed a flexible, original four-year curriculum plus an industrial research project. Lessons began in January 1916 with 30 students. In 1917, the project evolved towards the *Institut de Química Aplicada* (IQA). It held a school for managers, the laboratory and a provincial hygiene lab – Agell had a pharmacy background and played an important role in developing analytical techniques and quality control. Industrialists, very often acting as sponsors of the school, were to furnish the museum with chemicals, instruments, machines and chemical samples. Beyond the 'textbook' classical training in the official 'schools', Agell's IQA provided experimental chemical training to solve modern, industrially rooted laboratory research problems.[92]

Other private initiatives are worth mentioning. For example, the case of Francesc Novellas (1874–1940), a Catalan chemist who, a few years after obtaining his university degree in 1898, launched Titan, a paint chemicals company (1907). In 1905, he became director of the journal *La industria química*, which targeted chemical industrialists, pharmacists, engineers, dyers and 'anyone interested in the progress of chemical science and its applications to industry, pharmacy, agriculture and trade'.[93] The journal included translations of foreign chemistry papers,[94] book reviews, prices of drugs and chemicals, chemical industry advertisements, a commercial section and a space for dry-salters and industrialists to ask for technical advice.[95] Novellas founded the *Instituto Físico-químico* (1899–1904), which later became the *Instituto Químico-técnico* (1905–20), a private firm for chemical analysis and testing for industry and for chemistry teaching by correspondence. Novellas covered several fronts: private business, technical training, industrial assessment, patents and even popular chemistry.

In 1924, Agell's IQA took a leading role in the publication of a new journal, *Química e Industria*, which enjoyed a wide readership. Academic chemists, industrial engineers and pharmacists – many of them closely

---

[91] Galí, *Història de les institucions*, p. 208.
[92] This was Alexandre Galí's view in the 1980s. Galí, *Història de les institucions*, p. 214.
[93] Gaziella Artís, Mireia Artís, 'Francesc Novellas i Roig (1874–1940). La química industrial', in Antoni Roca-Rosell, Josep Maria Camarasa (eds.), *Ciència i Tècnica al Països Catalans. Una aproximació biogràfica*. 2 vols. (Barcelona: Fundació Catalana per a la Recerca, 1995), pp. 937–66, 945. See also: Gerardo Monge, *Figuras de la industria química española* (Barcelona, 1955), pp. 23–5. *La industria química* was founded in 1903 by Antonio Mora.
[94] From journals such as *La revue des produits chimiques*, *Zeitschrift für Elektrochemie*, *Journal of the Society of Chemical Industry* and *Revue de chimie industrielle*. *La industria química*, VII, 141, 1 January 1910.
[95] *La industria química*, VII, 142, 16 January 1910, p. vii.

related to the IQA – pioneered the new journal in a new attempt to bring chemical knowledge closer to industrial needs. The journal published academic and industrial papers, the latter covering topics such as synthetic products, tanning, pigments and explosives. It also included short articles by industrial experts, books reviews and journals abstracts, together with updated price lists for chemical products.[96] Under the direction of Uhthoff, one of the leaders of the backwardness campaign, *Química e Industria* became the official voice of the National Board of Chemical Industry (*Cámara Nacional de Industrias Químicas*, founded in 1919 as an entrepreneurial board of chemical companies).[97] The whole project drew its inspiration from the French journal *Chimie et industrie*, which was published under the banner of the *Société de chimie industrielle*. Industrial advertising covered international firms such as Pirelli, Goodyear and Krupp, but also local firms and pharmaceutical labs, including a Spanish version of Muspratt's industrial chemistry encyclopaedia.[98] *Química e Industria* also counted on the support of important pharmaceutical entrepreneurs and the *Unión de laboratorios químico farmacéuticos de España*.[99] In fact, the leading group of *Química e Industria* managed to get by with a form of industrial modernity through chemistry, and they adapted their economic and professionals interests to Primo de Rivera's policies.

There were, however, other kinds of dreams of modernity in industrial chemistry, such as the case of the *Instituto Químico de Sarriá* (IQS). In 1905, initially as an appendix to the astronomical *Observatorio del Ebro*, the Society of Jesus founded the *Laboratorio Químico del Ebro*.[100] The project was led by Eduardo Vitoria, who trained in organic chemistry at the Catholic University of Leuven. In the early years of the twentieth century, Vitoria had visited the most prominent chemistry labs in Germany, Belgium, Holland and France and strengthened the links among Catholic chemists.[101] The early laboratory fitted very well within the framework of the Jesuits' early twentieth century agenda to create new teaching and research institutions in order to strengthen the compatibility

---

[96] *Química e Industria*, I(1), February 1924.
[97] *Química e Industria*, VI(67), August 1929.
[98] Friedrich Stohmann et al. (eds.), *Gran Enciclopedia de Química Industrial. Teórica, práctica y analítica. 12 Tomos + 1 Suplemento* (Barcelona: Francisco Seix, 1930).
[99] The *Unión de laboratorios químico farmacéuticos de España* counted on successful entrepreneurs such as Santiago Pagés, Francisco Mandri, Francisco Uriach, Enrique Puig and Pabro Borrell. *Química e Industria*, IV(41), June 1927, p. 174.
[100] Eduardo Vitoria, *Conferencias de química moderna dadas en el Laboratorio químico del Ebro de la Compañía de Jesús* (Tortosa: Biarnés, Foguet, 1907); Eduardo Vitoria, *La ciencia química y la vida social. Conferencias de vulgarización científica ... por Eduardo Vitoria* (Barcelona: Tipografía Católica Pontificia, 1916).
[101] Vitoria, *Conferencias de Química Moderna*, p. 33.

between science and religion.[102] The chemical lab was very close to the Faculty of Theology of Tortosa, where young Jesuits received their early training and where they now had to complete their education with courses in 'positive' science. However, that chemistry education, originally restricted to future Jesuits, soon attracted the attention of secular chemists and pharmacists who wanted to improve their chemistry skills with Professor Vitoria. Vitoria's *Manual de química moderna* was published in 1910, and it later went into its 14th edition and sold more the 100,000 copies (including in the Latin American market). *Catálisis química* (1912) and *Prácticas químicas para cátedras y laboratorios* (1914) were pragmatic reflections of Vitoria's own experiments and lectures.[103]

In 1915, at the beginning of his career in leading Jesuit chemistry in Spain, Vitoria had already expressed his commitment to reconciling chemistry and religion in public lectures. At the University of Valencia that year, Vitoria's popular addresses were entitled 'Chemical Science and Social Life', and he talked of the nation, the region and the family. For the sake of the nation, chemistry had to explore life and make drugs and explosives; for the prosperity of the Valencian region, chemistry had to serve agriculture and industry; for family well-being, chemistry had to be applied to domestic life and housewives' needs.[104] In 1916, when the Faculty of Theology moved to Barcelona, the chemical lab followed. The new IQS attracted chemists and pharmacists to its three-year graduate programme based on mineral, analytical and carbon chemistry, with a ratio of one hour of theory to six hours of experiments in the lab per day.[105] The IQS students, very often belonging to families of Catalan chemical entrepreneurs, sought a practical chemical education with a view to later joining their own firms. The IQS postgraduate programme progressively evolved into an undergraduate syllabus for the training of future chemists

---

[102] Miguel M. Valera S. J., 'Eduardo Vitoria S.J. A Contemporary Leader in the Spanish Chemical World', *Journal of Chemical Education*, 33(4) (1956) 161–6. Lluís Victori, 'Eduard Vitoria i Miralles: pioner de la química al nostre país', *Revista de la Societat Catalana de Química*, 7 (2006), 60–4.

[103] For further biographical details on Vitoria, see: Eduardo Vitoria, *Autobiografía, 1864–1958* (Barcelona: IQS, 2005). See also: Joaquín Pérez Pariente, 'El Jesuita Eduardo Vitoria: la química como apostolado', in Pedro Bosch Giral, Juan Francisco García de la Banda, Joaquín Pérez Pariente, Manoel Toural Quiroga, *Protagonistas de la química en España: los orígenes de la catálisis* (Madrid: CSIC, 2010), pp. 61–116.

[104] Ibid.

[105] Núria Puig, Santiago López, *Ciencia e industria en España. El Instituto Químico de Sarriá, 1916–1992* (Barcelona: IQS, 1992); Núria Puig, 'The Frustrated Rise of Spanish Chemical Industry between the Wars', in Anthony S. Travis et al. (eds.), *Determinants in the Evolution of the European Chemical Industry, 1900–1939* (Dordrecht: Kluwer, 1998), pp. 301–20; Núria Puig, 'El crecimiento asistido de la industria química en España: Fabricación Nacional de Colorantes y Explosivos, 1922–1965', *Revista de Historia Industrial*, 15 (1999), 105–36.

with a strong applied, experimental emphasis. In 1921, the journal *Afinidad* appeared originally as a newsletter for former IQS students, but soon contributed substantially to consolidating the identity of the Jesuit chemists. In 1927, Paul Sabatier visited the IQS, and in the same year Vitoria's *La química del carbono* appeared, which became a key textbook for the teaching of organic chemistry.[106] Due to his ideological affinity with Vitoria, Sabatier welcomed IQS graduates, who could study for a degree in chemical engineering at the University of Toulouse after one year in a French University.[107]

Also from a conservative position, other industrial projects emerged and developed from the heart of the university. This was the case for Emilio Jimeno, who was trained in physical chemistry and did research work on particular aspects of the electrochemical decomposition of hyposulphates. In 1913, after a short collaboration at the LIF with Moles, Jimeno obtained a *pensionado* to work on kinetics at Ostwald's Physical Chemistry Institute in Leipzig. In 1916, Jimeno became professor of inorganic chemistry at the University of Oviedo, at the heart of an important mining region. In 1919, he went on to a second stage of international scientific training in the United States with a JAE grant at the School of Mining of Columbia University (New York) and the Bureau of Standards (Washington, DC). He later travelled to several metallurgical labs and industries in France, Belgium and Britain.[108] Back in Spain, in 1924, Jimeno developed a highly dynamic metallurgy project combining teaching and research applied to industry, and he was very active in publishing research papers and textbooks. This was an ideal environment for developing his ambitious project for the growth of high-level academic metallurgy closely linked with local industrial demands. During that decade, Jimeno increased his metallurgy lectures and courses and progressively equipped a new lab for the same purpose.

In tune with Primo de Rivera's goals of industrial growth and social order, industrial chemistry rose to prominence in the public sphere in the late 1920s. At the 1929 International Exhibition in Barcelona, foreign but

---

[106] Eduardo Vitoria, *Química del carbono. Teoría y práctica con vistas muy especiales a la síntesis en el laboratorio y en la industria* (Barcelona: Tipografía Católica Casals, 1940).
[107] Miguel M. Valera S. J., 'Eduardo Vitoria S.J.', p. 165.
[108] Felipe Calvo, José María Gulemany, 'Contribución del profesor Emilio Jimeno Gil al prestigio de la Universidad de Barcelona: Rector de 1939 a 1941', in *Història de la Universitat de Barcelona. I Simposium del 150 Aniversari de la Restauració. Barcelona 1988* (Barcelona: Publicacions de la Universitat de Barcelona, 1990), pp. 471–83. See also: Felipe Calvo, *Emilio Jimeno Gil. Semblanza de un maestro* (Santander: Amigos de la Cultura científica, 1984); María Ignacia Moya, *La enseñanza de la química inorgánica en la Universidad de Barcelona (1900–1936)*. PhD thesis (Barcelona: Universitat de Barcelona, 1980), pp. 160–233; Archivo General de la Adminsitración (AGA). *Emilio Jimeno Expediente personal (21–20358-001)*. JAE Archive. JAE/82-66.

Figure 1.5 The building of the Chemistry Pavilion (*Palacio de la Química*) at the 1929 Barcelona International Exhibition.
Private collection of the author.

also local industrial chemistry made their mark in the Chemistry Pavilion (*Palacio de la Química*), which displayed an impressive number of stands and chemicals (Figure 1.5).[109] There were special halls for Italian, French, Swiss and Spanish products. The latter were particularly numerous: oxygenated water, dyestuffs, borax, soap, coal tar derivatives, oils, waxes, paintings, soda, lead salts and drugs, among many others. In the Italian hall, there was a display of the *Fabrica Reunnita Agricultori Italiani*, dyestuffs for printing processes and stands of the *Istituto Seroterapico Milaneseese* and the *Istituto Nazionale Medico Farmacologico Serono*. The French section held dyed and printed textile samples, tanned and dyed animal skins, synthetic perfumes, pharmaceuticals and scientific instruments. Drugs and food were the main representatives of the Swiss section. Journalist S. de Llinás presented figures concerning Spanish chemical imports and expressed his hope for a promising future for the Spanish chemical industry in the following terms:

It cannot be denied that the Spanish Section in the Chemistry Pavilion represents the hope for the industrial tomorrow of our homeland, which in this field is strongly indebted to foreign countries. [...] [It] reveal[s] the nation's capacity to

[109] On 18 July 1929, S. de Llinás, a journalist from the Barcelona newspaper *La Vanguardia*, published a detailed article on a visit to the Chemistry Pavilion. See also: Susana Torrellas, *Química i indústria a l'Exposició Internacional de Barcelona de 1929*. Master's thesis (Bellaterra: Universitat Autònoma de Barcelona, 2013).

consume materials which, when transformed, find their proper application in several of our industries. Other products, however, for which we still depend on receiving from abroad, will very soon be produced by our homeland in such quantities that they will fulfil the demand for domestic consumption and even exportation.[110]

Outside the Pavilion, chemical firms also built several corporate pavilions to showcase their products. The German I. G. Farben displayed its chemicals in several pavilions (electricity, chemistry, agriculture and textiles).[111] Local firms such as Cros chemical company and the cement company Asland put a lot of thought and money into their displays. Cros built its own pavilion to showcase its factories and chemical products.[112]

The Barcelona 1929 International Exhibition was also an excellent setting for organising international conferences on a great number of subjects. It hosted chemical-related meetings on dyestuffs, energy, industrial chemistry, wood, pharmacy, wine, iron, forestry, medical presses, technical presses and astronomy.[113] At the opening session of the 9th International Conference on Industrial Chemistry,[114] held on the premises of the exhibition, Moles, who was at that time the president of the RSEFQ and the main representative of the Barcelona section of the *Sociedad*,[115] gave the inaugural speech, 'La Universidad y la Industria'.[116] Tackling once again the pure–applied chemistry issue, Moles stressed the need to introduce good research into a weak Spanish university system that had been mainly focusing only on teaching for decades and had neglected experimental equipment and labs for potential research schools. Quoting great luminaries such as William Ramsay, John Tyndall, Louis Pasteur and Ernst Rutherford, Moles again strongly dismissed any divide between pure and applied chemistry. In his view, a solid university culture (with a balance between research and teaching) would sooner or later bring about industrial prosperity, and he used the case of the French *Société de chimie industrielle* as a good example of fruitful collaboration between great academic names such as Moissan, Berthelot, Pasteur, Sabatier, Grignard, Matignon and Fourneau and successful

---

[110] S. De Llinás, 'Notas del día. La Exposición Internacional. Apuntes para mi archivo. Palacio de la Química', *La Vanguardia*, 18 July 1929, p. 6.
[111] *Diario Oficial de la Exposición Internacional de Barcelona 1929*. 19 October 1929.
[112] Josep Uhthoff, 'La indústria química a l'Exposició', *Ciència: Revista catalana de ciència i tecnologia*, 5(36) (1930), 627. See also: Torrellas, *Química i indústria*.
[113] *Diario Oficial de la Exposición Internacional de Barcelona* (Barcelona: P. Yuste, 1929–1930).
[114] *La Vanguardia*, 17 October 1929, p. 9. Ulrike Fell (ed.), *Chimie et industrie en Europe. L'apport des sociétés savantes industrielles du XIXè sièvle à nos jours* (Paris: Editions des archives contemporaines, 2003).
[115] 'Acta de la constitución de la Sección de Barcelona'. *Anales*, 27 (1929), 765–8.
[116] Enrique Moles, 'La Universidad y la Industria', *Química e Industria*, 7 (1930), 148–53.

industrial projects. Moles presented chemistry as a whole – even including very specialised areas such as his own research field on the experimental calculation of atomic weights – as a highly basic, academic science that had to build bridges to industry, since industrial applications were the main professional prospects for many of the members of his own research school.[117]

In order to develop a similar pattern in Spain – and, obviously, also taking into account the virtues of the German 'science-based industry' – Moles advocated an ambitious reform of chemical education and robust university research to support industrial research. In terms of training, Moles had introduced a research project at the end of the undergraduate studies at the Faculty of Science in Madrid, which would better train young chemists to adapt to future industrial problems. Therefore, a young graduate with good research skills was, in Moles view, an excellent technician for industry. Moles used figures from a 1928 report of the *Consejo de la Economía Nacional* to demonstrate the trade deficit in several chemicals that the country was unable to produce and needed to import: ammonium sulphate (75.5 million pesetas), superphosphates (11.4 million pesetas), calcium cyanamide (320 million pesetas), insecticides made from copper sulphate, arsenic salts and alkaline cyanides (4 million pesetas).[118]

Among the invited speakers, the Conference on Industrial Chemistry welcomed Nobel Prize winner Paul Sabatier (Toulouse), Henry Armstrong (Imperial College London), M. Louis Hackspill (Strasbourg) and Rocasolano (Zaragoza).[119] Also in 1929, in the context of the Barcelona International Exhibition, Masriera, at that time lecturer of 'Technical Chemistry' at the university, again discussed the utility of chemistry in the following terms:

> Many of our minerals are sent abroad. Our iron has to leave Spain to return in the form of machinery. While chemical industries worldwide start out with the distillation of bituminous coal, in Spain there is hardly a single national product that goes beyond the distillation of coal tar. And if there is any exception, in textiles for instance, it is under a shameful regime of tariffs.[120]

Masriera called for a more liberal free trade economy to improve the chemical industry and Moles advocated for compatibility between pure and applied chemistry, but both men expressed their hope for a modern

---

[117] Ángel Juliá, 'El IX Congreso de Química Industrial', *Química e Industria*, 67 (1929), 202–3.
[118] Moles, 'La Universidad y la Industria', p. 152
[119] Juliá, 'El IX Congreso de Química Industrial'.
[120] Miguel Masriera, *La Enseñanza de la Química Técnica* (Barcelona: Uriach, 1929), p. 5.

chemical industry in Spain that could make the dream of modernity come true.

***

In 1917, a report on the chemistry of alkaloids by García Banús – at that time Professor of Organic Chemistry at the University of Barcelona – epitomised a good part of the chemical dream of modernity and the material culture of modernisation in the early twentieth century.[121] García Banús used his international experience at the ETH in Zurich in the training of new experimental chemists who could devote unlimited time to lab experiments, learn foreign languages (particularly German), become acquainted with foreign international chemistry journals and books and adapt their skills to new glassware, instruments and chemicals. He chose alkaloids as a family of natural products that allowed discussions on pure and applied science, including the industrial dimension of chemistry as part of its modernity. But alkaloids also required mastery of the international literature and international standards on nomenclature. In his training abroad as a *pensionado* and after settling down in his new lab in Barcelona, García Banús amassed a valuable collection of alkaloids, which he made available to students for analytical tests, potential synthetic approaches and further medical applications.

In fact, García Banús's programme on alkaloids can be taken as a good illustration of the struggle against backwardness by the seminal community of Spanish chemists who had been active since the foundation of the RSEFQ. During the final decades of the Bourbon Restoration and Primo de Rivera's dictatorship, chemistry gained prominence and authority in the universities, technical schools, new research labs and chemical industries. The new generation of chemists of the early decades of the twentieth century created a material discourse and modernisation project. Many shared the liberal values of the JAE, but others, from more conservative positions, also used chemistry to strengthen their professional ambitions, creating strong alliances with industry. Despite the social tensions of the 1910s and 1920s, chemical 'modernity' basically meant the fight against 'backwardness' through reforms in chemical training, cosmopolitan and internationally connected research projects and an ambitious renewal of the material culture of chemistry in terms of instruments, reagents and laboratories.

---

[121] Antonio García Banús, 'Memoria-Informe que presenta al "Consell d'Investigació Pedagògica" el Profesor A. García Banús ', 20 January 1917. *Diputació de Barcelona. Arxiu General. Consell d'Investigació Pedegògica.* Expedient 53.

Borrowing again from Mary Jo Nye's pattern for assessing the identity of a particular scientific discipline, we can conclude that Spanish chemists created a genuine scientific community.[122] *Anales* played a key role as the community's core literature. Sites for teaching and research grew substantially, in spite of public laments of backwardness and a lack of experimental equipment and labs. Members of competing scientific communities at home and abroad recognised Spanish chemists' new dynamism and contributions and shared research priorities at the beginning of the Second Spanish Republic.

[122] Mary Jo Nye, *From Chemical Philosophy to Theoretical Chemistry*.

2 A Republican Science

In 1931, just after the advent of the Second Spanish Republic in April, chemist and pharmacist José Giral was elected as a member of parliament and soon became Rector of the University of Madrid and Minister of the Navy in one of the first cabinets of the new Republic. During the 1920s, J. Giral had been active in the republican opposition to Primo de Rivera's dictatorship. His personal contact with Manuel Azaña – the future President of the Republic – and the gatherings in the back room of the family pharmacy led J. Giral towards his calling for chemistry and liberal thought.[1] In the same year, his son, Francisco Giral, translated into Spanish a book by British science journalist James Gerald Crowther (1899–1983),[2] *Science in Soviet Russia*, which described a recent visit to research centres, universities and scientific institutions in Leningrad and Moscow and praised the scientific project of the USSR by comparing it to those of Western countries. Crowther had successfully also published *An Outline of the Universe* (1931), and *The ABC of Chemistry* (1932), a widely read popular chemistry book. After the Second World War, he remained a Marxist activist for years, strongly criticising the use of science in capitalist countries, seeing it as an ideologically biased project with regard to society.[3]

In that founding year of the new Spanish Republic, which opened up new opportunities for the modernisation of the scientific culture of the

---

[1] Julián Chaves, 'Oposición política a la monarquía de Alfonso XIII. José Giral y los republicanos en la Dictadura de Primo de Rivera', *Hispania*, 252 (2016), 159–87; María Fernanda Mancebo, *La Universidad de Valencia. De la Monarquía a la República (1919–1939)* (Valencia: Universitat de València/Instituto de Cultura Juan Gil-Albert, 1994); Javier Puerto, *Ciencia y política. José Giral Pereira* (Madrid: Real Academia de la Historia/BOE, 2016).
[2] James G. Crowther, *La ciencia en el país de los Soviets. Traducción directa del inglés de Francisco Giral* (Madrid: Cenit, 1931). See also: James G. Crowther, *Science in Soviet Russia* (London: Williams & Norgate, 1930).
[3] See Agustí Nieto-Galan, *Science in the Public Sphere* (London: Routledge, 2016). On Crowther, see: Jane Gregory, 'Crowther, James Gerald (1899–1983), Science Journalist and Administrator', *Oxford Dictionary of National Biography* (2006) http://oxfordindex.oup.com/view/10.1093/ref:odnb/94975 (last accessed 18 October 2017).

country, F. Giral stressed the need to appropriate some of the features of the Soviet scientific model from a non-dogmatic position. He envisaged chances in new and perhaps utopian assemblages between chemistry, industry, education and the state, a project that lay very much at the heart of the values that had won over the majority in the local elections that preceded the fall of the monarchy and the dictatorial period. In F. Giral's words:

> Without wanting to fully transfer the communist regime from Moscow to Madrid, but it holds many things that can we can take advantage of and use in order to give our country a new organisation. We do not need to strictly adopt Marx and Lenin's doctrines, but we shall design a new order in which *dirty business, selfishness, incivility, laziness, unproductive capital and clericalism do not prevail* [...][4]

Since the late nineteenth century, Spanish republican political thought had its roots in the values of the French Third Republic; that is, in the struggle for universal access to a secular education and for real influence on cultural elites with regard to the leadership of the country. Spanish republicans shared a certain ethos based on anticlerical freethinking and trust in positive science. In their view, it was this rational study of nature that could provide useful ideas for the improvement of society, as well as a sort of moral refuge in times of national decline and social instability.[5] But nineteenth-century republican values had clashed at the heart of a backward university system, as the so-called '*cuestión universitaria*' of 1875, with the suppression of academic freedom, clearly epitomised. Free thinking, the rationalist philosophy of Krausism (rooted in the nineteenth-century idealist texts of the German philosopher Friedrich Krause [1781–1832] that favoured secular readings of nature)[6] and the long-standing attempts at modernisation and educational reform led to the foundation of the ILE in 1876 and the JAE in 1907, with the Second Republic of 1931 being another key landmark in that process.[7] F. Giral's appeal to the Soviets should therefore be understood in that context.

Liberal chemists such as José Giral and Francisco Giral, leftist socialists such as Antonio Madinaveitia and powerful figures such as Enrique

---

[4] Crowther, *La ciencia en el país de los Soviets*, p. 9 (emphasis added).
[5] Agustí Nieto-Galan, 'A Republican Natural History in Spain around 1900: Odón de Buen (1863–1945) and His Audiences', *Historical Studies in the Natural Sciences*, 42(3) (2012), 159–89.
[6] Otto Carlos Stoetzer, *Karl Christian Friedrich Krause and his Influence in the Hispanic World* (Vienna: Bohlau, 1998).
[7] José M. Cobos, 'Luces y sombras del apoyo de la II^a República española a la ciencia', in Elena Ausejo (ed.), *La ciencia española en la II^a República* (Madrid: FIM, 2008), pp. 9–39. See also: Julián Casanova, *República y Guerra Civil*. Vol. 8. Josep Fontana, Ramón Villares (eds.), *Historia de España* (Barcelona/Madrid: Crítica/Marcial Pons, 2007).

Moles – with a more ambiguous political profile, but committed to the scientific project of the Second Republic – contributed to the making of a 'republican' chemistry, which again strengthened chemical cosmopolitanism as a key value of the modernisation process of the profession. At the same time, bridges between academia and industry were frequent in a sort of Silver Age of Spanish industrial chemistry. Along with the League of Nations policies, Spanish republicanism promoted the idea of pacific, useful chemistry internationally committed to the progress of mankind, providing a new image of the profession far removed from the pernicious uses of chemical gases that issued from the First World War.[8] In spite of public pacifist rhetoric, fears of a new chemical war were associated with fears about Mussolini's fascist Italy and Nazi Germany attacks against Western democracies, which then included the Spanish Republic. Beyond the international context, at a domestic level, the freethinking rationalism of the JAE and even the panentheistic tradition of the Krausist philosophy seemed to place chemistry in a comfortable position to highlight material progress and a positivistic scientific culture as solid allies of the new regime. Krausism had inspired the cultural policies of the ILE and the JAE and associated God with matter and the universe. It facilitated the notion of chemistry as a secular science, a science of matter, without explicit links to religion and divinities, as well as the hope that the material progress of synthetic production would produce the social progress embedded in the republican dream.[9]

The new Republic was, in F. Giral's mind, a golden opportunity to make the old dreams of modernity come true. Cosmopolitanism, Europe and ambitious reforms in research and education made up the JAE policies that, in the preceding decades, had brought a generation of young chemists to the core of international networks. Chemical progress was therefore associated with new improvements in training, laboratories, research schools and industries, again in a context in which chemistry became a solid ally of the policies of the new Republic. In fact, the happy 1930s that the Girals were looking forward to had three main pillars in chemistry, to be examined in detail in this chapter: the opening in 1932 of the *Instituto Nacional de Física y Química* (INFQ), its leading role in chemical research and the academic and political context that surrounded it; the 9th International Chemistry Conference on Pure and Applied

---

[8] For the foreign policies of the Spanish Republic, see: Sebastian Balfour, Paul Preston (eds.), *Spain and the Great Powers in the Twentieth Century* (London: Routledge, 1999).

[9] On Krausist philosophy in Spain, see: Gonzalo Capellán, *La España armónica. El proyecto del krausismo español para una sociedad en conflicto* (Madrid: Biblioteca Nueva, 2006).

Chemistry in Madrid in 1934; and finally, the new attempts to link academic chemistry to industrial growth.[10]

A New Enlightenment

In 1934, Moles's address, *Del momento científico español*, resulted from his admission as new fellow of the Royal Academy of Sciences (RACEFN) in Madrid.[11] He referred to a number of eighteenth-century founding heroes of Spanish chemistry who dated back to the Enlightenment: Antoni de Martí Franquès (1750–1832) for his accurate analysis of the composition of atmospheric air in the times of Lavoisier; Fausto de Elhúyar (1755–1833) and his brother Juan José de Elhúyar (1754–96) for the isolation of tungsten on their trips to Latin America; and Andrés del Río (1764–1849) for the discovery of vanadium in New Spain. The three were eminent in the Enlightenment, a period considered as a good precedent for the 'modern' scientific policies of the JAE in the 1930s.

Moles perceived the period of 1775–1825 as a key moment of cosmopolitanism in Spanish chemical science, which followed the decadence of the nineteenth century and would be reborn in the 'glorious' 1930s. Acting as a chemist–historian,[12] he attempted to assess the results of the policies of modernisation since the beginning of the twentieth century, and he compared the policies of the new Republic with the golden years of the Spanish chemical Enlightenment. Moles's Enlightenment voyage seemed therefore to have found its harbour in the 1930s. The *Sociedad Española de Física y Química* – now it had formally lost its 'Royal' status – already had sections in Barcelona, Valencia, Seville, Asturias, Granada and Bilbao, published an average of 100 papers a year in Spanish and had 1200 fellows.[13]

Since 1925, in a period in which the economic nationalism of Primo de Rivera's dictatorship conceived of science and industry as part of the promotion of the regime, the International Education Board (IEB) of

---

[10] Helen Graham, *The Spanish Republic at War* (Cambridge: Cambridge University Press, 2002).
[11] Enrique Moles, *Del momento científico español, 1775–1825* (Madrid: C. Bermejo, 1934).
[12] Colin A. Russell, '"Rude and Disgraceful Beginnings": A View of the History of Chemistry from the Nineteenth Century', *The British Journal for the History of Science*, 21 (1988), 273–94; Agustí Nieto-Galan, 'Seeking an Identity for Chemistry in Spain: Medicine, Industry, University, the Liberal State and the New "Professionals"', in David Knight, Helge Kragh (eds.), *The Making of the Chemist* (Cambridge: Cambridge University Press, 1998), pp. 177–90.
[13] Carmen Magallón, *Pioneras españolas en las ciencias. Las mujeres del Instituto Nacional de Física y Química* (Madrid: CSIC, 1998).

the Rockefeller Foundation had been planning the creation of a research institute for physics and chemistry in Spain. In 1927, Moles, Catalán and the architects in charge of the construction of the INFQ building travelled for two months to France, Germany, Denmark and the Netherlands to observe different models for the institute.[14] Augustus Trowbridge, a representative of the IEB, visited Spain that same year.[15] The general impression that he reported encouraged local chemists to overcome old laments of backwardness and focus on new lines of research and productivity. After several interviews with chemists in Madrid and Barcelona, Trowbridge stated:

> Conditions do not appear to an outsider to be anywhere so bad as they make them out to be and A. Trowbridge has a strong 'hunch' that if several of the scientists he has met here would stop pitying themselves and use their leisure time which they seem to have in doing something instead of using the lack of facilities as an excuse for doing very little, they would be less discouraged than they are and might win for themselves the distinction even in their own country of which they deplore the absence.[16]

Beyond the bemoaning of backwardness and isolation, dreams of modernity and the material conditions of modernisation through chemistry seemed to an outsider to be more feasible than the Spaniards' expectations. A contribution of $420,000 from the Rockefeller IEB and the support of the Republican Government provided an ideal environment for around 100 elite researchers.[17] The JAE initially planned to offer the directorship of the INFQ to a prestigious international luminary, but physicist Blas Cabrera finally took up the position. Willstätter, one of the leading mentors of the JAE chemists, described in his autobiography how the new INFQ had offered him the directorship. In 1924, Willstätter resigned from his position of head of the Chemical Institute of the

---

[14] José Manuel Sánchez Ron, *Miguel Catalán. Su obra y su mundo* (Madrid, CSIC/ Fundación Ramón Menéndez Pidal, 1994).

[15] Thomas Glick, 'La Fundación Rockefeller en España: Augustus Trowbridge y la negociación para el Instituto Nacional de Física y Química, 1923–1927', in José Manuel Sánchez Ron (ed.), *La Junta para Ampliación de Estudios e Investigaciones Científicas 80 años después: 1907–1987.* 2 vols. (Madrid: CSIC, 1988), Vol. II, pp. 281–300. See also: Agustí Nieto-Galan, 'Free Radicals in the European Periphery: "Translating" Organic Chemistry from Zurich to Barcelona in the Early Twentieth Century', *The British Journal for the History of Science*, 37 (2004), 167–91.

[16] Augustus Trowbridge's travel report to Spain for 1927. Rockefeller Archive Center. International Education Board (IEB). Box 41, Folder 580, p. 30. Quoted in Nieto-Galan, 'Free Radicals in the European Periphery', p. 176.

[17] José Manuel Sánchez Ron, Antoni Roca-Rosell, 'Spain's First School of Physics: Blas Cabrera's Laboratorio de Investigaciones Físicas', in Gerald L. Geison and Frederic L. Holmes (eds.), *Research Schools. Historical Reappraisals. Osiris. Second Series*, 8 (1993), pp. 127–55.

University of Munich as a result of his disagreement with the University's anti-Semitic policy.[18] For several professional reasons – he was at that time tempted by other research institutions and universities for different positions – he did not accept the leadership of the Rockefeller, but maintained close contact with Madinaveitia and some of the Silver Age chemists.[19] Within a framework of ideological complicity, Willstätter was sympathetic to the scientific style of that emerging community. In his own words: 'It was during the years following the fall of the Spanish Monarchy, when a group of professors, all of them *liberal political thinkers* and passionate patriots, held positions of great influence in the Government'.[20]

On 6 February 1932, the Minister for Public Instruction, Fernando de los Ríos, officially opened the new INFQ, and the ceremony was attended by outstanding figures such as the physicists Arnold Sommerfeld (1868–1951), Paul Scherrer (1890–1969) and Pierre Weiss (1865–1940) and chemists Otto Hönigschmid (1878–1945) and Willstätter, whose work and scientific collaboration were close to the research lines of the new groups in Madrid.[21]

The INFQ was conceived as a continuation of the research lines of the LIF and organised into six main research groups: electricity and magnetism, under the direction of physicist Cabrera and with a special interest in the study of diamagnetism following Weiss's work; spectroscopy, under the leadership of Miguel Catalán, who, with his collaboration with Ángel del Campo and his discovery of multiplets, provided enormous international prestige to the Institute; electrochemistry, under the direction of Julio Guzmán, who had been trained in Leipzig at Ostwald's prestigious physical chemistry institute; X-rays, led by physicist Julio Palacios (1891–1970), working on crystal structures in collaboration with Scherrer at the ETH Zurich; organic chemistry, with the direction of Madinaveitia, who had close links with Willstätter and

---

[18] 'From Madrid I had a letter, on behalf of the Government, from the renowned and admired physiologist, Ramon y Cajal (in his function as president of the *Junta para Ampliación de Estudios*) offering me the office of president of the *Instituto Nacional de Física y Química* endowed by the Rockefeller Foundation. I visited the handsome institute and its dedication, and again later at the request of the Minister of Education to advise the chemical laboratory headed by my pupil Professor Antonio Madinaveitia'. Richard Willstätter, *From My Life: The Memoirs of Richard Willstätter* (New York: W. A. Benjamin, 1965), p. 368.

[19] In fact, he received three invitations from the Spanish government through his 'friend and pupil' Madinaveitia. Antonio Madinaveitia, 'El Profesor Richard Willstätter', *Ciencia. Revista hispano-americana de ciencias puras y aplicadas*, 10–11 (1942), 320–2.

[20] Willstätter, *From My Life*, p. 412 (emphasis added).

[21] Luis Enrique Otero Carvajal, José María López Sánchez, *La lucha por la modernidad. Las ciencias naturales y la Junta para Ampliación de Estudios* (Madrid: Residencia de Estudiantes, 2012).

Fourneau; and physical chemistry, under Moles' leadership, whose group became the largest, conducting outstanding research and gaining international recognition in new methods for the experimental determination of atomic weights (Figure 2.1).[22]

Madinaveitia, leader of the organic chemistry section,[23] benefitted from earlier teamwork at his chair of biological chemistry at the Faculty of Pharmacy in Madrid, which produced a wide range of collaborations and important national and international research papers.[24] The laboratory culture of the group used his textbook on drug synthesis extensively, a treatise published as a joint version with the French chemist Ernest Fourneau (see Chapter 1).[25] Adolfo González, another of Staudinger's pupils in Zurich, was Madinaveitia's right hand in the lab,[26] but other important names constituted his school: Ignacio Ribas (1901–96) – a pupil of Fourneau[27] – and Andrés León (1893–1976),[28] among others,[29] the latter having been trained in Oxford with Robert Robinson, a Nobel

---

[22] Sánchez Ron, Roca-Roselet, 'Spain's First School of Physics'. See also: Glick, 'La Fundación Rockefeller en España'; Sánchez Ron, *Miguel Catalán*.

[23] Rosario E. Fernández Terán, Francisco A. González Redondo, 'Antonio Madinaveitia y el Laboratorio de Química Biológica de la Junta para Ampliación de Estudios', in José Manuel Sánchez Ron, José García-Velasco (eds.), *100 años de la JAE. La Junta para Ampliación de Estudios e Investigaciones Científicas en su Centenario* (Madrid: Publicaciones de la Residencia de Estudiantes, 2010), pp. 743–61. Francisco A. González Redondo, Rosario E. Fernández Terán, 'Cajal y la nueva senda de la química orgánica en España. En torno a Antonio Madinaveitia Tabuyo', in *Actas del III Simposio 'Ciencia y Técnica en España de 1898 a 1945: Cabrera, Cajal, Torres Quevedo'* (Madrid: Amigos de la Cultura Científica, 2004), pp. 127–42.

[24] Madinaveitia's early research team at the pharmacy faculty can be associated with names such as José Rodríguez Carracido, José Sureda, Adolfo González, Joaquín Sorolla, José Puyal Gil, José Ranedo, Fernando Díaz Aguirreche, Ignacio Ribas, Natividad González, M. Gallego, Jesús Sáenz de Buruaga, Carmen Gómez, Francisco Giral, etc. JAE Archive. JAE/90-26.

[25] Ernest Fourneau, Antonio Madinaveita, *Síntesis de medicamentos orgánicos* (Madrid: Calpe, 1921). See also: Francisco Giral, 'Química orgánica (1932–1939)', in *50 años de investigación en física y química en el edificio Rockefeller de Madrid 1932–1982. Noviembre de 1982. Recuerdos históricos* (Madrid: CSIC, 1982), pp. 39–42, 40.

[26] JAE Archive. JAE/73-788. Adolfo González stayed at the ETH with Staudinger in 1917–18.

[27] See, for example: Ernest Fourneau, Ignacio Ribas, 'Stéréoisomérie et action anesthésique locale: Séparation du diméthylaminodiméthyléthylcarbinol en ses isomères optiques et préparation des deux stovaïnes actives', *Bulletin des sciences pharmacologiques*, 30(5) (1928), 273–8.

[28] *Diccionario de Profesores de Instituto vinculados a la JAE*. http://ceies.cchs.csic.es/?q=content/le%C3%B3n-maroto-andr%C3%A9s (last accessed 26 January 2018).

[29] Francisco García González, his son Juan Madinaveitia, but also his female pupils Carmen Gómez, Natividad González and Petra Barnés, the latter becoming F. Giral's wife and collaborator. Carmen Magallón, 'Mujeres de las ciències físico-químiques en España. El Instituto Nacional de Ciencias y el INFQ (1910–1936)', *Llull*, 20 (1997), 529–74.

2. = I. N. F. Q. = Laboratorio de Espectrografia.

3. = I. N. F. Q. = Laboratorio de Electroquímica.

Figure 2.1 Chemistry laboratories at the *Instituto Nacional de Física y Química* (INFQ), Madrid 1932. (a) spectroscopy (Catalán); (b) electrochemistry (Guzmán); (c) physical chemistry (Moles) and (d) organic chemistry (Madinaveitia).
De l'Espagne chimique (bref aperçu). Notes sur l'enseignement et la pratique de la chimie en Espagne. Avril 1934 (Madrid: IX Congreso Internacional de Química, 1934), p. 16 (2, 3, 4, 7). Reproduced with permission of the *Biblioteca de Catalunya*, Barcelona.

60   A Republican Science

c

4. = I. N. F. Q. = Laboratorio de Química física.

d

7. = I. N. F. Q. = Laboratorio de Química orgánica.

Figure 2.1 (cont.)

A New Enlightenment 61

laureate in chemistry in 1947 for his work on dyestuffs and alkaloids.[30] Research focused mainly on topics such as catalytic hydrogenation – one of the key fields imported from Willstätter's lab – the synthesis of natural products, naphthoquinones ('plumbagine'), vitamins, photochemical dimerization and joint work with physicist Julio Palacios on X-ray techniques. As was the case with other members of his family, Madinaveitia's committed socialism reinforced his network of professional and personal alliances.[31]

But perhaps the most prominent research school that flourished at the INFQ was Moles' physical chemistry lab. In the 1930s, already well settled in his university chair and committed to the new project of the Rockefeller, Moles became a prestigious and powerful group leader. Although physical chemistry was established in several science faculties, in terms of basic, new research, it was based on his own research school at the new institute. Moles's international reputation peaked during the Second Republic, and the productivity of his group at the INFQ was remarkable. Papers in Spanish in *Anales* did not hinder the appearance of other articles in prestigious international journals.[32] They were signed by Moles and his co-workers, many of them subsequently pursuing independent professional careers. The main research efforts focused on the experimental determination of atomic weights, gas adsorption, constants, gas density, viscosity, solubility and the study of hydrates and drying agents.[33] Although Cabrera was the director of the INFQ, Moles played his cards right in the new regime and, through his chemistry, came to epitomise the goals of the new republican chemistry: international standards, top-level basic research with potential industrial applications, new disciplinary boundaries and further university reforms.[34]

---

[30] Andrés León, Robert Robinson, 'Comparación de la estabilidad de diésteres', *Anales*, 30 (1932), 31–6. The research was carried out at the Dyson Perrins Laboratory, Oxford. León also worked at University College London and Manchester, and published other joint papers with Robinson.

[31] Indalecio Prieto, 'De cómo un Madinaveitia me lanzó a la política', *El Socialista*, 14 December 1938, 1–2; Manoel Toural, 'Antonio Madinaveitia. Un científico republicano', in Pedro Bosch Giral et al., *Protagonistas de la química en España: Los orígenes de la catálisis* (Madrid: CSIC, 2010), pp. 223–85.

[32] *Zeitschrift für Anorganische and Algemeine Chemie*, *Transactions of the Faraday Society*, *Monatshefte für Chemie* and *Comtes Rendues*.

[33] Among the members of Moles' research school at the INFQ, it is worth mentioning Miguel Crespí, Tomás Batuecas, Augusto Pérez Vitoria, Luis Solana, María Teresa Salazar, Gregorio Herrero, Juan Sancho, María Teresa Toral, P. Villán, Maximiliano Gutiérrez de Celis, Carlota Rodríguez de Robles, Carlos Nogareda, Ángel Vián and Fernando González Núñez.

[34] Siro Arribas, *Introducción a la historia de la química analítica en España* (Oviedo: Universidad de Oviedo, 1985), pp. 21–2.

The electrochemistry section was in the hands of Julio Guzmán, who had also begun his research work at the LIF some years earlier.[35] In 1912, Guzmán had worked with Carl Drucker (1876–1959) as a *pensionado* at Wilhelm Ostwald's Physical Chemistry Institute in Leipzig, another key node of the international chemical network. He took the *Physikalisch-Chemisches-Praktikum*, as well as working on scientific photography, electrolysis and liquid viscosity.[36] Back in Spain, Guzmán led the series *Trabajos de Electroquímica y Electroanálisis*.[37] Adolfo Rancaño also arrived at the INFQ to work with Julio Guzmán in the 1930s and later become a professor of physical chemistry at the Faculty of Science in Granada. At the INFQ, Guzmán collaborated with Moles on the *Cursos prácticos de Química-Física, Electroquímica y Electroanálisis* and later obtained a chair in electrochemistry at the University of Madrid. He also became the secretary of the institute, but unlike Madinaveitia, he had no defined political position in public.

Spectroscopy became a golden section of the INFQ, in which Catalán generated positive feedback among chemists and physicists. Catalán had already become one of the brightest Spanish scientists in the 1920s. After some early work in collaboration with del Campo, in 1922, he worked with the distinguished spectroscopist Alfred Fowler and published papers on multiplets for the interpretation of complex spectra. That same year, Sommerfeld had applied quantum chemistry to spectroscopy. Catalán complemented his work at the INFQ with his 1934 ad hoc chair in atomic molecular structure and spectroscopy at the University of Madrid. His international reputation also had a strong republican character, since he was a member of the political party *Izquierda Republicana*,[38] and he became one of the flagbearers of the republican JAE policies in the 1930s. Catalán's group at the INFQ worked on specific studies of the spectral lines of magnesium.

As in Moles's and Catalán's research groups, republican science soon meant an increase in the numbers of women in universities, research centres and scientific societies. At the INFQ, in the period of 1931–37, around 20 per cent of the academic staff were women, with 1934 being the strongest year for the recruitment of scientific staff, with a total of

---

[35] On Julio Guzmán, see: Rosario E. Fernández Terán, *El profesorado del 'Instituto Nacional de Física y Química' ante la Guerra Civil. El proceso de depuración y el drama del exilio*. PhD thesis (Madrid: Universidad Complutense de Madrid, 2005).
[36] In collaboration with Professor Carl Drucker, Guzmán published joint papers at the *Zeitschrift für Elecktrochemie*.
[37] José María Fernández-Ladreda, Emilio Jimeno, Mariano Marquina, Andrés León and Tomás Batuecas, among many others, published their research in that series.
[38] Republican Left Party.

A New Enlightenment 63

over 130.[39] Moles' research group, for instance, hosted 14 women in his lab,[40] and at the INFQ as a whole, the number of women chemists was double that of physicists. Also from a gender perspective, the case of the American chemist Mary Louise Foster (1865–1960) was particularly important.[41] After obtaining a PhD in chemistry in Chicago in 1914, in the 1920s, Foster had directed the International Institute for Girls in Spain (IIGS), in collaboration with the JAE, which put a special emphasis on the training of women and prepared its students for a professional career mainly as laboratory assistants in chemistry and pharmacy, but also in some cases for higher research training to enter the INFQ in the 1930s.[42] Foster described the training in experimental sciences in Spain in the following terms:

> Education in the sciences, based on practice in the laboratory, has been slow to develop in Spain. Not only were the ancient methods of instruction of memorised texts opposed to the modern methods but the excessive cost of laboratories maintenance was prohibitive. However, changes are coming rapidly, and have, in fact, already come in the University of Madrid, due to the influence and unremitting efforts of a splendid group of men and women who are inspired by the highest ideals for education.[43]

Foster was intellectually committed to the history of chemistry and chemical education as two main pillars for steering a new chemical culture internationally, but also in Spain, and these were, in her view, particularly important for women. She even attended the famous 1931 International Conference of the History of Science in London, a landmark for the introduction of Marxist ideology and methodology into what was at that time a young, emerging discipline.[44] From a leftist perspective very much

---

[39] Academic staff numbers at the INFQ were as follows: 1931 (63), 1932 (92), 1933 (121), 1934 (130), 1935 (79), 1936 (72), 1937 (28), 1938 (17) and 1939 (17). Magallón, *Pioneras españolas*, p. 237.

[40] Pura Barbero, Adela Barnés, Asunción Fernández, María Aragón García, María Luisa Garrayabal, Carmen Herrero, Narcisa Martín, Carmen Pardo, Carlota Rodríguez, Concepción Rof Carballo, María Teresa Salazar, María Teresa Toral and Pilar Villán. Magallón, *Pioneras españolas*, p. 226. See also: Antonina Rodrigo, *Una mujer silenciada, María Teresa Toral. Ciencia, compromiso y exilio* (Barcelona: Ariel, 2010).

[41] Foster attended the 1934 Chemistry Conference in Madrid. See: Carmen Magallón, 'El laboratorio Foster de la Residencia de Señoritas. Las relaciones de la JAE con el "International Institute for Girls in Spain", y la formación de las jóvenes científicas españolas', *Asclepio*, 59(2) (2007), 37–62; Margaret W. Rossiter, *Women Scientists in America before Affirmative Action, 1940–1972* (Baltimore/London: Johns Hopkins University Press, 1995).

[42] Mary Louise Foster, 'The Education of Spanish Women in Chemistry', *Journal of Chemical Education*, 8(1) (1931), 30–4.

[43] Ibid., p. 30.

[44] For the 1931 London International Conference of the History of Science, see: Nikolai I. Bukharin (ed.), *Science at the Cross Roads: Papers from the Second International Congress of the History of Science and Technology 1931* (London: Routledge, 2014).

in tune with the values of the internationalist atmosphere of the Second Republic, Foster also campaigned against the use of chemical weapons in wartime.[45]

In a political regime that had generally promoted academic autonomy and new regional political power – at least in the cases of the Basque Country and Catalonia – several chemists contributed to the new liberal values and became active in the co-construction between chemistry and politics. In 1933, an 'autonomous' University of Barcelona (UAB) took shape in the context of the new Autonomy Act (*Estatut d'Autonomia*), passed by the Republican parliament in Madrid a year earlier. From the beginning, García Banús played an important role in the governing body of the new UAB. The Governing Council was composed of five members from the Spanish government (García Banús being one of them), five members representing the Catalan government and the rector. In 1933, García Banús was also among the five members who wrote the new legal framework (*estatutos*) of the university. In 1931, with the establishment of the new Republic, he had already written a report on the weaknesses of the Spanish university system and on the need to appoint full-time academic staff and to create strong research schools with better international standards, as well as new university research institutes.[46]

In 1933, in the context of the Republican UAB,[47] and following similar developments in Madrid, Seville, Zaragoza and Oviedo, García Banús sketched the first plans for the creation of a new chemistry institute – the *Institut de Química* (IQ) – in Barcelona. It aimed to get chemistry professors from the science and pharmacy faculties to work together. It wanted to intensify research work from the previously established labs for physical chemistry, inorganic chemistry, organic chemistry and technical chemistry at the Faculty of Science and the lab of therapeutic chemistry at the Faculty of Pharmacy. The IQ had to attract the most brilliant students for further research training, which again ideally combined basic research and industrial applications. The IQ '*Patronat*' included the rector of the university, a representative of the *Generalitat* – the autonomous Catalan government – another from the Faculty of Science and the director of the institute. The budget, which had to rely on extra funding from the central government in Madrid, was devoted to research fellowships, invitations to foreign chemistry professors, books, journals and new lab

---

[45] Magallón, 'El laboratorio Foster', p. 48.
[46] Universitat de Barcelona. Archive. 'Univestiat Autònoma de Barcelona. Institut de Química' (15 November 1937) CAT AUB 02 C-I. I am indebted to Joaquim Sales for this information.
[47] Albert Ribas, *La Universitat Autònoma de Barcelona (1933–1939)* (Barcelona: Edicions 62, 1976).

equipment, as well as additional salaries for the chemistry professors, assistants and the director.[48] In spite of the political tensions of the Second Republic – in October 1934, under the conservative government of the *Confederación Española de Derechas Autónomas* (CEDA), the UAB and the Catalan autonomous administration were abrogated and not re-established until 1936 with the election victory of the leftist 'Front Popular' – the IQ aimed to optimise university research and produce a skilled workforce, to overcome the rigidities and bureaucracy of traditional faculties and to create a critical mass of professional chemists with international standards, all within the political framework of the Catalanist republicanism of the 1930s.

Notwithstanding the degree of internationalisation and Spain's growing scientific reputation at home and abroad, the fact is that the republican 1930s saw a proliferation of chemistry research schools. The INFQ strengthened the role of prestigious group leaders such as Moles and Madinaveitia, and university chairs and institutes reshaped the relationship between teaching and research. Although teaching, lab equipment, book translations and publications in local journals such as *Anales* occupied almost all of the daily working hours of many chemists, the cosmopolitan dream now enjoyed stronger institutional support and funding, all under a public rhetoric of chemistry as a tool for peace and solidarity among nations. It was a way to erase the nightmares of the chemical First World War and the new fears of approaching fascism from Italy and Germany.

Nobel Visitors

On 25 April 1931, just a few days after the proclamation of the Second Republic following the elections on 14 April, Marie Skłodowska-Curie (1867–1934) was welcomed at the *Sala Rectoral* of the University of Madrid. On 24 April, Curie gave a lecture on α, β and γ rays, chaired by Moles, who was at that time president of the SEFQ.[49] Her visit epitomised the quest for scientific internationalisation that, following earlier JAE policies, the new regime focused on.

In that context, Willstätter praised the efforts of the Spanish Republican Government for the internationalisation of chemistry. He openly expressed his political affinity with Madinaveitia, Moles and the other republican chemists in their success at involving the government in the promotion of

---

[48] Universitat de Barcelona. Archive. 'Univestiat Autònoma de Barcelona. Institut de Química' (15 November 1937) CAT AUB 02 C-I.
[49] *Anales*, 29 (1931), p. 187.

basic research instead of making the war effort a priority.[50] This was probably a utopian dream, but chemists in their public addresses and political positions were keen to present chemistry as a new, modern, peaceful science that could potentially contribute to international collaborations instead of the national rivalries that had 'corrupted' this science in the First World War. Staudinger's lectures in the spring of 1933 represented another example of the cosmopolitan chemistry that the republican elites wanted to promote. Staudinger spoke at the University of Madrid on the constitution of rubber and cellulose.[51] In fact, in the early years of the Second Republic, public funds went to the INFQ for the improvement of science faculties, but also for the hosting of foreign luminaries and emblematic events such as summer schools and international conferences.[52]

This was the case for the Santander *Universidad Internacional de Verano* (International Summer School), which acted as a site for international academic exchange. From 9 to 20 August 1933, the *Reunión de Ciencias Químicas* (Chemical Sciences Meeting) brought together a group of prestigious foreign chemists at the *Palacio de la Magdalena* (Figure 2.2).[53] Nobel Prize winners such as Fritz Haber and Willstätter mingled with Spanish names such as del Campo, J. Giral, O. Fernández, Madinaveitia, Moles, Calvet and Ribas.[54] Keynote speeches and some of the lectures were later translated and printed in Spanish.[55] Willstätter acknowledged that 'it was expensive for the government, but how much cheaper than even a single cannon. The audiences were very small, since none of us were fluent in Spanish'.[56] In spite of its limitations in terms of reaching a wide

[50] Willstätter, *From My Life*, p. 413.
[51] Hermann Staudinger, 'Sobre la constitución del caucho. Conferencia explicada en la Universidad de Madrid el 23 de marzo de 1933 (traducción de I. Jiménez Herrero)', *Anales*, 32 (1934), 426–35; Hermann Staudinger, 'Sobre la constitución de la celulosa. Conferencia pronunciada en la Universidad de Madrid el 23 de marzo de 1933 (traducción de J. Sureda)', *Anales*, 32 (1935), 74–6.
[52] Cobos, 'Luces y sombras'.
[53] The list of foreign luminaries included G. Barger (Edinburgh), E. Billmann (Copenhagen), E. Cohen (Utrecht), H. von Euler (Stockholm), F. Fichter (Basel), C. Matignon (Paris), N. Parravano (Rome), E. Späth (Vienna), W. Schlenck (Berlin), Seidell (Washington, DC), N. Zelinsky (Moscow) and J. Gérard (general secretary of the International Union of Chemistry).
[54] As well as engineers José Antonio Artigas and Enrique Hauser. Benito Madariaga, Celia Valbuena, *La Universidad Internacional de Verano en Santander (1933–1936)* (Guadalajara: Universidad Internacional Menéndez Pelayo/Ministerio de Universidades e Investigación, 1981), pp. 87, 91. Haber lectured on auto-oxidation, Zelinsky on the organic origins of oil, Cohen on the instability of matter and Matignon on high-temperature chemistry. Enrique Moles, 'La 1ª Reunión internacional de la Universidad Internacional de Verano de Santander', *Anales*, 31 (1933), 352–5, p. 352.
[55] By Haber, Matignon, Parravano, von Euler, Barger and Willstätter. See: Willstätter, *From My Life*, p. 413.
[56] Ibid., p. 413.

Figure 2.2 The 1933 Summer School in Santander. Front row (seated): F. Haber, R. Willstätter, H. von Euler, E. Billmann, E. Cohen, N. Parravano, C. Matignon, E. Hauser, F. Fichte; second row: Ms Cohen, Ms Ribas, Ms Seidell, Ms Calvet, Ms del Campo, G. Barger, Ms del Fresno, J. Gérard, P. E. Barredo; third row: F. Calvet, A. del Campo, A. Pérez Vitoria, E. Moles, C. Del Fresno, A. Madinaveitia, I. Ribas, A. Seidell.
Enrique Moles, 'La 1ª Reunión internacional de la Universidad Internacional de Verano de Santander', *Anales*, 31 (1933), 352–5, p. 352. Reproduced with permission of the *Real Sociedad Española de Química*.

audience, the 1933 summer school acted as a marketing strategy for the republican chemists of the INFQ, for the renewed science faculties and for government foreign policy.

A year later, in 1934, an international chemistry conference in Madrid became one of the main paradigms of the republican dream to overcome post-war international conflicts and the 1929 economic crash. On 5 April 1934, O. Fernández was ready for the opening session of the 9th International Congress of the International Union of Pure and Applied Chemistry (IUPAC).[57] Fernández presented the state of the art of

---

[57] Mary Louise Foster, 'Chemistry in Spain', *Journal of Chemical Education*, 11(7) (1934), 426–7.

chemistry that year as compared to 1912. The consequences of the First World War, the damage of the 1929 Wall Street Crash and some reluctance from the political line of the new Republic had delayed the organisation of the event for years. O. Fernández acted as a qualified representative of a worldwide generation of chemists that had achieved a lot in terms of chemical progress. They had developed an electronic interpretation of the chemical bond, discovered new elements and isotopes and synthesised new natural products and artificial molecules and polymers, but also made enormous advances in the synthesis and industrial production of new drugs, heralding the start of the biological turn, in which carbohydrates, enzymes, steroids, vitamins and alkaloids paved the road towards biochemistry.[58] In his speech, Fernández introduced himself as a privileged witness to the blossoming of chemistry in Spain.[59] Before a crowded, distinguished audience, which included first-rate political authorities and Nobel Prize winners, he presented a modern, optimistic and cosmopolitan image of chemistry. Perhaps for the first time, the Spanish community, led by the liberal chemists in tune with the values of the Republic, talked face to face with international luminaries about the past, present and future prospects of the discipline. That optimistic, modern message was in tune with the achievements of the Spanish chemical community in the early decades of the century, which Fernández aimed to stress during that unique opportunity (Figure 2.3).

Despite notable absences from Latin America and from the United States, the conference attracted around 1,200 delegates, mainly from Italy, France and Germany. Also at the opening ceremony at the lecture hall of the *Residencia de Estudiantes*, the President of the Spanish Republic, Niceto Alcalá Zamora, chaired the session, together with the Minister for Public Instruction and Fine Art, Salvador de Madariaga, and the Minister for Trade and Industry, Ricardo Samper. Alcalá Zamora spoke to all delegates with great enthusiasm about the contribution of chemistry to progress and, again, made a particular reference to the use of chemical weapons. Alcalá Zamora:

sympathized with that philosophy of chemistry that sought to benefit humanity in every way, and [...] he wished every success to result from the research of those interested in this aspect, but [...] on the day that a government demanded the production of catastrophic bombs and asphyxiating gases, he

---

[58] Carsten Reinhardt (ed.), *Chemical Science in the Twentieth Century: Bridging Boundaries* (Weinheim: Wiley-VCH, 2001).

[59] Obdulio Fernández, 'La evolución de la química desde el VIII Congreso Internacional', in *IX Congreso Internacional de química pura y aplicada. Madrid 5–11 Abril 1934* (Madrid: C. Bermejo, 1934), pp. 3–48.

Nobel Visitors 69

Figure 2.3 (a) The opening ceremony of the 9th International Conference of the International Union of Pure and Applied Chemistry (IUPAC), Madrid, April 1934, and (b) the emblem of the conference. (a) Augusto Pérez Vitoria, 'El IX Congreso Internacional de Química pura y aplicada y la XI Conferencia de la Unión Internacional de Química', *Anales*, 32(2) (1934), 195–207, p. 196. Reproduced with permission of the *Real Sociedad Española de Química*; (b) *De l'Espagne chimique* (front cover). Reproduced with permission of the *Biblioteca de Catalunya*, Barcelona.

Figure 2.4 Gilbert Newton Lewis during his plenary lecture at the IUPAC Conference, Madrid, 1934: 'On the different kinds of water'. Augusto Pérez Vitoria, 'El IX Congreso Internacional de Química pura y aplicada y la XI Conferencia de la Unión Internacional de Química', *Anales*, 32(2) (1934), 195–207, p. 200. Reproduced with permission of *Real Sociedad Española de Química*.

could do no less than hope that their minds would refuse to function on such research.[60]

After the formalities, the plenary lectures began. The distinguished keynote speakers included Gilbert Newton Lewis (Berkeley) on the isotopes of water (Figure 2.4), Giovanni Battista Bonino (Bologna) on the Raman spectrum and F. Giral's mentor, Richard Kuhn (Heidelberg), on flavine. Camille Matignon (Paris) had also been scheduled to talk about phosphorus derivatives, but he had died in March 1934.[61] The conference also offered 19 introductory lectures and 274 papers, 96 of which were presented by Spanish chemists – many of them young students – and it hosted eight scientific sections: (1) theoretical and physical chemistry; (2) inorganic chemistry; (3) organic chemistry; (4) biological chemistry; (5) analytical chemistry; (6) agricultural chemistry; (7) chemical education and chemical economy; and (8) combustible chemistry.[62]

[60] Quoted by Foster, 'Chemistry in Spain', pp. 426–7.
[61] Finally, engineer Augusto Gálvez gave a plenary lecture on the history of quicksilver and silver metallurgy, two emblematic topics of the Spanish chemical colonial tradition.
[62] Augusto Pérez Vitoria, 'El IX Congreso Internacional de Química pura y aplicada y la XI Conferencia de la Unión Internacional de Química', *Anales de Física y Química*, 32(2) (1934), 195–207, p. 199.

A keenness to anchor the Spanish chemical community in international networks completed the ceremony. Henry E. Armstrong, Le Chatelier, Lewis, Nicola Parravano and Paul Walden received honorary doctorate degrees at the Faculty of Science, and Fourneau, Paul Karrer and Robert Robinson obtained honorary doctorate degrees at the Faculty of Pharmacy.[63] The conference ran in parallel with the meeting of the International Union of Chemistry, at which some Spanish delegates took an active part in the inorganic, organic and biological nomenclature sections, as well as in the section of international constants and patrons.[64] The final banquet was attended by a considerable political representation that, in public speeches, emphasised once again the universal fraternity of chemistry, its contribution to peace and its role as a cosmopolitan discipline.[65]

A booklet entitled *De l'Espagne chimique* was available to all conference delegates, aiming to provide a general overview of the state of the art of chemistry in Spain in the mid-1930s in academic and industrial settings.[66] Although the subtitle focused on chemistry education (*Notes sur l'enseignement de la chimie en Espagne*), and despite its centralised bias, which basically only presented chemistry schools and labs in Madrid, the booklet provided a useful overview. The SEFQ, the INFQ, the JAE and, to a lesser extent, the Academy of Sciences

---

[63] Javier Puerto, *Giral: El domador de tormentas, la sombra de Manuel Azaña* (Madrid: Corona Borealis, 2003). Del Campo introduced the candidates to receive the awards at the science faculty, and J. Giral did the same for the pharmacy faculty. The Republican 'chemical foreign office' went even further. At the Madrid Academy of Science (*Academia de Ciencias, Exactas, Físicas y Naturales*), other distinguished delegates were formally nominated as corresponding members, with the famous Spanish inventor Leonardo Torres Quevedo being the president of the Academy. Enrique Hauser introduced the nominees: Barger (Edinburgh), Bartow (Iowa), Bertrand (Paris), Bonino (Bologna), Cohen (Utrecht), Fichter (Basel), Ruff (Breslau), Sörensens (Copenhagen), Späth (Vienna), Votolek (Prague), who spoke on behalf of all nominees, and Zelinksy (Moscow). Pérez Vitoria, 'El IX Congreso Internacional'.
[64] Obdulio Fernández, Enrique Hauser, D. C. López-Sánchez and Carlos del Fresno worked in the inorganic nomenclature section, Antonio Madinaveitia in the organic nomenclature section, José Giral in the biological nomenclature section, Blas Cabrera, del Fresno, D. E. Hauser and Moles in the international constants section and del Campo at the International Office of Chemistry and Physics patrons. Ibid., p. 204. On del Campo, see: Francisco González de Posada et al., *Ángel del Campo y Cerdán. Eminente químico español. Cuenca 1881; Madrid 1944* (Madrid: Amigos de la Cultura Científica. Academia de Ciencias e Ingenierías de Lanzarote, 2006).
[65] Alejandro Lifchuz, 'El IX Congreso Internacional de Química Pura y Aplicada', *Química e Industria*, 124 (1934), 105–11.
[66] *De l'Espagne chimique (brief aperçu). Notes sur l'enseignement de la chimie en Espagne. Avril 1934* (Madrid: IX Congreso Internacional de Química, 1934).

appeared as the only 'pure'-research institutions. Chemistry teaching (often combined with some research) took place in science faculties (Madrid, Barcelona, Zaragoza, Oviedo, Seville, Salamanca, Valencia, Granada),[67] pharmacy faculties (Madrid, Barcelona, Santiago, Granada) and industrial engineering schools (Madrid, Barcelona, Bilbao), whereas chemistry teaching was also common at civil engineering schools, faculties of medicine, trade and customs and veterinary schools. Chemists also had a say in other labs and centres such as the institutes of oceanography, geology, physiology and legal medicine, in the Museum of Natural History and at the national institutes for hygiene, pharmacology, military health and animal biology. Other military installations and state industries and labs added new names to a long list of institutions. Since Spain in the 1930s was not a leading country in terms of science, the list aimed to impress delegates at the conference. Profusely illustrated, the booklet included references to beer, perfume, gas and concrete chemical industries in Madrid and other industrial areas in which chemistry played an important role, such as Barcelona, Flix, Sabiñánigo, Almadén, Puertollano, Huelva, Bilbao, Santander and Oviedo (Figure 2.5).

L'Espagne chimique, the honorary doctorate degrees, all of the events and the impact of the 1934 conference, along with the cosmopolitan atmosphere of the summer schools in Santander, contributed to situating Spanish chemistry as a standardised profession at the international level. Perhaps not by chance, in 1935, Bermejo, Madinaveitia and J. Giral acted for the first time as nominators of the Nobel Prize for Chemistry.[68] Equally, the flow of Nobel visitors did not stop until the beginning of the Civil War in July 1936. Some months earlier, for instance, George Barger (1878–1939), the director of the medical chemistry department of the University of Edinburgh and one of the acclaimed guests in Madrid in 1934, visited Barcelona to give three public lectures (in French): one on ergometrine at the Catalan Society of Biology, one on the chemistry of the thyroid hormone at the UAB and a third lecture on cholesterol at the Barcelona branch of the SEFQ.[69]

---

[67] As well as partially in Valladolid, La Laguna, Cádiz and Murcia.
[68] 'Nomination Database'. Nobelprize.org. Nobel Media AB 2014. www.nobelprize.org/nomination/archive/show_people.php?id=5441 (last accessed 29 January 2018). See also: Elisabeth T. Crawford, John L. Heilbron, Rebecca Ullrich, *The Nobel Population 1901–1937: A Census of the Nominators and Nominees for the Prizes in Physics and Chemistry* (Berkeley: Office for History of Science and Technology/University of California, 1987).
[69] *La Vanguardia*, 13 May 1936.

The Silver Age of Industry

Late nineteenth-century science-based industry, and what historian Jeffrey Allan Johnson has called the academic–industrial symbiosis in German chemistry in the early decades of the twentieth century,[70] influenced Spanish chemists, especially after their frequent training as *pensionados* in close contact with several German research schools and industries. The academic–industrial partnership was therefore an important challenge for republican chemistry. Recent research has unveiled, for instance, J. Giral's industrial entrepreneurship in the areas of food, drugs and chemical analysis, in tune with his prestigious academic profile and his strong political commitment.[71]

As Foster emphasised in her report of the 1934 conference in Madrid, the potential for the Spanish chemical industry seemed quite promising:

> Spain is rich in mineral wealth. Its mines of pyrite are the richest in the world, having yielded during the past year 1,831,000 tons for export. Mercury, zinc, galena, manganese, phosphate, and potash deposits of extraordinary abundance are only a few of the mineral deposits. Mines exploitation originated in early Roman times. Spaniards have been in no sense unaware of the existence of these potential riches, but they have lacked initiative in the past, and the mines have generally been in the hands of foreigners. The situation is changing. That there is change is largely due to the constant effort, pressure, and initiative of the *Junta para Ampliación de Estudios*.[72]

That contemporary, rather optimistic vision opposes the received view of chemistry in the JAE years, which supposedly only focused on elite basic research, mainly centralised in Madrid, neglecting the application of chemistry to local industrial needs across the territory. This was one of the core theses of chemists in Franco's regime, their aim being to legitimise an abrupt institutional discontinuity after the Civil War (see Chapter 4). Nevertheless, the industrial key was also at the heart of the JAE reforms. It became a basic pillar of the modernisation project, in tune with the republican spirit. There is evidence of the 1930s growth of the Spanish chemical industry, with more efficient exploitation of raw materials and notable international cooperation. The capacity for innovation, however,

---

[70] Jeffrey Allan Johnson, 'The Academic–Industrial Symbiosis in German Chemical Research, 1905–1939', in John Lesch (ed.), *The German Chemical Industry in the Twentieth Century* (Dordrecht: Kluwer, 2000), pp. 15–56.
[71] Javier Puerto, 'José Giral Pereira (1879–1962) profesor de bioquímica', *Panacea*, Junio (2015), 21–5. Puerto refers to several of J. Giral's papers at the *Archivo Histórico Nacional*, Madrid.
[72] Foster, 'Chemistry in Spain', p. 427.

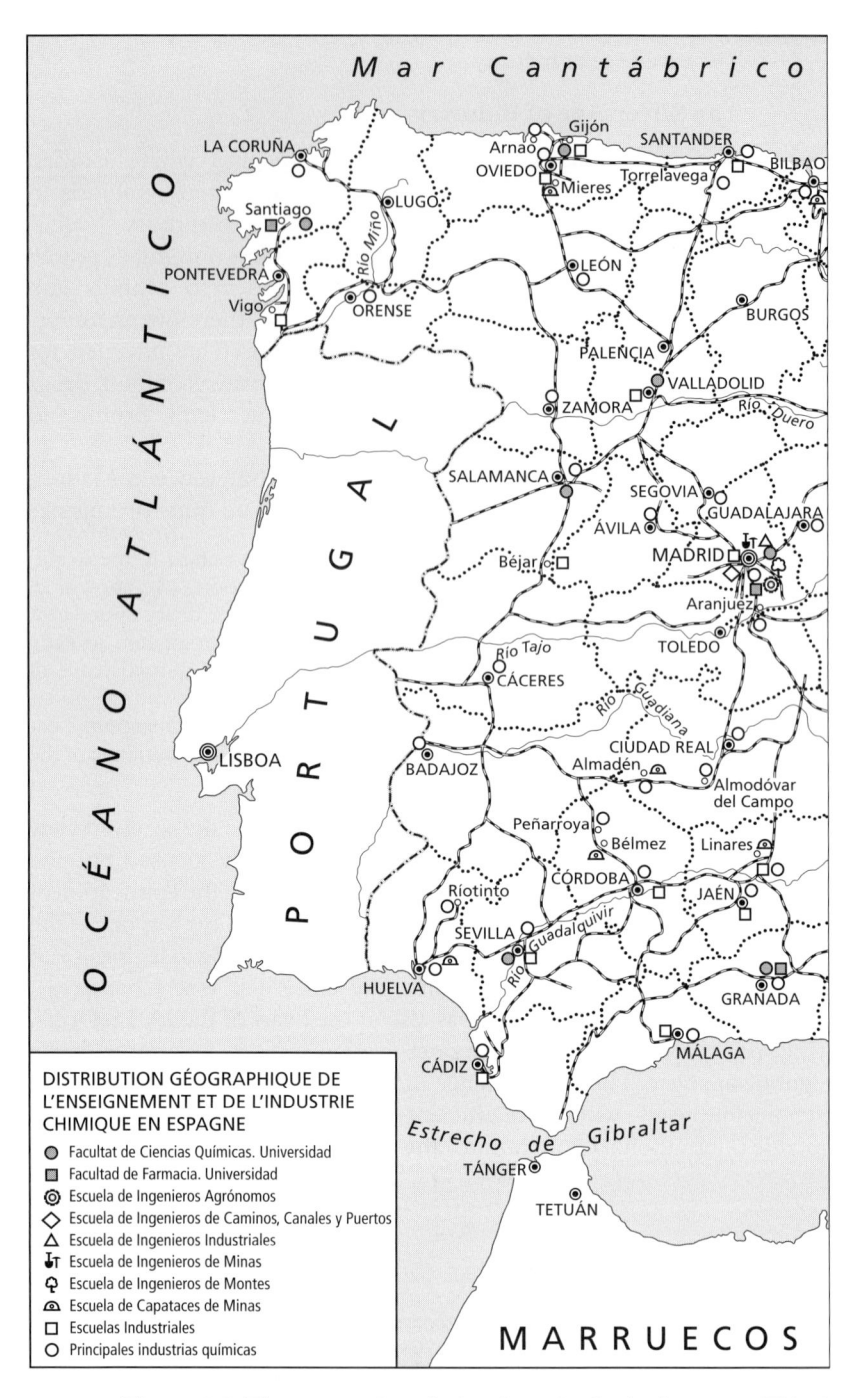

Figure 2.5 The geography of chemistry in Spain in 1934. This is a recreation of a 1934 map from the leaflet *De l'Espagne chimique* (1934). Names appear as they were published in 1934.
Reproduced with permission of the *Biblioteca de Catalunya*, Barcelona.

FRANCIA
SAN SEBASTIÁN
VITORIA
PAMPLONA
LOGROÑO
Sabiñánigo
ANDORRA
HUESCA
GERONA
Cascante
Tudela
Río Ebro
Cardona
SORIA
LÉRIDA
ZARAGOZA
Tarrasa
Épila
BARCELONA
Villanueva y Geltrú
Flix
TARRAGONA
TERUEL
ISLAS BALEARES
I. MENORCA
CUENCA
CASTELLÓN
PALMA
Sagunto
I. MALLORCA
VALENCIA
Río Júcar
I. IBIZA
ALBACETE
Alcoy
Hellín
Río Segura
ALICANTE
MURCIA
Cartagena
Mar Mediterráneo
ALMERÍA
OCÉANO ATLÁNTICO
I. ALEGRANZA
I. GRACIOSA
I. LANZAROTE
I. LA PALMA
I. TENERIFE
I. FUERTEVENTURA
La Laguna
SANTA CRUZ DE TENERIFE
I. GOMERA
LAS PALMAS
I. HIERRO
I. GRAN CANARIA
OCCIDENTALES
ORIENTALES
ISLAS CANARIAS

remained weak, particularly in small firms, so debate on the lack of the scientific, technical and managerial capacities of industrial chemists as experts was still in its infancy. Technical schools, universities, professional associations and industrial policy-makers obviously had a lot to say in this long-standing debate – as reflected in the journal *Química e Industria* – which would go beyond the liberal period of the early 1930s.[73]

Economic historians place the early dynamism of the Spanish chemical industry mainly in the 1910s, and particularly during the First World War,[74] when the JAE had already launched its scientific policies through *pensionados*, new labs and training reforms, whereas the old industries of explosives, mineral acids and alcohol progressively shifted towards dyes, fertilisers, perfumes and pharmaceuticals. Although growth trends seemed to weaken after the war, chemists had new opportunities beyond the strict academic arena. In the mid-1930s, more than 300 chemical firms represented nine per cent of the whole economy.[75] Mineral acids and alkalis remained key compounds for use in several industrial processes thanks to local production by prosperous firms such as Cros, but also to the progressive introduction of foreign investors such as Solvay.[76] In 1914, the country produced 33,000 metric tonnes of sulphuric acid, whereas by 1935 figures had risen to 334,000. Equally, the making of explosives experienced significant diversification in terms of production: gunpowder for the navy, phosphates in Morocco, potash in Catalonia and pyrites in Huelva.[77]

Antonio Llopis, President of the *Cámara Nacional de Industrias Químicas* (CNIQ),[78] gave a plenary lecture on the chemical industry

---

[73] For the evolution of the pharmaceutical industry, see: Raúl Rodríguez Nozal, 'Orígenes, desarrollo y consolidación de la industria farmacéutica española (ca. 1850–1936)', *Asclepio*, 52(1) (2000), 127–59.

[74] Nuria Puig, 'The Frustrated Rise of Spanish Chemical Industry between the Wars', in Anthony S. Travis et al. (ed.), *Determinants in the Evolution of the European Chemical Industry, 1900–1939* (Dordrecht: Kluwer, 1998), pp. 301–20.

[75] Ibid., p. 304.

[76] Ángel Toca, *La introducción de la gran industria química en España. Solvay y su planta de Torrelavega (1887–1935)* (Santander: Universidad de Cantabria, 2005).

[77] Puig, 'The Frustrated Rise'.

[78] The *Cámara Nacional de Industrias Químicas* (CNIQ) was created in 1919. For an earlier address by its president, Antonio Llopis, see: 'Conferencia del Presidente de la Cámara Nacional de Industrias Químicas', *La Voz de la Farmacia*, 23 (1932), 24. See also: Antonio M. Llopis, 'La industria química en España', in D. J. Péres Unzueta (dir.), *Anuario de industrias químicas* (Barcelona: Cámara Nacional de Industrias Químicas, 1932), pp. xiv–xxiv. Raul Rodríguez Nozal, 'Sanidad, farmacia y medicamento industrial durante la segunda República', *Papeles de la Fundación de Investigacions Marxistas*, 28(2) (2008), 163–85.

in Spain at the 1934 International Conference.[79] Organisers introduced Llopis as a key expert in Spanish raw materials and chemicals so as to improve their rational exploitation. He pointed out the leading position of the country in terms of pyrite ores as raw materials for the production of sulphuric acid and phosphates.[80] Spain was also rich in nitrogen, sphalerite, calamine, galena, manganese, sodium and potassium chloride, as well as other minerals used in the chemical industry, covering internal demand and allowing for some exportation. Internal demand for nitric acid and hydrochloric acid was also covered. In addition, there were three factories for the production of ammonia fertilisers, which at that time competed well for the internal market with the foreign Haber–Bosch process.[81] Llopis stressed the positive results of the olive oil industry and the use of carbon sulphide and trichloroethane as extraction solvents. He remarked on Spain's extraction of natural products, but also on the weaknesses of its organic industrial synthesis. Only a few factories produced perfumes and artificial dyestuffs, and all needed state support. There was also small-scale production of nitrocellulose paint, artificial silk and wood distillation products (acetic acid, methanol, formaldehyde and acetone). Dyes and colorants and protectionist policies that were intended to defend Spanish industry from the German hegemony up to the First World War soon gave way to the *Fábrica Nacional de Colorantes y Explosivos* (FNCE). The FNCE dominated the market from 1922 onwards, under the partial control of I. G. Farben in 1926, who led the quest for synthetic fuel in the late 1920s. Growth in fertilisers was also significant during that period.[82]

In the 1930s, some landmarks along the transition between artisanal drugs and industrial production should also be taken into account. One, for example, was the case of the production of 'neo-salvarsan' by *Laboratorios Esteve* and synthetic sulphonamides by *Laboratorios*

---

[79] Antonio M. Llopis, 'Primeras materias y productos químicos españoles', in *IX Congreso Interacional de Química Pura y Aplicada. Tomo IX, Grupo VIII. Conferencias, memorias y comunicaciones de enseñanza y economía* (Madrid: C. Bermejo, 1934), pp. 5–12. I thank Joaquim Sales for providing the original text of Llopis' lecture.
[80] Llopis, 'Primeras materias', p. 8.
[81] The production of inorganic salts also flourished. Llopis, 'Primeras materias', p. 10.
[82] From the production of 76,000 metric tonnes of superphosphates in 1905, figures rose to 1,076,000 in 1935. In addition, whereas in 1900 all nitrogen fertilisers came from imports, in 1935 the country was already producing 3,800 metric tonnes. The *Sociedad Española de Industrias Aragonesas* produced the first nitrogen fertilisers in the period of 1920–30 (cyanamide, ammonia, ammonia sulphate); in 1930, the *Sociedad Española del Nitrógeno* in Asturias used the Haber–Bosch process for the first time.

Andreu.[83] Similarly, in 1936 – just before the Civil War – there were three factories of artificial fibres: *Sociedad Anónima de Fibras Artificiales* (SAFA), *La Seda de Barcelona* and *Sociedad Española de Seda Artificial*, importing cellulose as a raw material.[84] Similarly, since 1930, chemist José Manuel Pertierra had led the *Instituto del Carbón* at the University of Oviedo in the mining district of Asturias and had worked on methods of hydrogenation of coal and oil to obtain fuels for combustion. In 1933, in close collaboration with the Department of Inorganic Chemistry led by Carlos del Fresno and the *Istituto di chimica industriale di Milano*, it became the *Instituto de Química Aplicada*.[85]

As Moles had already stated in 1929, basic research was not incompatible with industrial applications (see Chapter 2). At the INFQ, Madinaveitia had clear ideas on how to link academic research with industry. He established collaborations with sugar firms, and some of his co-workers ran joint projects with pharmaceutical laboratories.[86] Private chemical firms backed the activities of the Barcelona academic chair in technical chemistry, and student visits to factories were frequent.[87] But on a more institutional level, the creation of the *Fundación Nacional para Investigaciones Científicas y Ensayos de Reformas* (FNICER) can be considered a landmark of the republican policies towards the industrial Silver Age. The FNICER aimed to complement JAE policies by establishing industrial links with public and private firms, funding new labs in research centres and universities (again, the laboratory emerged here as a crucial space for the modernity of chemistry), decentralising the Madrid-orientated structure of science policies and promoting and piloting small-scale, well-focused projects (*ensayos*).

---

[83] Raul Rodríguez Nozal, Antonio González Bueno (eds.), *El medicamento de fabricación industrial en la España contemporánea* (Madrid: CERSA, 2008).
[84] Emilio de Diego, *Historia de la industria en España. La Química* (Madrid: Actas Editorial, 1996), p. 99.
[85] José Manuel Pertierra, 'Disolución coloidal e hidrogenación de un lignito', *Anales*, 31 (1933), 779–809.
[86] Laboratories Abelló, Urigoiti, and Ibys. F. Giral, 'Química orgánica (1932–1939)', p. 42 (see Chapter 7).
[87] 1934–35: *Subproductos del alquitrán*, B. Bardinas, *Catalana de Gas y Electricidad*, *Vidriería Barcelona*, FNCE, CAMPSA, *Sociedad electroquímica Flix*; 1935–36: FNCE, *Catalana de Gas y Electricidad*, *Nacional Pirelli* (*Vilanova i la Geltrú*), *Damm*, *Myrurgia*, *Potasas Ibéricas* (*Suria*), *Sociedad Española de Carburos Metálicos* (*Berga*). It also included phosphates, paper, sulphuric acid and '*La industrial química*' in Zaragoza, the electrochemical plant at Flix (Tarragona) and the chemical plant in Sabinánigo (Huesca). *La Vanguardia*, 12 July 1936.

In 1934, the FNICER achieved its first tangible results.[88] In spite of its fragile status and brief period of existence, the Foundation opened new university laboratories in Zaragoza (haematology), Salamanca and Oviedo (chemistry), Santiago (geochemistry) and Valencia (metallurgy). Ignacio Ribas, from Madinaveitia's research school at the INFQ, led the organic chemistry laboratory in Salamanca. Despite resistance from farming groups, in 1933 the *Centro de Investigaciones Vinícolas* (CIV) regulated a good part of the alcohol market, and a major cartel of alcohol industries was established in 1935. Led by the agricultural engineer Juan Marcilla (1886–1950), but in collaboration with a wider network of wine producers, it was devoted to improving the quality of Spanish wines through chemical research into their composition and fermentation processes.[89] Marcilla led a research school focusing on the chemistry of Spanish wines, particularly their chemical analysis in Andalusia, and he created the journal *Anales del Centro de Investigaciones Vinícolas*. The CIV endorsed the republican agricultural reforms, favouring small-scale farms and limited land ownership – one of the landmarks of the republican policies – which was linked to the liberal left.[90]

Other projects underlining the symbiosis between academic and industrial chemistry paradoxically came again from the opposition to the republican government. The dissolution of the Society of Jesus by the government in 1932 interrupted the IQS project, but it had already left its mark in strong alliances with chemical entrepreneurs and a particular style of chemistry education that was very much focused on laboratory experiments and industrial needs, which had to be revived after the Civil War (see Chapter 4).[91] The director of the IQS, Eduardo Vitoria, was deeply conservative, very averse to fashion and novelties and fully removed from the liberal, republican values of the 1930s, but he was a reputable experimental chemist who devoted his entire life to

---

[88] Santiago López, 'El Patronato Juan de la Cierva (1939–1960), I. Parte: Las Instituciones precedentes', *Arbor*, 619 (1997), 201–38. See also: Rosario Fernández Terán, Francisco González Redondo, 'Blas Cabrera y la Física Española durante la República', in Elena Ausejo (ed.), *La ciencia española en la II^a República*, pp. 67–105, 83; Justo Formentín, Esther Rodríguez, *La Fundación Nacional para Investigaciones Científicas (1931–1939)* (Madrid: CSIC, 2009); Santiago López, 'Las ciencias aplicadas y las técnicas: La Fundación Nacional de Investigaciones Científicas y Ensayos de Reformas y el Patronato Juan de la Cierva del CSIC (1931–1961)', in Ana Romero and María Jesús Santesmases (eds.), *Cien años de política científica en España* (Madrid: Fundación BBVA, 2008), pp. 79–106.

[89] Santiago López, 'El Patronato Juan de la Cierva (1939–1960), I Parte'.

[90] Ibid., p. 223. See also: Manuel Lora-Tamayo, *La investigación química española* (Madrid: Alhambra, 1981), p. 315.

[91] Núria Puig, Santiago López, *Ciencia e industria en España: El Instituto Químico de Sarriá, 1916–1992* (Barcelona: Institut Químic de Sarrià, 1992), pp. 77–100.

his work. His textbooks and publications had a great impact internationally, particularly in Latin America (see Chapter 1).[92] In disagreement with the Republican government, Llopis and the CNIQ – which united most chemical entrepreneurs – regretted the closure of the IQS and continued to praise its capacity to drive chemical innovation in industry.[93]

In 1933, in the university context, conservative chemist Emilio Jimeno created the *Instituto de Metalurgia y Mecánica* (IMM), which worked in close collaboration with Catalan firms.[94] In 1935, he began publishing the journal *Metalurgia y Construcción Mecánica* (*MCM*), which attempted to spread knowledge of the main international innovations in the field, together with his own views on the relationship between science and industry.[95] In *La ciencia y la industria*,[96] Jimeno denounced the lack of an industrial research tradition and the need to further stimulate new science-based industry patterns. He listed the names of the earliest prominent twentieth-century American academic chemists – many of them influenced by Ostwald's physical chemistry school – and their specific links to powerful firms such as General Electric, Union Carbide, Bakelite Corporation, DuPont and Dow Chemical.[97] For the actual making of a successful link between academic chemistry and industry, Jimeno believed that mediators such as technicians, popularisers, teachers and entrepreneurs played a crucial role that needed to be imitated in Spain with utilitarian, protectionist aims and by reinforcing the 'scientific–industrial teaching of chemistry'.[98] In October 1935, the first issue of *MCM* clearly demonstrated Jimeno's project, his network of collaborators and his intended audience. The 1930s became a crucial period for the creation of Jimeno's research school, which enjoyed a degree of continuity despite the Civil War, and its mark remained until the 1960s, the peak of Franco's dictatorship (See Chapter 4).[99]

---

[92] Lluis Victori, 'Eduard Vitoria i Miralles: Pioner de la química al nostre país', *Revista de la Societat Catalana de Química*, 7 (2006), 60–4, p. 63. See also Joaquín Pérez Pariente, 'El jesuita Eduardo Vitoria: La química como apostolado', in Pedro Bosch Giral et al., *Protagonistas de la química en España*, pp. 61–116

[93] *La Vanguardia*, 10 February 1932, p. 7.

[94] *La Maquinista Terrestre y Marítima, Altos Hornos de Cataluña* and *Fundación Escorza*.

[95] Felipe Calvo, José María Gulemany, 'Contribución del profesor Emilio Jimeno Gil al prestigio de la Universidad de Barcelona: Rector de 1939 a 1941', in *Història de la Universitat de Barcelona. I Simposium del 150 Aniversari de la Restauració* (Barcelona: Publicacions de la Universitat de Barcelona, 1990), pp. 471–83, 473–4.

[96] Emilio Jimeno, Facundo Rolf Morral, *La ciencia y la industria* (Barcelona: Publicaciones del Instituto de la Metalurgia y la Mecánica. Universidad de Barcelona, 1936) (articles published in *Metalurgia y Construcción Mecánica* in 1935 and 1936).

[97] Ibid., p. 12.   [98] Ibid., p. 26.

[99] Emilio Jimeno, *Antecedentes del Instituto Tecnológico Metalúrgico 'Emilio Jimeno'. Consideraciones sobre educación* (Barcelona: Facultad de Ciencias. ITMEJ. Universidad de Barcelona, 1970). Among Jimeno's pupils, José Ibarz is worth noting (1905–72). He

Chemistry and its industrial dimension acted as a sort of wild card to be appropriated by different groups for the sake of their own professional interests and ideological positions in society. Socialist Madinaveitia's links with industry, the FNICER wine chemistry programme in tune with agricultural reform and the broad interest in Spanish raw materials and their exploitation as expressed by the political authorities at the 1934 Madrid conference, among other alliances, coexisted with conservative approaches to industrial chemistry such as those supported by Vitoria and Jimeno. In one way or another, what all of these examples show is the unquestionable vitality of the Silver Age of industry during the Second Republic. They challenge the frequent criticism of a supposedly overly academic chemistry at the height of a troubled and finally frustrated republican dream.

***

Obdulio Fernández's bombastic address to the delegates of the IUPAC Conference in Madrid in 1934 stressed the importance of the republican dream of progress and cosmopolitanism in the following terms:

If our satisfaction as chemists is great, for we share the pleasure of this invitation to our country with our companions from all parts, that, which we as Spaniards feel, is much more intense. Chemistry in Spain is beginning to respond to the impulse of glorious memory and to the necessity of developing the sources of our wealth. The stimulus which your presence here communicates to us will give breath to more difficult enterprises and result in fruitful research.[100]

Beyond the official rhetoric, however, the fact is that 'republican' chemistry continued and intensified former policies for the internationalisation of research, educational reforms and new laboratories and university research institutes from the early twentieth century. Utopian dreams of material progress opposed any use of chemicals for war purposes as a sign of the internationalism of the new times, which supported the invitation of foreign chemists to visit Spain, as well as the sending of large numbers of young chemists for training abroad. Under the banner of the Krausist philosophy of the JAE, chemistry seemed to fit very well with the republican ideal of positivism, anti-clericalism and secular educational reforms. Along similar lines to the DuPont campaign

---

obtained his PhD in 1933 under Jimeno's supervision on the influence of cadmium in several metallic alloys. He joined Jimeno's labs at the University of Barcelona and soon began teaching electrochemistry and electrometallurgy.

[100] Quoted by Foster, 'Chemistry in Spain', p. 426.

'Better Things for Better Living through Chemistry' in the 1930s,[101] Spanish chemists placed explicit emphasis on material progress and endorsed a more utilitarian, industrial appropriation of the Silver Age. The intersection between pure and applied chemistry occurred more frequently than expected, and is probably still insufficiently explored.

Pacifist, internationalist rhetoric, such as the way in which chemical progress was presented at the 1934 IUPAC Conference in Madrid as a victory against the pernicious uses of chemicals for war purposes, also had its counterbalance in conservative proposals such as those of Vitoria and Jimeno who, also in the republican 1930s, opposed the main values of the regime and created their own training routes, international connections through religious networks and an ambitious plan for industrial chemistry projects allied with entrepreneurial elites. Unlike in Krausism, they did not see any contradiction between God and matter in their scientific study. Conversely, Madinaveitia and the Girals stressed the materialist tradition of Marxism as a useful tool to be employed with moderation in Spain, but one that contributed to chemical and social progress according to the leftist values of the Republic. In that sense, the republican chemistry of wine, as a tool for the agricultural reform plans of the government, clashed with the anti-republican Jesuit alliance with industry.

In 1936, just at the outbreak of the Spanish Civil War, political, ideological and religious tensions aside, the SEFQ had reached a significant membership of more than 1,300, until Franco's coup d'état seriously jeopardised the republican dream.[102]

---

[101] David J. Rhees, 'Corporate Advertising, Public Relations and Popular Exhibits: The Case of DuPont', *History and Technology*, 10 (1993), 65–75.

[102] José Manuel Sánchez Ron, 'La física en España durante el primer tercio del siglo XX', in José Manuel Sánchez Ron (ed.), *Ciencia y sociedad en España: De la Ilustración a la Guerra Civil* (Madrid: El Arquero/CSIC, 1988), pp. 283–306, 291; Manuel Lora-Tamayo, *Cincuenta años de Física y Química en España, 1903–1953. Bodas de Oro de la Real Sociedad Española de Física y Química* (Madrid: C. Bermejo, 1954), pp. 34–5. At that commemorative event in 1953, Lora-Tamayo was President of the RSEFQ.

3   War Weapons

Historians, and historians of chemistry in particular, have had a special interest in the role of chemistry in warfare.[1] Bitter controversies about Fritz Haber's receipt of the Nobel Prize for Chemistry in 1918 after his commitment to the weapons of the First World War led to the debate on the moral responsibility of chemists as experts when contributing to national, common endeavours. Equally, the collaboration of chemical companies in Nazi Germany during the Second World War and the dramatic split in several scientific communities under that totalitarian regime provided excellent examples for further analysis of the complex intersection between science and ideology.[2] Many questions have therefore arisen about the production of chemical weapons on a laboratory scale, at the industrial level and regarding their use in the battlefield, as well as on poisonous chemicals as agents of mass destruction.[3] Nevertheless, chemists also played a crucial role in the care of the sick and wounded and in the health of the troops (the humanitarian side of the war), and they developed gas masks and other devices for protection. Coal powered the war, but it also became a raw material for chemicals that were crucial in the production of explosives, dyes, medicines and antiseptics. As David Edgerton pointed out in his history of Britain as a warfare state between 1920 and 1970, any account of twentieth-century science should include academic and industrial chemistry, but also military applications and medical research as the only way in which 'to address the key historical questions we must have about the

---

[1] Jeffrey Allan Johnson, 'Chemical Warfare in the Great War', *Minerva*, 40(1) (2002), 93–106; Jeffrey Allan Johnson, 'Crisis, Change and Creativity in Science and Technology. Chemists in the Aftermath of Twentieth-Century Global Wars', *Ambix*, 58 (2) (2011), 101–15.
[2] Peter Hayes, *Industry and Ideology. IG Farben and the Nazi Era* (Cambridge: Cambridge University Press, 1989).
[3] Ludwik Fritz Haber, *The Poisonous Cloud. Chemical Warfare in the First World War* (Oxford: Clarendon Press, 1986); Michael Freemantle, *The Chemists' War, 1914–1918* (London: Royal Society of Chemistry, 2015). See also: David Aubin, Patrice Bret, *Le sabre et l'éprouvette. L'invention d'une science de guerre, 1914–1939* (Paris: Editions Noésis/Agnès Viénot, 2003).

relations of science, technology and society'.[4] The role of chemists in the Spanish Civil War is an excellent case for evaluating the co-construction of chemistry and politics.

From July 1936 onwards, the dreams of the 1934 International Conference faded when the Spanish Civil War broke out after General Franco's coup d'état against the Republican order. Given that the country soon split between the fascist, Francoist side and the defenders of the liberal Republic – the latter including left-wing anarchist and communist revolutionaries – chemists became involved, throughout the three years of war, in a highly polarised political clash tinged with deep ideological undercurrents. Collectivisation of chemical industries by anarchist groups on one side and the leading role of experts on chemical weapons of the Francoist side on the other brought chemists to an ideological clash in a context in which public rhetoric about the neutrality, objectivity and universalism of chemistry seemed particularly untenable. The Civil War dramatically changed daily practices, and the chemical community suffered cruel divisions and polarisation. Skilled chemists who had opened up new research fields in the 1930s were, in many cases, condemned to marginalisation and mediocrity, whereas other prominent chemists became active figures in the repression of many of their colleagues. In the following sections, I will describe how, in practice, chemistry became a war weapon and at the same time a dangerous 'poison' for the scientific community itself, as pre-war skills became seriously jeopardised and reshaped.

A Chemical Civil War

In the Rif War (1921–27) in Morocco, the Spanish army used chemical weapons against the Berber rebellion. In 1924, at the beginning of Primo de Rivera's dictatorship, air force squadrons fired mustard gas (yperite) over the rebels in retaliation for the 1921 Battle of Annual that had caused thousands of Spanish casualties.[5] The *Fábrica Nacional de Productos Químicos* in La Marañosa (near Madrid) produced the gas, under the supervision of chemist Hugo Stolzenberg (1883–1973), who had worked with Fritz Haber during the First World War on the production of mustard gas and who was later to become a member of the Nazi party.

---

[4] David Edgerton, *Warfare State: Britain, 1920–1970* (Cambridge: Cambridge University Press, 2006), p. 330.
[5] Sebastian Balfour, *Deadly Embrace: Morocco and the Road to the Spanish Civil War* (Oxford: Oxford University Press, 2002).

La Marañosa also produced chlorine, phosgene and ethyldichlorarsine.[6] The Rif events took place just a year before the signing of the Geneva Protocol, which formally prohibited chemical weapons. Spain ratified the protocol in 1929, but it only banned the use of toxic weapons, not their production and storage. Moreover, limitations of their effective use in 'internal' conflicts were unclear.[7]

Concerns about chemical weapons continued into the early 1930s. In April 1935, in the pacifist, pro-League of Nations context of the Republic, Barcelona City Council held a meeting 'against chemical warfare', with the support of a league for the protection of the population from chemical weapons (*Liga protectora de la población civil contra de la guerra química*)[8] and the endorsement of José Giral, who in 1936 became Minister of the Navy of the *Frente Popular* cabinet. Even in July 1936, a few days before Franco's coup d'état, J. Giral had created a special section of the navy for chemical warfare (*Dirección de Guerra Química de la Armada*).[9] All of those efforts focused on the correct uses of gas masks and shelters and the training of the population to resist chemical attack,[10] but the literature on chemical weapons proliferated widely. Many of these popular books revived memories of the damage and casualties that had been caused, mainly by chlorine gas, phosgene and mustard gas, during the First World War. They became relevant in a culture of chemistry that often hid expertise under a rhetoric of apolitical neutrality.[11]

In spite of an atmosphere of full-on violence and attacks against the civilian population, as seen in the bombings of Guernica and Barcelona, and despite the frequent use of bombs elsewhere, there is no consensus on the actual use of chemical weapons.[12] Historian Kim Coleman referred to a possible republican attack in Guadarrama using tear gases

---

[6] José María Manrique, Lucas Molina, *Las armas de la Guerra Civil española* (Madrid: La Esfera de los Libros, 2006), p. 526. See also: Manrique, Molina, *Guerra química en España, 1921–1945* (Valladolid: Galland Books, 2012).
[7] Balfour, *Deadly Embrace*. [8] *La Vanguardia*, 20 August 1935.
[9] *La Vanguardia*, 1 July 1936. [10] *La Vanguardia*, 18 April 1936.
[11] Among the most widespread books, it is worth mentioning: Julio Guerra, Mariano Ferrer, *Gases de Guerra* (Barcelona, 1935); José María Serra, *Los gases de guerra. Protección de la población civil. Vulgarización científica* (Barcelona, 1936); Juan Morata, *Guerra química y bacteriológica* (Barcelona, 1938); Vicente Reig, *Gases de Guerra. Fisiopatología, química, defensa individual y colectiva* (Valencia, 1939); Luis Blas, *Química de guerra* (Barcelona, 1940); Juan Izquierdo, *Manual del arma química* (Madrid, 1940); Sancho Ezama, *Química de los explosivos* (Madrid, 1941); Jesualdo Martínez, José Rojas, José María Fernández-Ladreda, *Pólvoras y explosivos modernos* (Madrid, 1944). See also: Juan Manuel Sarasua, *La protección colectiva de la población civil contra las armas químicas en el Guerra Civil española*. Master's thesis (Bellaterra: Universitat Autònoma de Barcelona, 2007).
[12] I thank Jaume Valentines-Álvarez for his information on the use of chemical weapons in the Civil War.

and a later Francoist revenge attack in Madrid using nerve gas.[13] During the siege of the Alcazar de Toledo, the Republican army seems to have used chemical weapons.[14] There are also hints of the production of certain gases at Solvay in Torrelavega, near Santander, by the Francoist side, as well as secret documents reporting on the use of chemical gases in the same city by the Republican army in 1937, but there is poor evidence of their actual use against the civilian population.[15] That controversy, however, does not throw doubt on the unquestionable fact of the production, commercialisation and storage of chemical weapons as propaganda on both sides, as well as their crucial role in shaping civil defence strategies.

Despite the 1930s rhetoric of pacifism and the alignment with the League of Nations, once the war broke out, F. Giral endorsed the chemists offering their services to the country, since, in his view, almost half of the war was based on chemistry.[16] In July 1936, F. Giral became director of the *Centro de Estudios y Experiencias* at La Marañosa for the production of chemical weapons and the protection of the civilian population from chemical threats and the supervision of several chemical factories.[17] In 1937, in an article addressed to communist and socialist readers, F. Giral encouraged chemists to get out of their labs and offer their chemical skills in the service of the Republic.[18] In addition, since the beginning of the war, Moles had acted as technical advisor for the Republican side at La Marañosa, and later became director of the *Servicio de Pólvoras y Explosivos* (Gunpowder and Explosives Service).[19] Madinaveitia also contributed to advising on the purchase of chemical

---

[13] Kim Coleman, *A History of Chemical Warfare* (Basingstoke: Palgrave/Macmillan, 2005).
[14] Manrique, Molina, *Las armas de la Guerra Civil española*; Manrique, Molina, *Guerra química en España*.
[15] Guillermo Lusa, Antoni Roca-Rosell (eds.), *Fem memòria per fer futur. Tècnica, Medicina i Guerra Civil. V Jornada Memorial Democràtic* (Barcelona: UPC, 2007), p. 127. See also: Ángel Toca, *La introducción de la gran industria química en España. Solvay y su planta de Torrelavega (1887–1935)* (Santander: Universidad de Cantabria, 2005); Jaume Valentines, 'La introducció de la gran industria de la química a Espanya i el nou règim de sabers a la perifèria', *Quaderns d'Història de l'Enginyeria*, 8 (2007), 361–75.
[16] Carl August Rojahn, *Preparación de productos químicos y químico-farmacéuticos* (Mexico City: Atlante, 1942), p. xiii. See also: Francisco Giral, 'Ciencia aplicada. Productos lacrimógenos de aplicación en guerra química', *Ciencia*, 1(5) (1940), 214–20.
[17] In Medina de Segura (Murcia) and Cocentaina (Alicante). F. Giral also became *Subsecretario de Armamento del Ministerio de Defensa Nacional* and *Ingeniero Director de la Fábrica 19 de la Subsecretaría de Armamento en Cocentaina*. María Fernanda Mancebo, 'Tres vivencias del exilio en México: Max Aub, Adolfo Sánchez Vázquez y Francisco Giral', *Migraciones y Exilios*, 5 (2004), 85–102, p. 95.
[18] Francisco Giral, 'La ciencia colaborará para crear una industria de guerra', *Ahora. Diario de la Juventud Socialista Unificada*, 24 January 1937, p. 15.
[19] Enrique Moles, *'Determinació de pesos moleculars de gasos pel mètode de les densitats límits'*, Enric Moles i Ormella. Traducció, introducció i comentaris de Joaquim Sales i Agustí Nieto-Galan (Barcelona: Clàssics de la Química. Societat Catalana de Química, 2013).

weapons, particularly tetraethyl lead (TEL), as an anti-knocking agent for plane fuel.[20] In Granada, Adolfo Rancaño, who in 1934 had obtained the first physical chemistry chair at the science faculty, also worked on strategies of civil defence against chemical weapons. He tested several reagents for the detection of small concentrations of phosgene, yperite (mustard gas) and ethyldichlorarsine. Rancaño designed a kit for quick tests at the front lines, active carbon filters and a protocol in case of an attack using chemical weapons.[21]

In Barcelona, the university had recovered its autonomy after the leftist victory of the Catalan *Front Popular* in February 1936, but with the war, the situation became exceptional. The *Institut de Química* (IQ), at that time directed by García Banús's pupil, Miguel Masriera,[22] fell under the supervision of the *Consell de Sanitat de Guerra* (Wartime Health Council) and contributed to the making of drugs and medicines, but its research capacity was limited (Figure 3.1). After the early attempts of industrial collectivisation by the anarchist worker's union *Confederación Nacional del Trabajo* (CNT), in August 1936, the Catalan government created the *Comissió d'Indústries de Guerra* (CIG) (War Industries Commission), which became in practice a form of ministry for the conversion of civil industries into producers of war weapons.

The CIG appointed experts from 17 commandeered metallurgical, steel production and chemical industries in 15 new firms designed for the war effort[23] and ended up having more or less effective control of around 200 factories and 60,000 workers. The hegemony of the big chemical firms in Catalonia[24] was only partially complemented by

---

[20] Rosario E. Fernández Terán, *El profesorado del 'Instituto Nacional de Física y Química' ante la Guerra Civil. El proceso de depuración y el drama del exilio*. PhD thesis (Madrid: Universidad Complutense de Madrid, 2005).

[21] José Manuel Cano Pavón, 'La investigación química en Granada en el siglo actual (1900–1975)', *Dynamis*, 16 (1996), 317–67, p. 336.

[22] Agustí Nieto-Galan, 'From Papers to Newspapers: Miguel Masriera (1901–1981) and the Role of Science Popularisation under the Franco Regime', *Science in Context*, 26(3) (2013), 527–49.

[23] The CIG took control of important firms such as *Sociedad Anónima Cros, Fabricación Nacional de Colorantes y Explosivos, Sociedad Electro-Química de Flix, Unión Española de Explosivos, Maquinista Terrestre y Marítima, Hispano-Suiza, Sociedad Anónima de Material para Ferrocarriles y Construcciones, Pirelli, Riviere, Elizalde, Metales y Platería Rivera, Francisco Lacambra, G. de Andreis Metalgraf Española*. Francisco de Madariaga, *Las industrias de guerra de Cataluña durante la Guerra Civil*. PhD thesis (Tarragona: Universitat Rovira i Virgili, 2003). See also Francisco Javier de Madariaga, 'La Comissió d'Indústries de Guerra de la Generalitat', in Guillermo Lusa, Antoni Roca-Rosell (eds.), *Fem memòria per fer futur. La tècnica i la Guerra Civil. IV Jornades Memorial Democràtic a la UPC. 30 de novembre de 2006* (Barcelona: UPC, 2006), pp. 123–35.

[24] *Sociedad Anónima Cros Fábrica Nacional de Colorantes y Explosivos, Sociedad Electro-Química de Flix* and *Unión Española de Explosivos*.

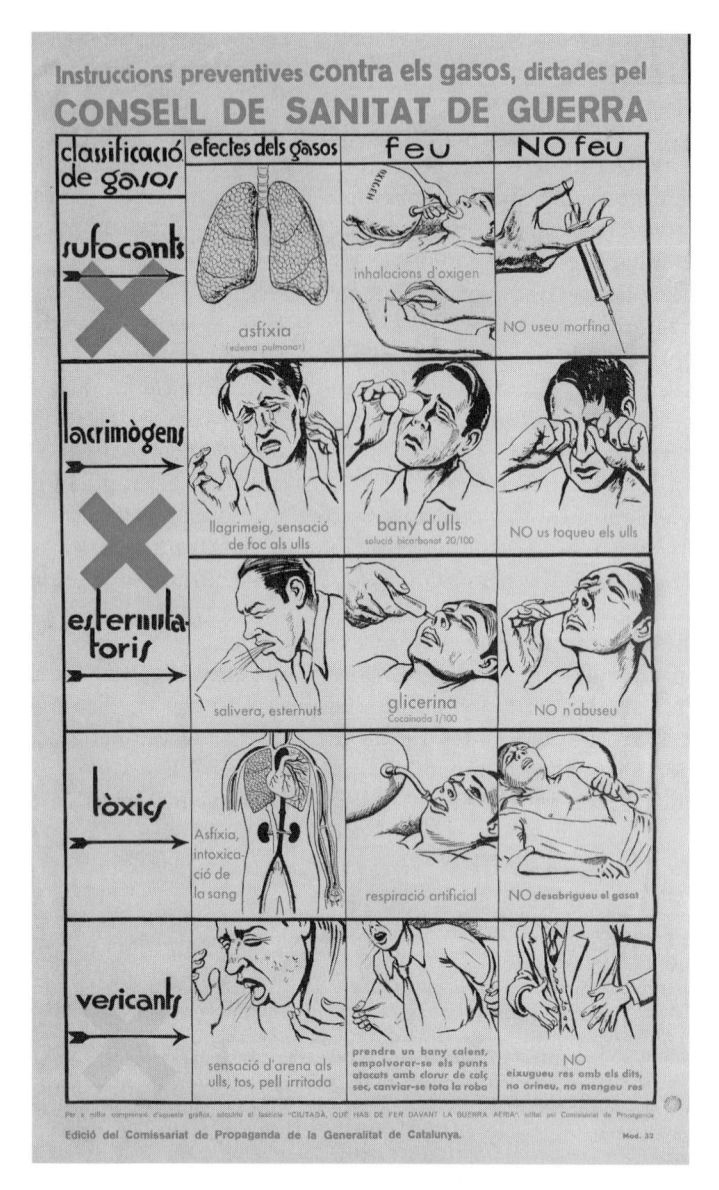

Figure 3.1 A poster produced by the Catalan *Consell de Sanitat de Guerra* (Wartime Health Council) for the protection of the civil population from chemical threats, c. 1938. It described suffocating, lacrimogenous, irritant, toxic and vesicant compounds and gave advice for counteracting their damaging effects.
http://mdc.csuc.cat/cdm/ref/collection/cartellsBC/id/28. Reproduced with permission of the *Biblioteca de Catalunya*, Barcelona.

A Chemical Civil War 89

small producers of fertilisers, pyrotechnics, pharmaceuticals and other chemicals for textiles and cleaning.[25] Under the banner of the CIG, the *Laboratori de Química Orgànica* (LQO) at the *Escola d'Enginyers Industrials* oversaw the analysis of raw materials and the synthesis of new chemicals to be produced on a large scale. Factory F-1 was set up in Badalona (near Barcelona) in the premises of the former *Unión Española de Explosivos* (UEE), producing TEL for aircraft fuel. F-2 produced gunpowder and other explosives. F-3 made trinitrotoluene (trilite), nitronaphthaline and pricic acid. F-4 produced cellulose from esparto as a raw material for trilite and gunpowder. After the transfer to Catalonia of a good part of the chemical weapons produced in La Marañosa, F-5 was fully devoted to the production of toxic gases in a factory of dyestuffs in the Pyrenees. F-6 was originally designed for the production of chemical gases, but there is no evidence of their actual production there. F-7 produced artillery cartridges and mortars. F-8 and F9 made rockets, bombs and fuses. F-10 made potassium chloride and bromine. F-11 was devoted to the making of several kinds of explosives after the rearrangement of a former dyestuffs factory. It produced mercury fulminate and lead nitride as raw materials. F-12 and F-13 were also devoted to the making of several kinds of explosives, whereas F-14 and F-15 focused on rifle production, repair and maintenance.[26]

The industrial war effort was politically supervised by the chemical branch of the anarchist CNT, but it also required academic support. In 1937, the main raw material weaknesses arose from a dependence on petroleum, a lack of new industrial hydrogenation processes for the production of 'artificial' gasolines and the need to further develop the electrochemical industry for the production of alkalis, fertilisers and metals.[27] Chemists such as José Pascual and Masriera acted as

---

[25] de Madariaga, 'La Comissió d'Indústries de Guerra', p. 135.
[26] de Madariaga, *Las industrias de guerra de Cataluña*, p. 13. See also: Guillermo Lusa, Antoni Roca-Rosell, Jaume Valentines (eds.), *Fem memòria per fer futur. Tècnica, medicina i Guerra Civil. V Jornada Memorial Democràtic a la UPC (29-11-2007)* (Barcelona: UPC, 2007): Document 2. 'Informe no. 140'. Laboratori de Química Orgànica EEIB. Comissió de la Indústria de Guerra, 9 de juny de 1938. Arxiu Montserrat Tarradellas i Macià, Poblet. García Banús's pupils Manuel Tremoleda, Francisco Boqué and José Pascual belonged to the laboratory team. For a general overview of the chemical industry in Catalonia, see: Josep Maria Bricall, *Política econòmica de la Generalitat (1936–1939)* (Barcelona: Edicions 62, 1979); Francesc Cabana, *Fàbriques i empresaris. Els protagonistes de la revolució industrial a Catalunya*. Vol. I. 'Metal·lúrgics. Químics' (Barcelona: Enciclopèdia Catalana, 1992).
[27] 'Conferència de l'Aprofitament Industrial de les Riqueses Naturals de Catalunya' (CAIRN).

experts and had their say on topics such as the raw inorganic and organic materials for the chemical industry.[28] Ramon Peypoch (1898–1984), editor of the Catalan science journal *Ciència* and a committed Catalanist, became responsible for chemical products at the *Consell d'Economia* of the Catalan government, supervising many of the collectivisation processes during the war, including those of the chemical industries.[29]

The LQO focused its work on anti-knocking agents.[30] In August 1938, Pascual became the director of the LQO and responsible for Factory F-1, producing TEL for plane fuel.[31] Pascual only stayed a few months in his post and seemed not to show loyalty to the republican cause.[32] In practice, the LQO most probably became a shelter lab for many Francoists who, still being on the Republican side of the war, used their chemical expertise to remain in Barcelona and avoid the front.[33]

On the Francoist side, chemical weapons leaflets also spread throughout the civilian population.[34] In 1936, the military hospital in Seville was ready to admit 8,500 casualties of mustard gas. Moreover, there is consistent evidence of Mussolini's delivery of chemical gases to Franco's army in 1937:[35] 176 tonnes of phosgene, 50 tonnes and 36,000 canisters of mustard gas, 19,500 arsine missiles, 2,000 tonnes of tear gas substances and 1,600 tonnes of incendiary products (Figure 3.2). Mustard

---

[28] Document nº. 1: C.A.I.R.N. Conferència de l'Aprofitament Industrial de les Riqueses Naturals de Catalunya. Publicacions de la Conselleria d'Economia de la Generalitat de Catalunya, 1937', and Document nº. 2: 'Comissió Organitzadora de la C.A.I.R.N. Full mecanografiat que forma part de la caixa núm. 441 de l'Arxiu de l'ETSEIB', in Guillermo Lusa, Antoni Roca-Rosell (eds.), *Fem memòria per fer futur. La tècnica i la Guerra Civil*, pp. 109–12.

[29] On Ramon Peypoch, see: Àngela Garcia Lledó, *La revista Ciència (1926–1933): Una eina al servei del catalanisme*. Master's thesis (Bellaterra: Universitat Autònoma de Barcelona, 2012).

[30] Bromine, sodium, ethylene, aluminium oxide, ethylene dibromide and chloronapthalene, as well as substances such as urotropine, diethylaniline and ethylenediamine.

[31] Francisco Javier de Madariaga, 'Las industrias de guerra de Cataluña durante la Guerra Civil', *Ebre*, 38 (2008), 43–64.

[32] Agustí Nieto-Galan, Joaquim Sales, 'Josep Pascual Vila (1895–1979): Una aproximación biográfica', *Anales de Química*, 113(1) (2017), 10–47.

[33] Lusa, Roca-Rosell (eds.), *Fem memòria per fer futur. Tècnica, medicina i Guerra Civil*; Guillermo Lusa, *La Escuela de ingenieros en guerra (1936–1939)* (Barcelona: Escola Tècnica Superior d'Enginyeria Industrial de Barcelona, UPC, 2007).

[34] See, for instance: Juan José Antequera, *Huelva: Los peligros del cielo. Prevenciones del bando faccioso frente a ataques republicanos en 1936* (Seville: Facediciones, 2008).

[35] Manrique, Molina, *Las armas de la Guerra Civil*, p. 526. See also: Alberto Rovighi, Filippo Stefani, *La partecipazione italiana alla Guerra civile spagnola* (Rome: Ufficio storico dello Stato maggiore dell'Esercito, 1993).

A Chemical Civil War 91

Figure 3.2 Mussolini offering poison gases to Franco. 'A Christmas present'. Mussolini to Franco: 'Here, son, you've been a good boy'. Cartoon in the communist journal *Ahora*, 2 January 1937.
Reproduced with permission of the *Biblioteca de Catalunya*, Barcelona.

gas and diphosgene also arrived in the same year from Nazi Germany.[36] Similarly, the *Fábrica Nacional de Colorantes y Explosivos* (FNCE) produced ammonia and explosives.[37] The company *Energía e Industrias Aragonesas* (EIA) produced phosgene in its chemical plant in Sabiñánigo.[38]

---

[36] Manrique, Molina, *Las armas de la Guerra Civil*, pp. 526–9.
[37] Núria Puig, 'The Frustrated Rise of Spanish Chemical Industry between the Wars', in Anthony S. Travis et al. (eds.), *Determinants in the Evolution of the European Chemical Industry, 1900–1939* (Dordrecht: Kluwer, 1998), pp. 301–20, 313.
[38] Manrique, Molina, *Las armas de la Guerra Civil*, p. 526.

Rocasolano's research school in the University of Zaragoza also contributed substantially to the war effort for the Francoist army.[39]

During the Civil War, Luis Blas (1906–67) became a professor of chemistry at the *Escuela de Oficiales de Guerra Química de Toledo y Salamanca* and published a widely read treatise on chemical war: *Química de guerra*.[40] The book summarised the main uses of chemical weapons in the First World War and provided a taxonomy of aggressive chemical agents for the general reader.[41] During his years as a PhD student, Blas had encountered José María Fernández-Ladreda,[42] who had worked with Julio Guzmán at the LIF on electrochemistry[43] and was an active member of the right-wing *Asociación Católica Nacional de Propagandistas* (ACNP). Fernández-Ladreda was an artillery general, former collaborator in Primo de Rivera's dictatorship and conservative Member of Parliament for the CEDA during the years of the Republic. He became a local hero of the Francoist side for his defence of Oviedo during the war. Blas and Fernández-Ladreda shared an intellectual passion for war chemicals and for Francoist right-wing ideology, so, in their view, chemical weapons became intrinsic tools in the making of the new regime and in victory against the Republic.

Fierce struggles for the making of chemical weapons as part of the war duties placed distinguished chemists such as F. Giral, Moles and Madinaveitia on one side and Rocasolano, Blas and Fernández-Ladreda on the other. The extreme political polarisation gave additional power to the anarchist CNT union. In one way or another, chemical weapons reified different political ideologies, often under a form of 'technical', 'apolitical' description of the synthesis of phosgene, yperite, chlorine and other gases, which appeared in textbooks and leaflets that circulated widely on both sides of the war. Analytical and synthetic procedures in the laboratory, large-scale production methods and strategies of resistance for the civilian

---

[39] José Luis Cebollada, 'Antonio de Gregorio Rocasolano y la Escuela Química de Zaragoza', *Llull*, 11 (1988), 181–216; Antonio de Diego, 'Movilización industrial. Algunos de sus aspectos', in Emilio Jimeno (ed.), *Universidad de Barcelona. Aspectos y problemas de la nueva organización de España. Ciclo de Conferencias organizado por la Universidad de Barcelona* (Barcelona: Universidad de Barcelona, 1939), pp. 263–80, 263.

[40] *Real Academia Nacional de Farmacia*. www.ranf.com/sesiones-y-actos/45-acad%C3%A9micos.html?start=140 (last accessed 15 March 2017).

[41] Suffocating (chlorine, phosgene, chlormethyls), toxic (hydrocyanic acid, arsenic compounds), lacrimogenous (benzylchloride), vesicant (mustard gas) and irritant (chlor arsines) agents.

[42] José María Fernández-Ladreda, *Discurso leído por el Excmo. Sr. D. José María Fernández-Ladreda y Menéndez-Valdés* (Madrid: Real Academia de Farmacia, 1947); Manuel Lora-Tamayo, *La investigación química española* (Madrid: Alhambra, 1981), pp. 160–1.

[43] Julio Guzmán, José María Fernández-Ladreda, 'Cátodo de cobre y ánodo de hierro en el electroanálisis de los latones', *Anales*, 11 (1915), 308.

A Damaged Community

population against chemical weapons appeared as a kind of standard information. However, behind that apparently ambiguous academic morality, the biographical profiles of many of the aforementioned chemists were profoundly determined by their active involvement in a 'political' war.

A Damaged Community

F. Giral's book *Ciencia española en el exilio (1939–1989)* provides biographical data on around 500 Spanish scientists – almost half of the university positions – who, for political reasons, left the country as a consequence of the Civil War between 1936 and 1939 (see Chapter 7). As in many other cases, F. Giral's promising career was dramatically interrupted.[44] He was, at that time, a young chemistry professor at the University of Santiago and a first-hand witness of how the Civil War brought fierce academic repression, exile and misery to many of his colleagues.[45] University rectors[46] and distinguished scholars[47] were executed and many others suffered repression, prison and the loss of their academic chairs.[48]

Perhaps his scientific career was not particularly brilliant when compared to those of Moles, Madinaveitia and García Banús, nor did his case have the same impact as the shocking execution of the poet Federico García Lorca on 18 August 1936, but, in the chemical community, Jesús Yoldi has become a symbol of the cruel repression by the Francoist troops in Andalusia (Figure 3.3). In 1922, Yoldi had obtained a chair in general chemistry at the University of Seville and moved to Granada in 1924. Like Miguel Catalán, he was a member of the *Izquierda Republicana* political party, alderman of the City Council in 1934 and mayor of Granada at the beginning of the Civil War. Yoldi was shot dead by Francoist troops on 23 October 1936, just a couple of months after the killing of Lorca.[49]

---

[44] Francisco Giral, *Ciencia española en el exilio (1939–1989). El exilio de los científicos españoles* (Barcelona: CIERE/Anthropos, 1994), p. 21.

[45] Giral, *Ciencia española en el exilio*; Jaume Claret, *El atroz desmoche* (Barcelona: Crítica, 2006); Luis Enrique Otero Carvajal et al. (dirs.), *La destrucción de la ciencia en España: depuración universitaria en el franquismo* (Madrid: Editorial Complutense, 2006).

[46] Leopoldo García Alas in Oviedo and Salvador Vila Hernández in Granada.

[47] Such as Juan Peset, Professor of Legal Medicine and former Rector of the University of Valencia; Arturo Pérez Martín, Professor of Theoretical Physics and Vice-Rector of the Universidad de Valladolid; and the naturalist Sadi de Buen.

[48] This was the case for some of Moles's pupils: Miguel Crespí in Madrid, Fernando González Núñez in Granada and Alberto Chalmeta, Professor of Practical Pharmacy in Madrid. On Chalmeta, see: Javier Puerto, 'Alberto Chalmeta Tomás (Catarroja, Valencia 1897–Paris 1987?) Professor of Practical Pharmacy at the Central University: A Broken Life', *Medicina e Historia*, 3 (2014), 4–16.

[49] Luis Fermín Capitán (ed.), *Un siglo de estudios de química en Granada (1913–2013)* (Granada: Universidad de Granada, 2014). I am indebted to Pedro Mateo for this information and for making me aware of the repression of the Granada chemical community during the war.

94   War Weapons

Figure 3.3 Jesús Yoldi in his laboratory at the Faculty of Science of the University of Granada in the 1920s. In 1936, Yoldi was executed by Franco's troops near Granada.
Courtesy of Pedro Luis Mateo.

The war also affected talented young chemists such as León Le Boucher (1904–37), who in the 1920s did excellent work on the chemistry of coordination complexes and managed to publish his first papers in *Anales* with Ángel del Campo.[50] George Urbain's course in Madrid in 1917 had

[50] Lora-Tamayo, *La investigación científica*, pp. 128–30; María Fernanda Mancebo, *La Universidad de Valencia. De la Monarquía a la República (1919–1939)* (Valencia: Universitat de València/Instituto de Cultura Juan Gil-Albert, 1994). See also: Ernest Sánchez Santiró, *Científics i professionals: La Facultat de Ciències de València (1857–1939)* (Valencia: Universitat de València, 1998).

been a rich source of inspiration for the young Le Boucher (see Chapter 1), but he also acquired great experience in the customs laboratory in Valencia and later obtained *pensionados* at the Polytechnic School of Hannover to work on molecular volumes at low temperatures and at the Technical Institute of Bremen. Part of his work appeared in the *Zeitschrift für Anorganische und Allgemeine Chemie* in 1932.[51] Le Boucher's brilliant start had placed him as a potential leader of coordination chemistry in Spain. Many years later, Manuel Lora-Tamayo, one of the key chemists of the new regime after the war, recognised Le Boucher as a man of integrity and of honest leftist ideology, who, living in Valencia as a customs chemist, did not come to terms with a passive, comfortable rearguard attitude and joined the Republican army, in the service of which he died at the front in 1937.[52]

The emerging Francoist elite took an active role in cleansing commissions after the war, which resulted in internal and external exile for many fellow colleagues.[53] Equally, many chemists who were viewed as active members of the old Silver Age research schools were systematically dismissed from selection committees (*oposiciones*) under the strict ideological control of names such as José María Albareda, Jimeno, Rocasolano, Lora-Tamayo, Antonio Rius Miró – a distinguished pupil of Rocasolano who acquired great academic power in the new regime – Bermejo, Fernández-Ladreda and Pascual. Rocasolano and Bermejo strongly expressed their views against the JAE project in public,[54] bitterly attacking what was conceived as the dangerous, powerful force of the secular, liberal, republican educational programme of the ILE, the JAE and the spirit of the INFQ in the 1930s (see Chapter 4). Rocasolano, in particular, accused the ILE of political sectarianism and deep aggression against the traditional, religious roots of Spanish culture.[55] He contrasted his supposedly 'good' research and his 'honest', 'apolitical' quest for truth with, in his view, the 'bad', Masonic, anticlerical and 'politically biased' Republican chemistry.[56]

---

[51] León Le Boucher, Werner Fischer, Wilhelm Biltz, 'Über Molekular- und Atomvolumina 41. Tieftemperaturdichten kristallisierten Fluorwasserstoffs und einiger kristallisierter Fluoride', *Zeitschrift für Anorganische und Allgemeine Chemie*, 207(1) (1932), 61–72; Otto Ruff, León le Boucher, 'Der Dampfdruck von ZnF_2, CdF_2, MgF_2, CaF_2, SrF_2, BaF_2, AlF_3', *Zeitschrift für Anorganische und Allgemeine Chemie*, 219(4) (1934), 376–81.

[52] Lora-Tamayo, *La investigación científica*, p. 129.

[53] Mariano Tomeo, 'Rocasolano', in Instituto Alonso Barba (ed.), *Trabajos del Laboratorio de Bioquímica y Química aplicada* (Zaragoza: Universidad de Zaragoza, 1941), pp. 3–22, 15–16. See also: Miguel Artigas et al., *Una poderosa fuerza secreta* (San Sebastian: Editorial Española, 1940). See also: Luis Enrique Otero Carvajal (dir.), *La Universidad Nacional Católica. La reacción antimoderna* (Madrid: Universidad Carlos III, 2014).

[54] Artigas, *Una poderosa fuerza secreta*.

[55] Antonio de Gregorio Rocasolano, 'La táctica de la Institutción', in Artigas, *Una poderosa fuerza secreta*, pp. 125–30.

[56] Rocasolano, 'La investigación científica', in Artigas, *Una poderosa fuerza secreta*, p. 160.

Jimeno's scientific reputation from the JAE years did not prevent him from taking part in the cleansing committees after the war. In the early 1930s, he had already tried to find a position outside Catalonia, since he felt uncomfortable with the new autonomy of the Catalan university system in the context of the Second Republic. Jimeno rejected the public use of the Catalan language, which contradicted his strong Spanish nationalism and conservative ideology.[57] Bermejo, one of the most important chemists of the pre-war period and former rector of the University of Madrid, expressed his own views in bitter public critiques against his former colleagues at the INFQ. He labelled the luminaries of the Silver Age as an isolated group of chemists who had supposedly endorsed the JAE ideology. In his own words:

> The custodians of the Rockefeller Institute are keen to live in isolation, in order to serve the sectarian goals of the *Institución Libre de Enseñanza*. They are mainly radical extremists, or collaborators with extremism wearing the mask of pious and illustrious believers, which allows them to divert the attention of the gullible away from their Masonic nature.[58]

Rocasolano followed the same line and despised the JAE chemists for their ideological bias, but also for their supposed lack of interest in applied chemistry.[59] As a founding member of the right-wing journal *Acción Española*, he felt disappointed with the JAE research policies for their seeming lack of support for his lab in Zaragoza and for their secular, anticlerical bias. *Acción Española* was a right-wing, Catholic, conservative journal that strongly opposed the policies and values of the Second Republic. It often held anti-Semitic, anti-Masonic positions. The journal inspired the creation of the political party *Acción Nacional*, which had close links to the Catholic ACNP – the chemist Fernández-Ladreda being another of its members – and the newspaper *El Debate*, led by Cardinal Ángel Herrera Oria. In 1937, in a book published from an earlier paper in *Acción Española*, Rocasolano reflected on the physical–chemical and biochemical transition from life to death, *De la Vida a la*

---

[57] Equally, Cándido Torres (1902–95), Professor of Organic Chemistry at the Faculty of Pharmacy in Barcelona (and a pupil of Obdulio Fernández), felt uncomfortable with the new trends of the Republican UAB. After spending the whole of the Civil War on the Francoist side in Toledo as a provincial chief of the *Falange Española* (the Spanish fascist party), he returned to his position in 1940. Joaquim Sales, *La química a la Universitat de Barcelona* (Barcelona: Publicacions i Edicions de la Universitat de Barcelona, 2011), pp. 138–44.

[58] Luis Bermejo, 'El Instituto Rockefeller', in Artigas, *Una poderosa fuerza secreta*, pp. 197–202, 201. See also: Otero Carvajal, *La Universidad Nacional Católica*, pp. 102–14.

[59] Mariano Tomeo, *Rocasolano. Notas bio-bibliográficas* (Zaragoza: La Académica, 1941).

Muerte,[60] with the desire to see Spain recover a leading place in 'Western Christian civilization'. Rocasolano aggressively criticised the spirited philosophical systems of the nineteenth century geared towards materialism and monism. He obviously opposed Krausism, which had inspired the JAE policies for decades with secular overtones and had supposedly driven scientific–religious tension as perceived from the conservative, Catholic side. For the sake of saving the country from serious danger, in 1937 Rocasolano praised the thousands of young people who were at the front fighting under the leadership of the 'Generalísimo' Franco.[61]

Madrid was besieged by the Francoist army for a long period, but remained in the territory of Republican Spain almost until the end of the war.[62] From more than 70 academic positions in the early 1936, figures for the INFQ as a whole dramatically fell to less than 20 by 1938.[63] Since the director of the INFQ, Blas Cabrera, had gone into exile in 1936, Moles took the lead in that hostile environment. Maintaining routine laboratory work in daily life was not an easy task, but Moles and his decreasing number of co-workers tried to maintain a certain normality. Some months later, Moles's pupils Fernando González Núñez (1896–1985) and Augusto Pérez Vitoria (1908–91) were the directors in practice of the INFQ under very difficult circumstances. González Núñez did his best to preserve the laboratory equipment, but his symbolic role of resistance led to fierce repression from the Francoist victors and a period of several years in prison after the war. In the late 1920s Pérez Vitoria had worked at the LIF and later at the INFQ with Moles up until 1938. After his stay with Arthur J. Allmand at University College London (UCL), he published major papers with Moles in *Anales* and *Angewandte Chemie*. Pérez Vitoria also played an important institutional role at the SEFQ, as well as in the editorial office of *Anales*.[64]

During wartime, Madinaveitia joined the governing body of the JAE and, together with Moles, followed the Republican government in its

---

[60] Antonio de Gregorio Rocasolano, *De la vida a la muerte. Trabajos del Laboratorios de Investigaciones Bioquímicas de la Facultat de Ciencias de Zaragoza* (Zaragoza: Editorial Gambón, 1937).

[61] Ibid., p. 11.

[62] For more details about the INFQ professors during the war, see: Fernández Terán, *El profesorado del 'Instituto Nacional de Física y Química'*.

[63] Carmen Magallón, *Pioneras españolas en las ciencias: Las mujeres del Instituto Nacional de Física y Química* (Madrid: CSIC, 1998), p. 227. See also: Antonio Canales, Amparo Gómez-Rodríguez, 'La depuración franquista de la Junta para la Ampliación de Estudios e Investigaciones Científicas (JAE): Una aproximación cuantitativa', *Dynamis*, 37(2) (2017), 459–88.

[64] Augusto Pérez Vitoria, Archive of the Society for the Protection of Science and Learning. SPSL 220/9 fl.381–99. Bodleian Library. Oxford.

internal exile, first to Valencia and later to Barcelona, as well as supporting antifascist manifestoes. At the beginning of the war, Madinaveitia was dean of the Faculty of Pharmacy in Madrid. One of his pupils, Andrés León, directed the new residence of the *Fundación del Amo* at the *Ciudad Universitaria*, only few months before the beginning of the war.[65] The project originated with a donation to the JAE by the Spanish émigré in California, Gregorio del Amo. After the war, León could only survive professionally outside academia, working in the pharmaceutical industry. Ignacio Ribas was another member of Madinaveitia's research school, working on the chemistry of vegetable alkaloids, and as a pupil of Fourneau in Paris in the 1920s at the therapeutic chemistry lab – one of the natural nodes of Madinaveitia's international network.[66] Ribas obtained a chair in organic chemistry in Salamanca and later in Valencia, but after the war, he was also punished for his former relationship with Madinaveitia.[67]

In the university context, García Banús's prestigious research school was, in practice, dismantled. García Banús came from a refined, liberal, bourgeois family. Influenced by his uncle, the Spanish painter Joaquín Sorolla, he inherited the family passion for freedom and creativity, which did not match the highly polarised, revolutionary atmosphere of July 1936 in Barcelona. In November 1934, he had supported former Prime Minister Manuel Azaña in his defence of the liberal values of the Republic, which were then under serious threat from conservative groups. In the manifesto, he appeared next to names such as García Lorca, Valle-Inclán and León Felipe. In 1936, after the coup d'état, Banús travelled to Paris, and from there straight to Bogotá, and later to Caracas, to establish a new professional life.[68] He then became one of the many chemists to begin an uncertain international journey to escape from the war, desperately seeking academic and personal shelter abroad (see Chapter 7).

In the 1930s, British aid for exiled scientists and intellectuals mainly came from the Society for the Protection of Science and Learning (SPSL), which raised funds to help scientists escaping from totalitarian

---

[65] F. Giral, *La ciencia española en el exilio*, p. 55. See also: http://ceies.cchs.csic.es/?q=content/le%C3%B3n-maroto-andr%C3%A9s (last accessed 12 November 2017).

[66] Ernest Fourneau, Ignacio Ribas, 'Stéréoisomérie et action anesthésique locale: Séparation du diméthylaminodiméthyléthylcarbinol en ses isomères optiques et préparation des deux stovaïnes actives', *Bulletin des Sciences Pharmacologiques*, 30(5) (1928), 273–8.

[67] But he finally resumed his career with an organic chemistry chair in Santiago in 1942.

[68] He later went to Venezuela and never returned to Francoist Spain. García Banús died in Caracas in 1955. Agusti Nieto-Galan, 'Free Radicals in the European Periphery: "Translating" Organic Chemistry from Zurich to Barcelona in the Early Twentieth Century', *The British Journal for the History of Science*, 37 (2004), 167–91. See also: José Royo Gómez, 'El professor Antonio García Banús', *Ciencia*, XVI (1956), 236.

regimes and war. Its efforts focused on German scientists as refugees from the Nazi terror, but also the most important Spanish chemists who had suffered exile and repression in the Civil War.[69] The SPSL allocated funds to Fernando Calvet, who at that time held a DPhil at Queen's College, Oxford, in the research group of Frederick D. Chattaway during the period of 1925–27 while working on mechanisms of phenol-aldehyde condensation, and he had won a Ramsay Memorial Fellowship.[70] In Paris, his mentor, García Banús, obtained a four-month SPSL fellowship before departing for Colombia to begin his new professional life.[71] Eduardo de Salas, an assistant of García Banús who was working at UCL at the beginning of the war in 1936, also asked for the support of the Society.[72] Finally, Moles himself obtained a short fellowship, which he was never able to enjoy since he obtained a position in Paris at the Joliot-Curie lab.[73]

De Salas was another of the chemical talents for whom the Civil War truncated a brilliant academic career. In 1935, he had obtained a fellowship to work with Christopher Ingold and Christopher L. Wilson at UCL, and soon published several joint papers with his mentors in the *Journal of the Chemical Society* on hydrogen exchange in isotopes. In 1937, Ingold declared to the SPSL that de Salas had impressed him 'for his outstanding intellectual powers and considerable culture'.[74] Ingold also commented on De Salas's political situation: '[...] his sympathies are so far removed from those of both parties to the conflict that he very naturally does not desire to go to Spain only to become embroiled in it'.[75] In fact, that apparent 'neutrality' made it more difficult for the SPSL to treat de Salas as a political refugee. After staying for the whole of the Civil War in England, in 1939 he finally obtained a job, but at Imperial Chemical Industries (ICI) in Argentina, with prior training in Manchester, and later in the United States at DuPont. There is no evidence that he ever returned to Spain.[76]

---

[69] Alfredo Baratas, Manuel Lucena, 'La "Society for the Protection of Science and Learning" y el exilio republicano español', *Arbor*, 588 (1994), 25–48. See also: N. J. Baldwin, 'The Archive of the Society for the Protection of Science and Learning', *History of Science*, 27 (1989), 103–5. For the political controversies that the SPSL raised in Britain, see: David Zimmerman, 'The Society for the Protection of Science and Learning and the Politicization of British Science in the 1930s', *Minerva*, 44(1) (2006), 25–45.

[70] Sales, *La química a la Universitat de Barcelona*, pp. 159–62. Queen's College. University Archives, Oxford, UR2/8. See also: F. D. Chattaway, F. Calvet, 'Condensation of Chloral with Substituted Phenols', *Journal of the Chemical Society* (1928), 1088–94; BL Mss. SPSL 201/4.

[71] BL Mss. SPSL 213/3.   [72] BL Mss. SPSL 223/5.   [73] BL Mss. SPSL 220/5.

[74] SPSL 223, fl. 250–330, fl. 254.   [75] Ibid., fl. 267.

[76] Christopher Ingold, Eduardo de Salas, Christopher Wilson, 'Prototropy in Relation to the Exchange of Hydrogen Isotopes. Part I. Rates of Isomerisation and of Hydrogen

Another sign of international support for Republican Spain through chemistry came from the United States. In a letter to President Roosevelt on 24 April 1938, a group of prestigious scientists, including, among others, the Harvard Professor of Physiology Walter Cannon, Albert Einstein – at that time already exiled from Nazi Germany and settled in Princeton – and the Columbia Professor of Chemistry and Nobel Prize winner in 1934 Harold C. Urey (1893–1981), demanded an end to the arms embargo on the Spanish Republic. Cannon, Einstein, Urey and many others did not accept the ideological and geostrategic reasons – fear of communism and leftist revolution from one side and appeasement policies with respect to Hitler on the other – that led liberal Western governments to offer no help to the Spanish Republic. They considered their lack of action to be a serious mistake that, in their view, would make fascism and Nazism in Europe even stronger.[77] Urey, who later became the director of the Manhattan Project's uranium isotope-separation programme, led the support of the North American Committee for Help to Democratic Spain, which, in December 1938, managed to ship 26 pounds and 270,000 doses of nicotinic acid to treat pellagra, as well as the support of the cabinet of physiologist Juan Negrín (1892–1956), at that time Prime Minister of the Republican government.[78] In a similar line, British biochemist and Marxist John B. S. Haldane (1892–1964) supported the antifascist resistance of Madrid during the Civil War and advised the Republican government on toxic gases.[79]

The war was also seriously damaging to women. Adela Barnés was one of Moles's PhD students at the INFQ, and she ended up in exile in Mexico. Her sister, Dorotea, worked at the spectroscopy section of the INFQ and went into exile in France with her family, where she taught in a secondary school, only later to quit her scientific career once back in Francoist Spain. The case of María Teresa Toral (1911–94) is particularly striking in terms of repression. Toral earned her PhD at the INFQ as

---

Isotope Exchange in Unsaturated Nitriles', *Journal of the Chemical Society* (1936), 1328–34. Acknowledgements included the following: 'We thank the Royal Society for a grant for materials, and the *Academia de Ciencias de Madrid*, and University College London for financial assistance accorded to one of us'. See also: Eduardo de Salas, Christopher L. Wilson, 'Prototropy in Relation to the Exchange of Hydrogen Isotopes. Part IV. Isomerisation and Exchange in Methyleneazomethines', *Journal of the Chemical Society* (1938), 319–21; Thomas P. Nevell, Eduardo de Salas, Christopher L. Wilson, 'Use of Isotopes in Chemical Reactions. Part I. The Mechanism of the Wagner–Meerwein Rearrangement. Exchange of Radioactive Chlorine and of Deuterium between Camphene Hydrochloride and Hydrogen Chloride', *Journal of the Chemical Society* (1939), 1188–99.

[77] Sebastian Balfour, Paul Preston (eds.), *Spain and the Great Powers in the Twentieth Century* (London: Routledge, 1999).
[78] *La Vanguardia*, 17 December 1939.   [79] *Ahora*, 10 January 1937, front page.

A Damaged Community

part of Moles's research group.[80] After the war, because of her academic affiliations and political position as a member of the Communist Party, she was excluded from several competitions for academic positions.[81] In 1936, she had obtained a *pension* to work with Friedrich Paneth (1887–1958) at Imperial College London on the detection and separation of isotopes.[82] Toral stayed at the INFQ during the war, keen to rescue papers and books, and she also contributed to the making of explosives. In 1939, she was arrested, tortured and condemned to 12 years of prison. In 1946, Toral was put before a war crimes court, but was finally able to escape into exile in Mexico in the 1950s (see Chapter 7).[83]

Despite the repression, some of Moles's pupils managed to continue their professional careers after the war. This was the case for Ángel Vián,[84] who reported on the state of the INFQ in the following terms:

> I went back to Madrid in April 1939, to get on with my business. Nothing remained from my professional life. Professor Enrique [Moles] and some co-workers were in exile; the rest of the school in prison or under questioning; instruments for the PhD, so painstakingly designed and built, had been destroyed. I did not suffer any penalty, but it was obvious that I had lost the war, three years' hard work, and worst of all, my mentor.[85]

In a context of general violence, other chemists also suffered repression from radical revolutionary groups of the left. This was the case, for instance, of Albareda, the future leader of the Francoist research policies, who witnessed the murder of his father and brother by anticlerical anarchist groups. Although his PhD supervisor, Rius Miró, protected him in Madrid during the war, the violence against his own family made a profound mark on Albareda's right-wing political position during the dictatorship (see Chapter 4).[86] Equally, in 1936, leftist groups executed Ramon Casamada (1874–1936), Professor of Analytical Chemistry at the Faculty of Pharmacy of Barcelona and dean of the faculty in the early 1930s. At the end of the war, his corpse was found together with those of two other faculty colleagues. Casamada's conservative, Catholic ideology

---

[80] Enrique Moles, María Teresa Toral, 'Nueva revisión de los pesos atómicos del carbono y del nitrógeno', *Anales*, 35 (1937), 42–71.
[81] Otero Carvajal, *La Universidad Nacional Católica*, pp. 208–11.
[82] Paneth had escaped from the Nazis and was naturalised as a British citizen in 1939.
[83] Antonina Rodrigo, *Una mujer silenciada. María Teresa Toral: Ciencia, compromiso y exilio* (Barcelona: Ariel, 2012).
[84] Ángel Vián, Joaquín Ocón, *Elementos de Ingeniería Química* (Madrid: Aguilar, 1952). Lora-Tamayo, *La investigaación química española*; Ángel Vián, *Publicaciones científicas (1936–1984). Homenaje académico* (Madrid: Técnicas Reunidas, 1985).
[85] Vián, *Publicaciones científicas*, p. 47.
[86] Antoni Malet, 'José María Albareda (1902–1966) and the Formation of the Spanish Consejo Superior de Investigaciones Científicas', *Annals of Science*, 66 (2009), 307–32.

probably was a target of the anarchist, revolutionary groups that controlled Barcelona during the first months of the war.[87]

Similarly, the Civil War also represented a serious upheaval for the Jesuit chemists. Troubles had already begun in 1932 with the dissolution of the Society of Jesus in Spain by the Republican government. In fact, Salvador Gil, the director of the IQS after the war, described some years later how 'the normal life of the institute was fatally destroyed with the *disgraceful implantation* in Spain of the former Republic'.[88] After 1932, teaching activity experienced some clandestine continuity at the premises of the *Centro de Estudios Químicos*, where a technical school for essential oils, perfumes, cosmetics and soaps opened its doors in 1935. During the war, chemistry books and instruments were kept at the science faculty of the University of Barcelona with Jimeno's help. In fact, at that time, Barcelona was not a safe place for the Jesuits, since leftist anarchists and communist brigades executed a considerable number of clergy, so Jesuit chemists sought refuge in Mallorca – which had been in the hands of the Francoist side from the beginning of the war – and Eduardo Vitoria, the IQS founder, escaped to Italy. In his comments from Mallorca, Gil's political position speaks for itself:

There we rendered our services [...] both at sea and on land, both in air force fleets and on submarines, with the fervent dream of held by any decent Spaniard: *the triumph of the national cause, which was that of religion, of order, of peace and the victory of the true Spain.*[89]

The ideological polarisation of the war placed other chemists in dark territories. Although O. Fernández did not hold a position at the Rockefeller, he was one of the main chemists of the Silver Age and represented the glory of the generation that managed to successfully organise the 1934 International Conference in Madrid (see Chapter 2).[90] At the end of the war, Fernández was inevitably associated with

---

[87] As was frequently the case at that time, Casamada held degrees in chemistry and pharmacy and obtained his PhD in pharmacy with a study of tannins in wine. I am indebted to Professor Joaquim Sales for this information. Sales, *La química a la Universitat de Barcelona*, pp. 139–40.

[88] Salvador Gil, *Discurso pronunciado por el R.P. Dr. Salvador Gil Quinzá, S.J. Director del Instituto Químico de Sarriá, en el homenaje al R.P. Eduardo Vitoria en su 90° aniversario y 50° de la fundación del Instituto Químico. Barcelona 7 de mayo de 1955* (Barcelona: Instituto Gráfico Oliva de Vilanova, 1955) p. 31 (emphasis added).

[89] Ibid., p. 27 (emphasis added).

[90] Obdulio Fernández, *Recuerdos de una vida* (Madrid: Vda. de C. Bermejo, 1973) (new edition: Burgos, 1992). See also: *Libro de Homenaje al Prof. Obdulio Fernández y Rodríguez* (Madrid: RACEFN, 1969); *Recuerdo del Homenaje ofrecido al Catedrático D. Obdulio Fernández Rodríguez com motivo de su jubilación en la Universidad de Madrid* (Madrid, 1954).

Moles, Madinaveitia and the other leaders of the 1930s cosmopolitanism and the liberal spirit of the JAE. The new Francoist authorities also questioned Fernández about his personal friendship with J. Giral, probably one of the main defenders of Republican values in the chemical community and a 'bête noire' of the new regime.[91] Nevertheless, his conservative political ideology led Fernández to defend Franco's coup d'état. In his autobiography, he often spoke about the 'liberation war' as a 'useful' movement for fighting communism as well as Catalan separatism, and also as a response to the violence of anarchist and communist groups. Similarly, del Campo resisted the Civil War in Madrid with his son, but unlike Moles, he did not follow the Republican government to Valencia. In public, he maintained an 'apolitical' rhetoric, with no indications of manifestoes against fascism.[92] At the end of the war in 1939, and in spite of his close contact with many members of the INFQ – particularly his joint work on spectroscopy with Catalán – del Campo was restored to his post and managed to ensure the continuity of his research school of analytical chemistry in Madrid.[93]

The war also profoundly changed the sociology of the SEFQ. In December 1940, at the first meeting of the *Sociedad* after the war and following the cleansing process of many of its members, names such as J. Giral, Moles, Pérez Vitoria and A. León were now replaced by Albareda, Rius Miró and Bermejo.[94] The turmoil was particularly painful in the cases of the brilliant figures who had managed to work creatively on the fringes of the ever-changing boundaries of chemistry in the early decades of the twentieth century. Distinguished names of the Silver Age who had obtained brilliant results at the edge of several chemical specialities also suffered repression: Fernando Calvet for his seminal works on biochemistry, Miguel Catalán for the value of his research on spectroscopy, Enrique Moles for his scientific leadership on physical chemistry

---

[91] O. Fernández, *Recuerdos de una vida*, p. 244.

[92] Francisco González de Posada et al., *Ángel del Campo y Cerdán. Eminente químico español. Cuenca 1881; Madrid 1944* (Madrid: Amigos de la Cultura Científica. Academia de Ciencias e Ingeniería de Lanzarote, 2006).

[93] Ibid., pp. 23–8. The main names in del Campo's research school were Francisco Sierra, Fernando Burriel, J. García Escolar, Carlos Barcia, J. Hernández Cañavate, J. Antonio Parera, Carlos Nogareda and Maximiliano Gutiérrez de Celis (the latter two joined Moles's group). Siro Arribas, *Introducción a la historia de la química analítica en España* (Oviedo: Universidad de Oviedo, 1985), p. 31.

[94] Volume 35 of *Anales*, published in 1939 – '*año de la victoria*' – included a picture of Franco on a cross in homage to the Francoist casualties in the war and a new editorial board led by the new Francoist chemists. *Anales*, 35 (1939), front page. Manuel Valera et al., 'La Guerra Civil española y la investigación científica en química. Estudio preliminar', in Javier Echeverría, Marisol de Mora (eds.), *Actas del III Congreso de la Sociedad Española de Historia de la Ciencia. San Sebastián, 1 al 6 de octubre 1984* (San Sebastian: Editorial Guipuzcoana, 1986), Vol. 3, pp. 395–407.

Figure 3.4 Fernando Calvet at the award ceremony of his DPhil in Oxford in 1928.
Anthropos, 35 (1984), p. 6. Reproduced with permission of Anthropos Editorial, Barcelona.

and atomic weights and Miguel Masriera for his original approach to cosmochemistry deserve particular attention.

Tortured Skills

After obtaining his PhD on the condensation products of phenolic substances with chloral under Chattaway in Oxford (Figure 3.4), Calvet began new research on alkaloids with Heinrich Wieland in Munich[95]

[95] H. Wieland, F. Calvet, W. W. Moyer, 'Strychnos Alkaloids. VI. Typical Color Reactions', *Justus Liebigs Annalen der Chemie*, 491 (1931), 107–16.

and won a chair in organic chemistry at the University of Santiago in 1930. Calvet's school of organic chemistry and biochemistry in Santiago attracted a large number of young students (and women), and he became a pioneer in basic research on dioxins and enzymes, producing some remarkable international papers.[96] He combined Oxford rituals such as tea seminars and sports practice – rugby – with high standards in teaching and research. Politically, he merged his Catalanism and Republican culture with his enthusiastic support for the national identity of Galicia and its right to preserve its own language and to achieve political autonomy. When the Civil War broke out, Calvet was in the middle of a research period as a Rockefeller fellow at the Biochemistry Institute of Hans von Euler (1873–1964) in Stockholm, working on the enzymatic mechanisms of alcohol fermentation.[97] His application to the SPSL was endorsed by Wieland, Willstätter, Robinson,[98] Chattaway, von Euler and Moles. His first mentor, García Banús, praised Calvet for his pioneering work at the University of Santiago in an academic context that, in García Banús's view, lacked a research tradition.[99] The report of Professor Barger from the Department of Medical Chemistry at the University of Edinburgh, a frequent visitor to Republican Spain, was even more striking, as he explained Calvet's dismissal from his chair 'on account of Republican sympathies (he is a Catalan). He has little hope to return to Santiago'.[100] Robinson himself, in a letter from 1937, considered Calvet 'the most brilliant Spanish organic chemist'.[101] In spite of this international support, Calvet had to survive for years working in the chemical industry (see Chapter 7) until he could finally recover his chair, and he later obtained a position in 'technical chemistry' in Barcelona in the 1950s. His potential leadership of the emerging biochemistry scene in Spain was unquestionably damaged, and he never recovered the scientific dynamism he had had at the university in the 1930s.

Miguel Catalán found himself on the Francoist side at the beginning of the war, and his scientific prestige as a paragon of the JAE policies did not help in that context. After the war, Catalán lost his INFQ position and

---

[96] Among the members of Calvet's school, it is worth mentioning Ernesto Seijo, Enrique Niño, María Concepción Carreño, María Natividad Mejuto, Leopoldo Mosquera and Carmen Zapata. See Sales, *La química a la Universitat de Barcelona*, pp. 161–3. See also: 'Fernando Calvet Prats', *Anthropos*, 35 (1984), 1–27.
[97] BL Mss. SPSL 201/4, fl. 216–81.
[98] Sir Robert Robinson, Nobel laureate in 1947 for his research on plant dyestuffs (anthocyanins) and alkaloids. He was Waynflete Professor of Chemistry at Oxford University from 1930 and a fellow of Magdalen College, Oxford.
[99] A place in which, in his view: '*la recherche scientifique était inconnue*'. BL Mss. SPSL 201/4, fl. 228.
[100] Ibid., fl. 235. [101] Ibid., fl. 243.

his prestigious university chair in spectroscopy and atomic structure, and he would never go back to his former research position.[102] Catalán had belonged to the *Izquierda Republicana* political party, and the new Francoist authorities perceived him as one of the most prominent protégés of the JAE, associating him with Moles, another 'bête noire' of the new regime.[103] In his biography, historian José Manuel Sánchez Ron quoted a report from the Francoist army in 1937 with charges against Catalán:

Miguel Catalán Sañudo: Before the *Movimiento*,[104] he was a member of *Izquierda Republicana*. Always protected by those of leftist ideas at the *Institución Libre de Enseñanza*, he achieved several positions, among them, a chair in spectroscopy and atomic structure, created for his profile by the *Junta de Ampliación de Estudios* [sic.]. The JAE was under the control of the ILE, which was very much influenced by the 'committed' leftist Professor of the University of Madrid Enrique Moles. His friendship with Mr. Moles and the support of the JAE brought [Catalán] the gift of the chair and the direction of a section [spectroscopy] with an annual remuneration 12,000 pesetas (*Instituto Rockefeller de Madrid*).[105]

At the end of the war, and despite numerous signs of international solidarity, Catalán had to survive by working in private companies and publishing and translating textbooks. In 1946, he was allowed to teach again at a graduate level, and later, in 1950, he finally obtained a research position at the *Instituto de Óptica* in Madrid after years of humiliation.[106] The period of his creative work on spectroscopy, his discovery of multiplets and his privileged position internationally seemed to have disappeared irreversibly after all those years of personal and professional suffering. Although Catalán's old results were still appreciated abroad, new quantum mechanics and astrophysics had already profoundly changed spectroscopy.[107] So, after paying a high price for his political position before the war, the momentum of his career seemed to be over.

Considered by his contemporaries as the founding father of physical chemistry in Spain and a natural leader of the INFQ, the case of Moles's repression was probably the most shocking. Like hundreds of thousands of other Spaniards, Moles went down the path of exile in January 1939,

---

[102] José Manuel Sánchez Ron, *Miguel Catalan, su obra y su mundo* (Madrid: CSIC/ Fundación Ramón Menéndez Pidal, 1994).
[103] Catalán was also influenced by his father-in-law, the mediaeval historian and philologist, Ramón Menéndez Pidal (1869–1968), who went into exile after the war.
[104] The movement associated with the Francoist side of the war after the coup d'état of July 1936.
[105] Secret report on the Menéndez Pidal family, *Segunda Sección (Información) del Estado Mayor del VII Cuerpo del Ejército*, sent to the 'Jefe del Servicio de Información Militar, en Burgos, el 24 de octubre de 1937'. Cited by Sánchez Ron, *Miguel Catalán*, p. 295.
[106] Ibid., p. 376. [107] Ibid., p. 371.

from Barcelona – where the Republican government had taken refuge after escaping from Madrid and later from Valencia – to the French border, reaching Paris in February. In his laboratory at the *Collège de France*, he was welcomed by Frédéric Joliot-Curie (1900–58), winner of the Nobel Prize for Chemistry in 1935. News of Moles's exile and the loss of his chair in Madrid produced a massive expression of international solidarity. Around 100 chemists from France, the Netherlands, Belgium and Switzerland, including four Nobel Prize winners, unsuccessfully requested of the new Francoist foreign office of the new regime the immediate reinstatement of Moles's academic position, which also included the directorship of the INFQ at the end of the war.[108] Moles's status in Paris as *Maître de recherche* allowed him to continue his research, until the Nazi occupation and the establishment of the Vichy government made things considerably harder.[109] In spite of his explicit support for the Republic, Moles wanted to come back to Spain with the hope that his chemistry could be easily accommodated in different political regimes, as had previously been the case during the years of the Restoration and Primo de Rivera's dictatorship. A report of the International Education Board (IEB) of the Rockefeller Foundation in May 1939 confirmed Moles's desire to return to Spain and his personal perception – which was probably mistaken – of being 'above' any political fight:

[Moles] is quite confident that [his return to Spain] will ultimately be possible as he has never had anything to do with politics [...] Moles said that he has just seen announced in the Spanish papers the plan to create an Institute of Spain, patterned to some extent along the lines of the *Institut de France*, which will include all of the activities formerly carried out by the JAE. This was apparently a pet project of Sainz Rodríguez, [but he] has recently resigned or has been removed from office and may or may not be included in Franco's cabinet when it is established.[110]

Moles aimed to recover his teaching and research positions at the university and the INFQ, the preservation of which he credited to his

---

[108] Archivo General de la Administración. Expediente de depuración de Enrique Moles. Leg. 92059, folio 20. Agustí Nieto-Galan, Joaquim Sales, 'Exilio y represión científica en el primer franquismo. El caso de Enrique Moles', *Ayer. Revista de Historia Contemporánea*, 114(2) (2019), 279–311.

[109] Raúl Berrojo, *Enrique Moles y su obra*. PhD thesis (Barcelona: Universitat de Barcelona, 1980), 3 vols., pp. 1064–7. See also: Enrique Moles, *'Determinació de pesos moleculars de gasos pel mètode de les densitats límits'*, Enric Moles i Ormella. Traducció, introducció i comentaris de Joaquim Sales i Agustí Nieto-Galan (Barcelona: Societat Catalana de Química, 2013).

[110] Rockefeller Archive Center, IEB, box 41, file 581, New York. Memorandum. 'Professor Moles, formerly of the Institute of Physics and Chemistry, Madrid, and now refugee in France'.

own contributions during the war.[111] Nevertheless, on 8 December 1941, upon arrival at the Spanish border, he was arrested and charged with having produced weapons during the war and having dismissed some of the members of the INFQ. Then came bitter accusations from former fellow chemists, stating that he had supposedly exerted full, endogamic and corrupt control of all chemistry chairs. They even suggested, with no evidence, the possibility of Moles's Masonry, which was perceived as a network of solidarity and sectarian professional influence.[112] Moles was also blamed as being responsible for using his international reputation in favour of the Republican government and against the Francoist cause, contributing to the continuity of the liberal scientific institutions, mainly the INFQ, during the war. The accusation of José María Albareda, at that time the powerful general secretary of the Francoist *Consejo Superior de Investigaciones Científicas* (CSIC) (see Chapter 4), was particularly poisonous.[113] More proof of Moles's prosecution was a paper he signed in 1937 in the Republican journal *Madrid Cuadernos de la Casa de Cultura*, in which he summarised his main scientific results. On the last page of the paper, Moles compared the treatment of Jewish scientists in Nazi Germany with the more favourable science policies of the Soviet Union in the 1930s. He was finally accused of belonging the political party *Izquierda Republicana*, but this was never proven.[114]

Despite his lengthy repression,[115] Moles never ceased to ask for reductions, pardons and reviews, and in December 1942, he obtained provisional liberty.[116] The collapse of the prisons in the years of the truly harsh repression of the dictatorship – in 1945, more than 300,000 people occupied Spanish prisons as political prisoners[117] – and Ernest Fourneau's visit to Madrid in July 1942 to request Moles's liberation all played a part.[118] Since the 1910s, Fourneau had established fruitful

---

[111] In a letter to the physicist Julio Palacios, Moles seemed to assume an 'unproblematic' return to Spain. Archivo Ministerio de Defensa, Madrid. Sumario 25334, folios 80, 81.
[112] Archivo del Ministerio de Defensa, Madrid, Sumario 25.334.
[113] Archivo del Ministerio de Defensa, Madrid, Sumario 25.334, folio 62 (Declaración de José María Albareda Herrera).
[114] Enrique Moles, 'Veinte años de investigacions acerca de las densidades gaseosas', *Madrid. Cuadernos de la Casa de Cultura*, 1 (1937), 33–51.
[115] In 1942, the *Consejo de Guerra* condemned Moles for support of a military rebellion to six years of reclusion, which were formally extended in further trails to 30 years. Archivo del Ministerio de Defensa, Madrid, Sumario 25.334, folios 145–6 (Sentencia).
[116] Archivo del Ministerio del Interior, Madrid, Expediente 25.334, folio 69.
[117] Borja de Riquer, *Historia de España (IX): La dictadura de Franco* (Barcelona/Madrid: Crítica/Marcial Pons, 2010), p. 136.
[118] Jean-Pierre Fourneau, 'Ernest Fourneau, fondateur de la chimie thérapeutique française: feuillets d'albums', *Revue d'histoire de la pharmacie*, 275 (1987), 335–55.

collaborations with Spanish chemists, and he had received an honorary doctorate in Madrid during the International Conference of 1934 (see Chapter 2). However, Fourneau supported the Vichy regime during the Second World War and became a collaborator in the German occupation. These were useful credentials in Spain in 1942 with the Francoist authorities. Although Moles did not return to jail, further persecution followed in the late 1940s. That severe humiliation was added to the irreversible loss of his chair in inorganic chemistry in Madrid and his position at the INFQ. His work on atomic weights, which had acquired international recognition in the 1930s, and his research creativity and laboratory skills became irrelevant in the face of the mechanisms of repression after the war[119] and the new rhetoric against pure, liberal science in favour of a more utilitarian, applied chemistry.

Unlike Moles, Miguel Masriera did not suffer imprisonment but, following his early exile in Paris, he was never able to recover his position at the Faculty of Science of the University of Barcelona.[120] In Paris, Frédéric and Irene Joliot-Curie helped him to work on spectroscopy and infrared photography at the Astrophysics Institute (*Institut d'Astrophysique*) with Professor Daniel Chalonge (1895–1977),[121] and later at the *École Normale* and the Sorbonne. Masriera's name appeared in Franco's police files under the category of '*rojo*' (red) for his services to the Republic before and during the Civil War.[122] In his correspondence with some former colleagues who had managed to accommodate themselves in the new regime,[123] Masriera lamented his marginalisation from the new academic order and insisted on having at least a research position, but he was never again able to obtain an academic post.[124] Back in Spain, Masriera conducted independent scholarly work on theoretical aspects of cosmochemistry. He established his own laboratory at home with some of the instruments he was able to bring with him from Paris,[125]

---

[119] Nieto-Galan, Sales, 'Exilio y represión científica en el primer franquismo'.
[120] Nieto-Galan, 'From Papers to Newspapers'.
[121] Chalonge worked on experimental and theoretical astrophysics, stellar spectroscopy and precision spectrophotometry. He was one of the founders of the *Institut d'Astrophysique de Paris*. http://chalonge.obspm.fr/curso.html (last accessed 10 January 2016).
[122] A letter to José Baltà Elías, Professor of Electricity and Director of the *Instituto Nacional de Radioelectrónica* at the CSIC (17 June 1953). Fons Masriera (FM). 3.2.C. Institut d'Estudis Catalan (ICE). Barcelona.
[123] Nieto-Galan, 'From Papers to Newspapers'.
[124] Fons Masriera. Letter to José María Albareda. 05 July 1952. 3.2.C. *Institut d'Estudis Catalans*. Barcelona.
[125] Masriera set up his own laboratory in his private apartment: 'M. Masriera, Dr. Ing. Laboratorio Espectroquímico, Provenza, 293, Barcelona, Tel. 84317'.

acted as a private consultant for several industrial companies,[126] and became a very active science populariser (see Chapter 7).

Under the protection of the Count of Godó, an upper-class Catalan aristocrat enjoying a comfortable position in the Franco regime and the owner of the newspaper *La Vanguardia Española*, Masriera was appointed as the newspaper's science populariser. He established useful contacts with local and foreign chemical entrepreneurs, which became useful for supplementing his income and spreading his popular science products among the economic elite. Although his collaboration with *La Vanguardia* had begun in the 1920s, his presence in the daily press became particularly important from 1948 onward. Masriera became a tolerated dissident, but the fact is that his brilliant academic career and his interdisciplinary background that allowed him to move from theoretical physics and spectroscopy to applied chemistry in industry was never officially recognised after the war.

***

The scientific prestige of early 1930s, involving names such as Albareda, Rocasolano, Bermejo, Jimeno, Lora-Tamayo, Moles, Catalán, Madinaveitia, García Banús, J. Giral and many others, dramatically split to serve opposing political agendas, ideologies and professional interests. A preliminary prosopography of the different cases ranges from execution (Yoldi), exile (J. Giral, F. Giral, García Banús, Madinaveitia), jail (Moles) and marginalisation (Calvet, Catalán, Ribas, Masriera), to collaboration with cleansing committees (Jimeno, Bermejo) and enthusiastic Francoism (Rocasolano, Bermejo, Albareda, Lora-Tamayo). The way in which the Spanish chemical community was tortured and profoundly wounded during the Civil War will almost certainly shock the reader. Nevertheless, it was probably a natural reflection of the way in which Spanish society as a whole was badly damaged. Political ideology played a crucial role in the ways in which power relations were dramatically reshaped and rearranged, but personal rivalries and professional tensions were probably more important than those expressed in the historical actors' own accounts. Those who could not conquer the INFQ in the early 1930s often felt marginalised, particularly during the Republic – with Rocasolano being a paradigmatic case – and added their conservative, anti-ILE ideas to their criticisms against scientific elitism and the supposedly useless, liberal, pure research of the Madrid chemists. The fierce aggression against Moles in court and

---

[126] Nieto-Galan, 'From Papers to Newspapers'.

contempt for his atomic weight experiments need to be explained beyond political ideologies as a brutal struggle for academic power and prestige.

Madinaveitia's and F. Giral's chemistry of natural products inevitably had to find new raw materials in exile. Members of 'cleansing committees' such as Rocasolano and Jimeno frequently argued in public that Catalán's multiplets and Moles's atomic weights had become a sort of expensive, useless luxury of elitist research, which was supposedly too far removed from the new 'applied chemistry' for the progress of the country. War damage was profound and long-lasting, and a new age of chemistry and of chemists was waiting.

As historian Jeffrey Allan Johnson discussed a few years ago, the aftermath of wars produces serious effects in the scientific culture of a chemical community. What Johnson said about the aftermath of the two World Wars can be also applied to the consequences of the Spanish Civil War for chemistry. There were obviously deep changes in individuals, research groups and institutions, the destruction of instruments and laboratories, restrictions of raw materials, disruption of research lines, reversions of research priorities and reorganisations of information flows and international links. As happened elsewhere, the Spanish conflict profoundly changed contemporary perceptions of science, politics, industry and the military; it built up a new 'social context of chemistry' (in Johnson's terms) that I will describe in Chapter 4.[127]

---

[127] Johnson, 'Crisis, Change and Creativity'.

4 Totalitarian Ambitions

The alliance between scientific knowledge and political power was under discussion at the birth of the new regime. Franco's minister Pedro Sainz Rodríguez (1897–1986) initially endorsed the creation of an institution that would preserve a certain scientific autonomy, keeping it separate from political structures. In Burgos, intellectuals from the Francoist headquarters during the war wanted to see an imitation of the *Institut de France*,[1] in which prestigious scientists could develop their careers with a certain independence from political turbulence in an ivory tower-like image of scientific research. Nevertheless, it was the new minister, José Ibáñez Martín (1896–1969), in close collaboration with José María Albareda, who tipped the balance towards a totalitarian organisation of science policies such as the new *Consejo Superior de Investigaciones Científicas* (CSIC), one that was deeply embedded in the ideology of the new regime.[2] In that new organisation, Albareda, a prestigious chemist and devoted Catholic, played a major role. With degrees in pharmacy and chemistry and two doctorates, Albareda combined his academic credentials with his position as a high school teacher from 1928 to 1935. He also benefitted from excellent training in soil chemistry, which he had acquired as a JAE *pensionado* in the 1920s and 1930s,[3] and he also

---

[1] This was, for instance, the original ideal of intellectuals such as Eugeni D'Ors (1881–1954) who, after being active in the building of political Catalanism in the early decades of the century, progressively turned to a more conservative, fascist-like ideology and finally joined the Francoist side. On Eugeni D'Ors, see: Josep Maria Terricabras (ed.), *El pensament d'Eugeni d'Ors* (Girona: Documenta Universitària, 2010).

[2] Antoni Malet, 'Las primeras décadas del CSIC: Investigación y ciencia para el franquismo', in Ana Romero and María Jesús Santesmases (eds.), *Cien años de política científica en España* (Madrid: Fundación BBVA, 2008), pp. 211–56.

[3] In 1928, Albareda worked on agricultural chemistry in Bonn with Hubert Kappen, and later at the ETH Zurich with Georg Wieger and in Könnigsberg with E. H. Mitscherlich. In 1932, he obtained a Ramsay Fellowship to work on the chemistry of clay in Wales and Scotland. For more details on Albareda's biography, see: Antoni Malet, 'José María Albareda (1902–1966) and the Formation of the Spanish Consejo Superior de Investigaciones Científicas', *Annals of Science*, 66 (2009), 307–32. For an apologetic approach to Albareda's thought, see: Pilar León-Sanz, 'Science, State and Society. José María Albareda's *Consideraciones sobre la investigación científica*', *Prose Studies*, 31(3)

belonged to the prestigious, ideologically conservative school of Rocasolano in Zaragoza. Escaping from the Republican zone, he had also accompanied Josemaría Escrivá de Balaguer (1902–75), the founder of Opus Dei, during the Civil War (Figure 4.1).

Albareda's original design for the CSIC assigned an important role to the *Patronato Juan de la Cierva* (PJC), a key institution for the promotion of a new 'applied science' after the war. The name aimed to honour the famous Spanish inventor of the gyroplane (*autogiro*), Juan de la Cierva (1895–1936), who was also Franco's friend, and who had died in a plane crash in England at the end of the war.[4] Chemistry appeared at the intersection of the PJC (with the bulk of applied chemistry in the institutes of fats, metallurgy, combustibles, coal, plastics, fermentation and plant chemistry) and the *Patronato Alfonso X el Sabio* (with departments of analytical and inorganic chemistry and institutes of organic and physical chemistry).[5] Largely under the control of Albareda, the CSIC also funded and promoted several chemistry sections in universities. During the period of 1939–54, 42 per cent of CSIC researchers were chemists.[6]

Albareda had very specific plans to establish chemistry in the new political order. In his confidential report to Minister Ibáñez Martín, '*La química en España*', Albareda described in detail the scientific and personal profiles of the most prominent chemists whose job it would be to lead the new historical period. One of the names raised was Rocasolano, who held great academic prestige and had an ideal right-wing ideological profile, but in Albareda's view he had some weaknesses on the industrial side. Moreover, Rocasolano died in 1941, so he could not play a decisive role in the new organisation of chemistry. However, Albareda also had other names in mind: Rius Miró, one of Rocasolano's pupils and Albareda's supervisor – a politically faithful ally who could substitute for the repressed Moles in the leadership of physical and technical chemistry;

(2009), 227–40. In a similar vein, see: Enrique Gutiérrez Ríos, *José María Albareda: Una época de la cultura española* (Madrid: CSIC, 1970). See also: 'Professor J.M. Albareda', *Nature*, 210 (1966), 1313.

[4] José Warleta, *Autogiro. Juan de la Cierva y su obra* (Madrid: Instituto de España, 1977); Arthur W. J. G. Ord-Hume, *Juan de la Cierva and His Autogiros* (Catrine: Stenlake Publishing, 2011).

[5] José María Albareda, 'Organization et tendances de la recherche scientifique en Espagne. Le Conseil supérieur de la recherche scientifique', *Impact. Science et Société*, 15(1) (1965), 43–57.

[6] Néstor Herrán, Xavier Roqué, 'Los físicos en el primer franquismo: Conocimiento, poder y memoria', in Néstor Herrán, Xavier Roqué (eds.), *La física en la dictadura. Físicos, cultura y poder en España 1939–1975* (Bellaterra: Universitat Autònoma de Barcelona, 2012), pp. 85–104, 95. See also: Xavier Roqué, Néstor Herrán, 'An Autarkic Science: Physics, Culture and Power in Franco's Spain', *Historical Studies in the Natural Sciences*, 43(2) (2013), 202–35.

DON JOSE MARIA ALBAREDA HERRERA

(15 de abril de 1902-† 27 de marzo de 1966)

Figure 4.1 Portrait of José María Albareda in the journal *Arbor* in 1966. Reproduced with permission of Editorial CSIC.

and Casares, who maintained his scientific authority from the early decades of the century with international connections and faithful adhesion to the dictatorship. In organic chemistry, Albareda considered Ignacio Ribas, but his old links with the now-exiled Madinaveitia frustrated his promotion to an academic position. Albareda had two other names in mind: Manuel Lora-Tamayo and José Pascual. Lora-Tamayo was a close friend of Albareda and a prestigious scientist – well connected internationally, with strong ideological and religious affinities. Pascual was the natural replacement for the exiled García Banús. He was a prestigious organic chemist, but also a devout Catholic – a detail that particularly pleased Albareda – and a very close friend of Lora-Tamayo due to the years they had spent together at the University of Seville.[7] These were, in practice, the main names that constituted the core of the new Francoist chemistry and took the lead in terms of research priorities, educational reforms and professional strategies. As reflected at the RSEFQ, the geography of the profession had been profoundly altered. In inorganic and physical chemistry, Albareda's and Rius Miró's groups dominated the scene; in organic chemistry, Lora-Tamayo's and Pascual's schools controlled a good number of the published papers.[8]

At a university level, the 1943 Universities Act (*Ley de Ordenación Universitaria*; LOU) was a good reflection of the abrupt break with the liberal past of the 1930s and the totalitarian values of the new academic culture.[9] The whole academic system was strongly centralised. PhDs could only be submitted at the central university in Madrid, and students suffered the rigid control of the Spanish Universities Union, the *Sindicato Español Universitario* (SEU). Franco and the Minister for Education appointed university rectors, who became members of the *Cortes* (the new pseudo-parliamentary regime). Mandatory, censored textbooks, neo-scholasticism and the myth of the Spanish imperial past made profound marks on the intellectual atmosphere of the early 1940s.[10]

---

[7] *Informe ('Confidencial') de Albareda a Ibáñez Martín, titulado 'La Química en España'*, cited by José Manuel Sánchez Ron, *Miguel Catalán. Su obra y su mundo* (Madrid: CSIC, 1994), pp. 372–3.
[8] *V Reunión Anual de la Real Sociedad Española de Física y Química y II de los Institutos de Física y Química del CSIC. Resúmenes de las comunicaciones científicas: Granada, 26 abril al 1 mayo de 1948* (Madrid: Real Sociedad Española de Física y Química, 1948).
[9] Luis Enique Otero Carvajal (dir.), *La Universidad Nacional Católica. La reacción antimoderna* (Madrid: Universidad Carlos III, 2014), pp. 191–258. See also: Gonzalo Redondo, *Política, cultura y sociedad en la España de Franco (1939–1975)* (Pamplona: Ediciones Universidad de Navarra, 1999) (3 vols.); Enrique Gutiérrez Ríos, 'Investigación y libertad', *Nuestro Tiempo*, 109 (1963), 8–20; Gutiérrez Ríos, *José María Albareda*.
[10] Borja de Riquer, *Historia de España (IX): La dictadura de Franco* (Barcelona/Madrid: Crítica/Marcial Pons, 2010). See also: Carme Molinero, Pere Ysàs, *El règim franquista: Feixisme, modernització i consens* (Vic/Girona: Eumo/Universitat de Girona, 1992).

In that new order of things, the core group of new Francoist chemists became crucial actors in the new political ambitions of the time. In public, an aggressive rhetoric against Republican values helped them to reinforce their own professional positions. Although, in practice, some research lines continued from the JAE era, they demanded a radical discontinuity with the past and, as powerful ideological agents, enthusiastically ascribed the moral, political and religious values of the dictatorship to the ethos of their profession. Paradoxically, the materialist tradition of chemistry was not in contradiction with the Catholic, conservative, fascist values of the new regime, or with economic growth, profit and industrial progress. The new leading chemists wisely combined religious values, political priorities and research endeavours. They appropriated chemicals and research projects in a context in which religion strengthened group cohesion and legitimised academic and industrial practices in a peculiar alliance between chemistry and National Catholicism that I shall discuss in the following sections.

Fascist Chemistry

In October 1940, a year after the foundation of the CSIC, its president and Minister for Education Ibáñez Martín highlighted in his speech the totalitarian and radically anti-liberal nature of the new science:

The catastrophe we have experienced has taught us to consider bankrupt a historical age that began with the anarchy of human freedom and the deification of individual reason, against eternal principles, and brought us as a result that integral liberalism, which authorised the debauchery of the sciences, consecrating its regime of unsociability and prostituting it until it was used against the substance of the nation itself [...] The science of the new Spain will never be at the service of individual interests.[11]

Although continuity in the scientific practices of some research lines in organic chemistry, physical chemistry and other fields can be traced back, the new identity of the Francoist chemists was based on a strong anti-JAE claim, on 'applied chemistry', on the ambition to control the whole 'national' geography and on the importance of religious values. Albareda, Lora-Tamayo and many others benefitted on many fronts from the scientific training they received before the Civil War, but neglected their masters' generation for the sake of legitimating the dictatorship. In Albareda's view, the Silver Age had been a time of painful secularism, foreign praise and antipathy towards university. The JAE

[11] José Ibáñez Martín, *La investigación española* (Madrid: Publicaciones Españolas, 1947), pp. I, 22.

supposedly only promoted research centres such as the LIF or the INFQ, and appointed researchers according to strong ideological bias. Moreover, academics tended to act crudely in defence of their own individual interests and not for those of the nation. Albareda also blamed them for their aversion to technology and economics; in other words, he criticised what he conceived of as the JAE's exaggerated obsession with pure, elitist chemistry, which was often associated with the liberal values of the Republic.[12]

Similarly, Lora-Tamayo emphasised in public the differences between the JAE and the CSIC. During the Civil War, and as a member of the conservative association *Acción Católica*, he had already felt comfortable with the political project of the military rebels, and he had witnessed General Queipo de Llano's fierce repression in Seville during the Civil War. In his autobiography, he did not hesitate to give public praise to the value of Franco's personality and legacy for the modernisation of Spain.[13] In 1942, Lora-Tamayo obtained the chair in organic chemistry at the Faculty of Science in Madrid,[14] and he was soon made director of the *Instituto de Química Orgánica* at the CSIC. In the 1950s, he accumulated an enormous level of academic power.[15] In his view, the weak and scattered research efforts of the JAE contrasted with the post-1939 quantitative growth of research projects, which were spread throughout the whole of Spain and aimed to be 'useful' to industry. He presented the research of the pre-war period as suffering from "[...] restricted research groups [...] full of enthusiasm but lacking resources [...] with an almost absolute neglect of applied research'[16] – a set of weaknesses that only the new CSIC could overcome. Lora-Tamayo conceived the CSIC as a real state policy,[17] as the 'first' institution to establish a systematic research plan for the whole country and a standardised, centralised professional career in science.[18]

[12] Gutiérrez Ríos, *José María Albareda*.
[13] 'La figura ya histórica de Franco y la positiva efectividad de su obra en el desarrollo de España me impresionaron siempre', in Manuel Lora-Tamayo, *Lo que yo he conocido (Recuerdos de un viejo catedrático que fue ministro)* (Puerto Real: Federico Joly y Cia-Ingrasa, 2002), p. 88.
[14] At that early stage of his career, Lora-Tamayo had already published on the problem of the Spanish university. Manuel Lora-Tamayo, 'Orientaciones para una posible reforma de la Facultad de Ciencias', *Revista Nacional de Educación*, 1(2) (1941), 40–50.
[15] At that time, Lora-Tamayo was vice-rector of the University of Madrid, a professor at the *Escuela Superior de Armas Navales*, a member of the *Academia Real de Farmacia*, the *Academia de Medicina de Sevilla*, the *Academia de Buenas Letras de Sevilla* and the *Academia de Ciencias y Artes de Barcelona* and Head of Organic Chemistry at the *Instituto Alonso Barba de Química* at the CSIC.
[16] Manuel Lora-Tamayo, *La investigación científica* (Madrid: Editora Nacional, 1963,) p. 7.
[17] Manuel Lora-Tamayo, *La investigación en química orgánica* (Madrid: CSIC, 1951), p. 39.
[18] Lora-Tamayo, *La investigación científica*, p. 8. See also: Lora-Tamayo, *Lo que yo he conocido*, p. 259.

In spite of his close links with the JAE laboratories and his academic duties in the Republican 1930s, the former rector of the University of Madrid, Luis Bermejo, also delegitimised the INFQ. In his view, the 'Rockefeller' was a Masonic institute, an exclusive enclosure in which research contracts did not follow the traditional method of government competitions or *oposiciones* and did not count on the university at all. Moreover, he did not approve of the allocation of so much public funding for expensive scientific instruments. In particular, Moles's demanding experiments for the calculation of atomic weights became the paradigm of Bermejo's criticism.[19] Although Rocasolano and Bermejo both died in the early 1940s, they had campaigned for the creation of a new identity for chemistry in tune with the new totalitarian values of the dictatorship. In fact, the old Rockefeller soon became the new *Instituto Alonso Barba* (CSIC) – in honour of the seventeenth-century Spanish metallurgist, Alvaro Alonso Barba, who invented several silver–mercury amalgamation processes in mines in Latin America.[20] In 1946, the institute was renamed the *Instituto de Química Física Rocasolano* – in honour of Antonio de Gregorio Rocasolano, who had contributed notably to agrochemical and biochemical projects with his research group in Zaragoza and enthusiastically supported Franco's coup d'état (Figure 4.2). In fact, Ibáñez Martín, in his speech at the CSIC in December 1941, stressed the fact that '[Rocasolano] extolled the tradition of Spanish science, remembering the great imperial days and showing that our current scientific effort was not a shot in the dark but was linked with the highest days of Spanish thought'.[21]

Like Albareda and Rocasolano, Emilio Jimeno had similar views about the JAE being supposedly 'too political'. In February 1939, only a few weeks after the conquest of Barcelona by Franco's army, the *Caudillo* appointed Jimeno as the new rector of the university. Jimeno's conservative ideology and his personal friendship with Franco during the years they had spent together in Oviedo made him a solid ally after the war. He soon produced a report on the so-called 'totalitarian' organisation of chemistry in the new Spain.[22] In 1939, in one of his first public addresses as rector, Jimeno emphasised the need for all research efforts to satisfy the most urgent needs. He asked for closer collaboration with 'allied'

---

[19] Luis Bermejo, 'El Instituto Rockefeller', in Miguel Artigas et al. (eds.), *Una poderosa fuerza secreta. La Institución Libre de Enseñanza* (San Sebastian: Editorial Española, 1940), pp. 197–202.
[20] Modesto Bargalló, *La minería y la metalurgia en la América española durante la época colonial* (Mexico City: Fondo de Cultura Económica, 1955).
[21] Ibáñez Martín, *La investigación española*, pp. 41–2.
[22] Emilio Jimeno, *Ciencia y Técnica* (Madrid: SAETA, 1940), p. 146.

Fascist Chemistry 119

D. Antonio de Gregorio Rocasolano, fallecido el 25 de abril de 1941.

Figure 4.2 A portrait of Antonio de Gregorio Rocasolano in an obituary following his death in 1941.
Ión, 1 (1941), p. 67. Reproduced with permission of the *Biblioteca de Catalunya*, Barcelona.

countries, particularly with Mussolini's Italy, Latin America and even with the Philippines, a former Spanish colony until 1898, with the main point being in the two latter cases the common language and race, crucial aspects of Jimeno's strong Spanish nationalism.[23] In 1940, Jimeno warned of the 'liberal' dangers of the chemical industry in the new regime, and pointed out that:

> [...] At this time, if the Spanish state does not organise the chemical industry, large sums of foreign capital will be invested in it, with no concern other than profit. We must avoid this possibility at all costs, for the sake of not rendering to foreign hands one of the most important industrial activities of our country.[24]

Jimeno expressed scepticism about the Spanish chemical industry at the end of the war. Following some of the pre-war laments of chemical backwardness, he felt that the lack of well-trained experts, an overly theoretical education in the science faculties and the poor scientific background of Spain's entrepreneurs made things particularly hard.[25] Placing further emphasis on the discontinuity and rupture with the JAE period, he listed the weaknesses of pre-war chemical culture. Broadly speaking, these were: poor training of young chemists; lack of patriotism and of interest in creating a genuine Spanish chemistry; limited efforts to exploit local raw materials; selfishness and greed; high-level education only being for a small group of intellectuals; and the lack of good training for craftsmen and technicians.[26]

Jimeno's lectures, which were compiled and published in 1940, also reflected his old concerns on pure–applied chemistry.[27] Under the title *Ciencia y Técnica*, he conceived his writings as a seminal attempt to create a new journal that aimed to disseminate international chemistry novelties to entrepreneurs, but also to craftsmen and industrial workers, so as to diffuse foreign methods of chemical education and to provide a comprehensive international bibliography.[28] He denounced the long-standing weak alliance between chemistry and industry in Spain and again stressed the need to appropriate German and American patterns.[29] Jimeno also aimed to improve 'technical chemistry', which

---

[23] Emilio Jimeno (ed.), *Universidad de Barcelona. Aspectos y problemas de la nueva organización de España. Ciclo de Conferencias organizado por la Universidad de Barcelona* (Barcelona: Universidad de Barcelona, 1939), p. 8.

[24] Jimeno, *Ciencia y Técnica*, p. 243.   [25] Ibid., p. 35.   [26] Ibid., p. 39.

[27] Jimeno, 'De ningún modo podemos exponernos a defraudar a nuestro Generalísimo, que ha logrado para nosotros la victoria más importante de nuestro siglo', *Ciencia y Técnica*, p. 33.

[28] Jimeno, *Ciencia y Técnica*, p. 55. There is no evidence that *Ciencia y Técnica* actually became a periodical.

[29] Jimeno, *Ciencia y Técnica*, pp. 35–9.

needed to include basic concepts in applied mathematics, applied electricity, heat, conductivity, large-scale chemical reactions, quality control, main operations, materials and installations.[30]

Fiercely opposed to economic liberalism, Jimeno considered that the national industry had to operate under the strict control of new 'totalitarian' organisations such as the National Board for the Chemical Industry (*Consejo Nacional de la Industria Química*) and the Institute for Research into Industrial Chemistry (*Instituto de Investigaciones Químico Industriales*).[31] Both institutions had to efficiently exploit Spanish raw materials in a period of severe isolation and economic autarchy (see Chapter 5). They had to focus on the standardisation of chemical products, inspection, coordination of chemical research and teaching at a national level, all of which came under very strict government control.[32] The *Consejo Nacional* – its members being appointed directly by Franco in person – was in charge of that organisation through several bodies devoted to statistics, education and training, libraries, laboratories and industrial chemistry research, all for the sake of the standardisation of all chemical products, as well as for the cooperation between state and industry in a supposedly 'objective', 'neutral' process.[33] It was the job of the *Consejo Nacional* to assess, stimulate and support all of the chemical activities of the nation.[34] A totalitarian organisation of the Spanish chemical industry was a fundamental strategy, in Jimeno's view, to avoid investments of foreign capital and firms that, in that atmosphere of autarchy, were perceived with suspicion (Figure 4.3).

Under the supervision of the *Sindicato Nacional de Industrias Químicas*, a vertical – supposedly classless – totalitarian trade union and the political control of the ministries of agriculture, industry, education and defence, Jimeno conceived the whole endeavour in three steps. First came the establishment of basic chemical industries (mineral acids, metals and alloys, cellulose and coal distillation), the search for new national raw materials and the training of industrial experts. In a second step, protectionist policies for the Spanish chemical companies had to be implemented, as did a useful patent law for new chemicals, new control laboratories and regular inspections of all industries across the country. Finally, a third

---

[30] Ibid., p. 101.   [31] Ibid., pp. 243–60.   [32] Ibid., pp. 119–69.
[33] The *Consejo Nacional* was in charge of several bodies: the *Servicio de Ordenación de la Industria Química, Servicio de Enseñanza de la Química Industrial, Servicio de Bibliotecas e Información, Servicio de Estadística, Laboratorios de Normalización y Ensayos* and *Instituto de Investigaciones Químicas Industriales.*
[34] Jimeno, *Ciencia y Técnica*, pp. 35–9.

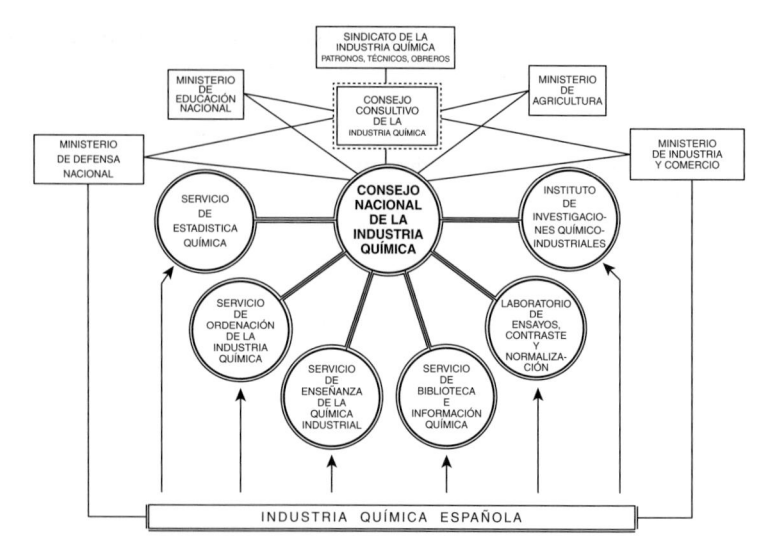

Figure 4.3 Emilio Jimeno's chart for a totalitarian chemistry in 1940.
Emilio Jimeno, *Ciencia y Técnica* (Madrid: Saeta, 1940), p. 245.
Reproduced with permission of the *Biblioteca de Catalunya*, Barcelona.

step needed to be the publication of statistical data, textbooks and popular chemistry texts, foreign translations and chemistry journals. He also envisaged the future creation of several institutes for industrial chemistry research on organic chemistry, inorganic chemistry, metals and alloys and electrochemistry.[35]

In spite of his privileged position at the end of the war and his collaboration with educational policies at the *Consejería de Educación Nacional*, in the 1950s Jimeno disagreed with Juan Antonio Suanzes (1891–1977), the president of the *Instituto Nacional de Industria* (INI) – a new official body for the promotion of industry – on the ways in which links between university and industry had to be established.[36] In 1945, the INI established a new research institute for iron and steel (*Instituto del Hierro y del Acero*), but did not count on Jimeno's earlier experience in metallurgy.

[35] Ibid., pp. 245–8.
[36] Felipe Calvo, *Emilio Jimeno Gil. Semblanza de un maestro* (Santander: Amigos de la Cultura Científica, 1984), p. 40; Felipe Calvo, José María Gulemany, 'Contribución del profesor Emilio Jimeno Gil al prestigio de la Universidad de Barcelona: Rector de 1939 a 1941', in *Història de la Universitat de Barcelona. I Simposium del 150 Aniversari de la Restauració. Barcelona 1988* (Barcelona: Publicacions de la Universitat de Barcelona, 1990), pp. 471–83; Joaquim Sales, *La química a la Universitat de Barcelona* (Barcelona: Publicacions i Edicions de la Universitat de Barcelona, 2011), pp. 129–37.

Fascist Chemistry   123

After a short period (1939–40) in which he contributed to the organisation of chemistry at academic and industrial levels, Jimeno moved to Madrid, where he obtained a chair in inorganic chemistry at the university in 1941. He replaced Moles, who at that time was suffering political repression. In spite of the isolation of the regime in those early years, Jimeno still enjoyed a great reputation abroad in teaching and applied research. In 1942, he obtained the *Premio Francisco Franco de Ciencias*, a new award that had been set up to strengthen the loyalty of the elite scientists to the new regime.

The fascist ideology of the *Falange Española* political party also included a chemical facet. In 1941, Gumersindo García became chief of the *Sindicato Nacional de Industrias Químicas*, which was placed at the top of Jimeno's chart for a totalitarian chemistry and occupied a key position in the organisation of the chemical industry (Figure 4.4). In the early 1930s, García was already a member of *Falange* and its general secretary in Madrid. After the war, he became one of the founders of the SEU, the fascist student organisation. In the early years of Francoism, he also had a significant political career as *Procurador en Cortes* (MP) during the period 1943–55.[37] The Falangist, vertical, totalitarian structure of the Francoist *Sindicatos* created an endless web of regulations, laws, groups and subgroups and a political–administrative framework that aimed to rigidly control all of the activities of the chemical industry.[38] The *Sindicato* reinforced the Falangist strong anti-liberalism ideology in politics but also in the economy, with the role of chemistry and chemicals being to exploit and produce ideal tools for the materialisation of the totalitarian dream in the early years of the regime. Moreover, under the banner of the *Sindicato*, experts were assigned different specialist commissions that worked in collaboration with the PJC in various applied chemistry projects.[39]

It was in that context that the journal *Ión* appeared, with a subtitle emphasising the insistent and renewed focus of applied chemistry (*Revista española de química aplicada*), which was very strongly reflected in sections for the 19 fields of industrial chemistry.[40] *Ión* aimed to show the vitality of the Spanish chemical industry, to establish new links between laboratory research and its industrial application, to keep readers informed about

---

[37] *Ión. Revista española de química aplicada*, 1(1) (1941).
[38] *Anuario de la industria química española* (Madrid: Sindicato Vertical de Industrias Químicas, 1949), pp. 11–16.
[39] Ibáñez Martín, *La investigación española*, pp. 41–2.
[40] (1) Fertilisers, (2) acids, (3) alcohols, (4) coal tar, (5) colourants, (6) candles, (7) fats, (8) soaps, (9) essences, (10) lubricants, (11) perfumes, (12) creams, (13) carbohydrates, (14) explosives, (15) rubber, (16) generic chemicals, (17) resins, (18) lactic products, and (19) pharmaceuticals. Colonel Juan G. Anleo was in charge of the standardisation of all chemical compounds.

Figure 4.4 Gumersindo García, the president of the *Sindicato Nacional de Industrias Químicas*, in 1941.
Ión, 13(2) (1942), p. 592. Reproduced with permission of the *Biblioteca de Catalunya*, Barcelona.

foreign journals and new developments and to closely collaborate with the *Sindicato*. *Ión* linked industrial chemistry with the Spanish economy in which, again, after strong and bitter criticism against liberalism and foreign capitalism, it listed the main chemical products that could ideally be produced in Spain.[41] In 1941, in the first issue of *Ión*, Rocasolano,

[41] Gasoline, oil, gas oil, lubricants, fuel oil, asphalt, paraffin, raw rubber, manufactured rubber, colourants, varnishes, perfumes, nitrates, potassium sulphate, potassium chloride, pharmaceutical products, paper and wood paste. Higinio París, 'La industria química y la economia española', *Ión*, 1(1) (1941), 27–9.

Lora-Tamayo and Julio Guzmán published short articles on the position of applied chemistry in the new regime. Similarly, politically loaded editorial notes under the banner of *Falange Española* and the totalitarian programme of the new *Sindicatos* – which strongly opposed the aims of class trade unions in liberal democracies – served as introductory articles. Later came 'technical' contributions devoted to chemical imports and to the exploitation of Spanish raw materials.[42] The editorial note of 1942 stressed the fact that 'before our *Movimiento* there was no efficient journal for the chemical industry. Today, if there is a journal that fulfils that mission, it was born from an organism of the Falange, from a *Sindicato*'.[43] Nevertheless, *Ión*'s protectionist, totalitarian agenda did not seem to be at odds with a section on foreign book reviews and foreign news and journals, which in practice became crucial for the profession. As a supplement to *Ión*, chemist José Barceló, a pupil of del Campo, acted as editor of the *Revista de Información de Química Analítica*. In 1948, Barceló became the new chief of the *Sindicato* and the director of all its chemistry publications.[44] In the early 1950s, the chemistry periodicals of the *Sindicato* sold more than 3,000 copies per month.[45]

We could obviously discuss in more abstract terms the meaning of totalitarianism: one-party rule, a leader with strong personal power and an elite determined to deeply transform society and to assimilate single individuals into state projects, strong nationalism and truly centralised control through repression and violence.[46] It is, however, through the

---

[42] See, for example: José María Bartolomé, 'La nueva estructura del Sindicato Nacional de Industrias Químicas', *Ión*, 2(6) (1942), 2–3.

[43] 'Editorial', *Ión*, 13(2) (1942), 594.

[44] *Los reactivos orgánicos en análisis inorgánica*, in CSIC series; *Monografías de Ciencia Moderna*; Siro Arribas, *Introducción a la historia de la química analítica en España* (Oviedo: Universidad de Oviedo, 1985), p. 31.

[45]

Journal	Sales
Ión (Mensual)	3,150 copies
Boletín Informativo (Bimensual)	1,500 copies
Química Analítica (Bimensual)	500 copies

María Silvia López, 'Aproximación al estudio de las publicaciones sindicales españolas desarrolladas durante el franquismo (1936–1975)', *Historia y Comunicación Social*, 8 (2003), 159–85. See also: José Barceló, 'Un siglo de espectroscopia: El centenario del descubrimiento del cesio por Bunsen y Kirchhoff', *Ión*, 20 (1960), 195–201. Trained by del Campo in spectroscopy, Barceló used a historical account to reflect on his own discipline and its potential as an ideal boundary work between physics and chemistry.

[46] Paul R. Josephson, *Totalitarian Science and Technology* (New York: Humanity Books, 2005); Abbott Gleason, *Totalitarianism: The Inner History of the Cold War* (Oxford:

establishment of that web of power and state control that Jimeno's chemical project, but also the Falangist chemistry of the *Sindicato* and the strong alliance with chemical companies and entrepreneurs, became major actors in the construction of early Francoism in itself. Chemicals, factories, chemists, production processes and social and political mechanisms of control intermingled profoundly in the 1940s. In addition, the chemical community was now embedded with a new powerful ethos: the religious values of National Catholicism, which merit more detailed analysis in the following section.

Chemistry and Religion

The research projects and industrial endeavours of the new chemical community found considerable legitimation in religious values that strongly supported the National Catholic ideology. To the Falangist chemists of the *Sindicato* aiming to design the whole range of industrial policies and Jimeno's plans for a totalitarian chemistry we must add the powerful Catholic right-wing biographies of the main leaders of the Francoist chemical community. The case of Albareda – one of the founding members of Opus Dei, a priest and the leader of the new science policies – is particularly relevant. Pascual and Lora-Tamayo – the latter also having a close relation to Opus Dei – were also active members of *Acción Católica*. Fernández-Ladreda was a distinguished member of the ACNP.[47] Chemists Eduardo Vitoria and Salvador Gil were important members of the Society of Jesus.

As described in detail by his biographers, Albareda merged his outstanding international chemistry training with a personal venture in close contact with Escrivá de Balaguer and the small core of the founding members of Opus Dei, who escaped from anticlerical persecution in the Republican camp during the Civil War and found comfortable shelter in Franco's provisional government in Burgos. Albareda therefore saw no contradiction between the totalitarian values of the regime, the organisation of scientific research at the CSIC and his radical Christian

---

Oxford University Press, 1995). See also: Richard Shorten, 'Rethinking Totalitarian Ideology: Insights from the Anti-Totalitarian Canon', *History of Political Thought*, 36(4) (2015), 716–61.

[47] Javier Tusell, *Franco y los católicos. La política interior española entre 1945 y 1957* (Madrid: Alianza, 1984), p. 69. See also: *Boletín de la Asociación Católica Nacional de Propagandistas (ACNP)*, 335, 1 August 1945. www.acdp.es/wp-content/uploads/boletines/1945/agosto-n355.pdf (last accessed 15 November 2016). Fernández-Ladreda was at that time director of a weapons factory in Oviedo, director of the Army Polytechnical School and vice-rector of the University of Oviedo.

worldview. Opus Dei's sanctification of this daily work fitted very well into a framework of utilitarian, applied chemistry and its industrial applications. A certain Weberian-like capitalist ethics, with an emphasis on asceticism, and the paradoxical combination of traditional conservatism and modernity made Opus Dei a 'totalising' organisation that was in tune with the totalitarian context of early Francoism.

Albareda's work on soil chemistry also grew in parallel with his religious beliefs. His field science placed him in close contact with nature, which at the same time he perceived as the result of divine creation. Through the *Instituto Nacional de Edafologia y Agrobiología*, Albareda set up a vast network of agrochemical stations across the country.[48] In his book, *El Suelo* (the soil), he described 'the soil of the fatherland' as a combination of hydraulic infrastructures, reforestation, agricultural improvements and, obviously, the power of fertilisers and soil chemistry as a whole.[49] It was therefore through soil chemistry – a deeply interdisciplinary field that included physical chemistry, geology, geobotany, colloid chemistry and biochemistry – that he managed to materially and spiritually unify the country as his personal method for approaching the 'truth', while, as he saw it, overcoming any simple anthropocentric system. At the *Instituto de Estudios Pirenáicos*, Albareda developed a scientific approach to the mountain as a unity, as a refraction of divine light, as a unifying object of study.[50] Equally, his concern for natural parks, in particular the case of Doñana, followed a similar path.

As a member of the *Academia Pontificia de Ciencias* – a Roman Catholic scientific society, which also counted Lora-Tamayo among its members – Albareda closely followed Pope Pius XII's reflections on the power of the experimental sciences to unveil God's science, as well as God's capacity to shape rocks, soils and matter as a whole.[51] In *Consideraciones sobre la investigación científica*,[52] Albareda stated his belief that only a real passion for 'truth' could inspire the right hypothesis to guide experimental work and lead to the Creator, so a real professional call was an extension of the Christian call and a road towards sanctification. Professional ambition

---

[48] *Centros de Edafología y Biología Vegetal* (Madrid, Salamanca, Santiago, Murcia, Seville, Tenerife), *Institutos de Orientación y Asistencia Técnica* (Murcia, Salamanca, Zaragoza, Badajoz, Valencia) *Aula Dei* (Zaragoza), *La Poveda* (Arganda del Rey), *La Mayosa* (Málaga), *Misión Biológica de Galicia* (Pontevedra), *Instituto de Aclimatación* (Almería). Universidad de Navarra. Archivo General. Fondo José María Albareda. Correspondence with Antonio Rius Miró (4 June 1966), 056/327.
[49] Ibáñez Martín, *La investigación española*, pp. 101–2.
[50] Gutierrez Ríos, *José María Albareda*, p. 231.   [51] Ibid., p. 328.
[52] José María Albareda, *Consideraciones sobre la investigación científica* (Madrid: CSIC, 1951), p. 183.

was closely linked to the Christian *summa perfectionis*.[53] In addition, scientific language and nomenclature helped, in his view, to unveil God's divine thought and design. In practical terms, Albareda presented his soil chemistry as a key contributor to agricultural progress, to modernity and to the ensuing industrialisation of the country.[54]

Other Opus Dei chemists also strove for 'personal perfection' through successful scientific careers and benefitted from religious contacts abroad. In 1955, Ángel Santos (1912–2005), professor of biological chemistry at the Faculty of Pharmacy in Madrid, organised the first international conference on oligo-elements in Rome, again under the banner of the *Academia Pontificia*. Santos played an important role in the emergence of biochemistry as a more established academic discipline in the 1960s.[55] Another case is that of José María González Barredo, who, after his training at the INFQ in the 1930s, did postdoctoral research in Göttingen and later obtained a physical chemistry chair in Zaragoza in 1942, just after Rocasolano's death. Like Albareda, González Barredo belonged to the early group who met Escrivá de Balaguer in the 1930s. In 1946, he moved to the United States on a Spanish fellowship, and soon obtained a contract at MIT to work on electrochemistry. Later, he went to Columbia, and then to the Washington Bureau of Standards – Jimeno's reference institution before the Civil War – and contacted several American scientific societies. González Barredo's chemistry played a crucial role in the founding of Opus Dei in the United States and in the international projection of Francoist Spain.[56] Similarly, Florencio Sánchez-Bella was another Opus Dei chemist who belonged to an influential family during the dictatorship – his brother Alfredo being Franco's Minister for Information and Tourism – and had important responsibilities in the congregation and at the University of Navarra. His early training in the pharmaceutical industry again seemed fully compatible with his religious commitments.[57]

---

[53] John Allen, *Opus Dei. An Objective Look behind the Myths and Reality of the Most Controversial Force in the Catholic Church* (New York: Doubleday, 2005). See also: Malet, 'José María Albareda'.

[54] Gutierrez Ríos, *José María Albareda*, p. 331; Pablo Pérez, 'José María Albareda. La ciencia al Servicio de Dios', *Nuestro Tiempo*, Novembre/Diciembre (2010), 52–7; Pablo Pérez, 'International Contacts in the First Years of the Spanish CSIC', in Fernando Clara, Cláudia Ninhos, Sasha Grishin (eds.), *Nazi Germany and Southern Europe, 1933–45: Science, Culture and Politics* (Basingstoke: Palgrave/Macmillan, 2016), pp. 68–83; Adolfo Castillo, *Albareda fue así: Semilla y surco* (Madrid: CSIC, 1971).

[55] María Jesús Santesmases, 'Severo Ochoa and the Biomedical Sciences in Spain under Franco, 1959-1975', *Isis*, 91(4) (2000), 706–34.

[56] John F. Coverdale, 'José María González Barredo. An American Pioneer', *Studia et Documenta Rivista dell'Istituto Storico San Josemaría Escrivá*, 10 (2016), 23–44.

[57] Francisco Ponz, 'En la muerte de Florencio Sánchez-Bella', *Diario de Navarra*, 2 October 2008.

In the 1940s, González Barredo published joint papers with Rius Miró in *Anales* as part of their research at the physical chemistry section of the *Instituto Alonso Barba*.[58] Similarly, once settled in his chair of biological chemistry at the Faculty of Pharmacy in Madrid, Santos became a prolific author for *Anales*, writing papers on hormones, steroids and vitamins, and even publishing joint papers with Albareda in collaboration between the *Instituto Cajal* and the *Instituto de Edafología*.[59] Moreover, in the same context, several landmark books appeared in 1940–1, published by Saeta. This was the case for Jimeno's lectures and papers on the totalitarian organisation of chemistry, which were published under the title *Ciencia y Técnica* (1940), but also Albareda's *El Suelo* (1940), Santos's *Hormonas* (1940; with a preface by Rocasolano) and Santos's *Vitaminas* (1941).[60] The Opus Dei chemists strengthened their academic power with new university and CSIC positions, but also with joint editorial ventures from research papers to textbooks, and even more popular books, reinforcing their cultural hegemony in the totalitarian values of the regime.[61]

After the Civil War, and thanks to Franco's victory, the Jesuits were able to return to Spain and managed to reopen their schools, including the IQS. In fact, by 1939 they had already recovered the pre-war buildings of the Institute, along with 49 students, but they mourned the 22 casualties caused by the war. These were hard times for the economy of the Jesuit chemists. Acceptance and academic prestige in the new regime were not necessarily associated with funds for research – which were mainly under the control of Albareda in the CSIC, as the leading member of the rival congregation. However, the IQS soon obtained private donations from the chemical industry and support from the Barcelona City Council and provincial board (*Diputación*). A new industrially orientated four-year experimental curriculum was set up in 1940. In spite of Eduardo Vitoria's fear of losing pietism and religious gravity, the new curriculum transformed the old school of chemistry into a professional college in which, in pragmatic terms, scientific training gained ground in

---

[58] Jesús Ynfante, *La prodigiosa aventura del Opus Dei* (Paris: Ruedo Ibérico, 1970). See the chapter devoted to the links between the CSIC and the Opus Dei.
[59] See, for example: J. M. Albareda, A. Santos, T. Albiñana, 'Materia orgánica en los suelos españoles', *Anales*, 39 (1943), 751–73.
[60] Emilio Jimeno, *Ciencia y Técnica* (Madrid: Saeta, 1940); José María Albareda, *El Suelo* (Madrid: Saeta, 1940); Ángel Santos, *Hormonas* (Madrid: Saeta, 1940); Ángel Santos, *Vitaminas* (Madrid: Saeta, 1941).
[61] Fernando García Naharro, *El papel de la ciencia. Publicaciones científico-técnicas durante el franquismo (1939–1966)*. PhD thesis (Madrid: Universidad Complutense de Madrid, 2017).

front of theology and humanism.[62] Even in the 1940s and the early 1950s, with the uncertain status of the IQS at the official level,[63] the experimental, practical character of its training and its strong links with the chemical industry soon bore fruit when compared to public universities.[64] It was in the 1950s that the IQS developed a service for assessing chemical industries. Its industrial samples lab became the seed for the future Department of Technical services, offering its academic staff and students a good reputation.[65]

In their search for elite calls to perfection in research, Opus Dei and the Jesuits competed for religious, institutional and political hegemony in the early 1940s in a complex relationship of collaboration and rivalry.[66] On the other side, members of the ACNP felt close to Opus Dei and often stressed the collaborative aspect.[67] Opus Dei progressively controlled the new university academic positions that needed to be filled after the enormous pruning following the Civil War, so religious links and affinities mattered for chemists when searching for new jobs in the regime. They even found solidarity in international Catholic networks abroad, as Albareda's honoris causa in Louvain and Toulouse and membership of the *Academia Pontificia* clearly show.[68]

Although historians have often interpreted the Society of Jesus in the nineteenth century as a representative of Spanish Catholic integrism,[69] the practice of experimental sciences brought them into a much stronger position in the early decades of the twentieth century, even competing with other secular discourses of modernity. The Society of Jesus stressed the compatibility between reason and God, between positivistic science and theological dogma. In the Jesuit tradition, values such as Christian

---

[62] In 1948, in a letter from Salvador Gil to the Provincial Jesuit Julian Sayós, he justified that strategic change. Letter to P. Julián Sayós, 5 September 1948. Attached report from P. Salvador Gil to P. Julián Sayós: 'IQS: Su historia, su vida actual, sus planes', in Lluís Victori, *P. Salvador Gil. Creador del IQS moderno (1934–1957)* (Barcelona: IQS, 2009), pp. 77–9.
[63] There is evidence of some economic support after 1953: '[...] the government has also lately offered to defray part of the running expenses of the Institute', Miguel M. Valera S. J., 'Eduardo Vitoria S.J. A Contemporary Leader in the Spanish Chemical World', *Journal of Chemical Education*, 33(4) (1956), 161–6, p. 165.
[64] Victori, *P. Salvador Gil*, p. 93.
[65] Vitoria became president of the Barcelona Academy of Science in the period of 1946–51.
[66] Josémaría Escrivá de Balaguer derived inspiration from spiritual exercises at the Society of Jesus. Joan Estruch, *Saints and Schemers: Opus Dei and Its Paradoxes* (Oxford: Oxford University Press, 1995).
[67] Ibid., pp. 182–99.
[68] *Universidad de Navarra. Archivo General*. Fondo José María Albareda. Correspondence with José Pascual (26 May 1953), 029/269.
[69] Carolyn P. Boyd, *Historia Patria: Politics, History and National Identity in Spain, 1875–1975* (Princeton: Princeton University Press, 1997), p. 100.

service beyond religious life, personal engagement with the world and the esteem for learning as a kind of sanctification process had reinforced the Society's interest in natural philosophy and later in modern experimental science since the sixteenth century.[70] As a counterbalance to William Draper's thesis, in the nineteenth century, Jesuit scientists became the exemplars of harmony between science and religion, promoting a personal approach to the frontier work with those outside the Church.[71] Ignatius's mystical experiences focused on the purpose of God's creation, so Jesuits sought the divine hidden power in nature through scientific work.[72] God's providence provided raw materials for chemists to study and exploit, in coherence between experimental science and theological dogma.[73]

In the early decades of the twentieth century, Vitoria perceived raw materials as being preserved by God's providence throughout the country, with chemists being called to extract and exploit them. His strong Spanish nationalism seemed to ideally merge with his Catholic faith. The soil, the subsoil, the rivers – all had to be exploited by chemical science through genuine Spanish companies and instruments.[74] In the context of the First World War, Vitoria presented a neutral, pure, basic chemistry that was supposedly not responsible for man's evils and perversities. It was a chemistry that came straight from God and went back to God, and 'cried' to see the cruelty and pain of bloodied Europe.[75]

Vitoria's early programme to reconcile chemistry and religion resisted the secularisation plans of the Republic, survived the Civil War and easily found its place in the new regime (Figure 4.5). In 1942, already in the Francoist context, he described acetylene and its chemical applications in the following theological terms:

---

[70] Steven J. Harris, 'Transposing Merton's Thesis: Apostolic Spirituality and the Establishment of the Jesuit Scientific Tradition', *Science in Context*, 3 (1989), 29–65.
[71] Agustín Udías, *Jesuit Contribution to Science: A History* (Dordrecht: Springer, 2015). See also: Thomas Worcester (ed.), *The Cambridge Companion to the Jesuits* (Cambridge: Cambridge University Press, 2008).
[72] As discussed by the Jesuit chemist Ernest G. Spittler in the 1960s, the 'priest–scientist' had to act as a mediator between God and man. E. G. Spittler, 'The Priest–Scientist in the Church', *Bulletin of the American Association of Jesuit Scientists (BAAJS)*, 39 (1962), 30–7. See also: William Meissner, 'The Jesuit as a Priest–Scientist', *BAAJS*, 40 (1963), 25–33, cited by Udías, *Jesuit Contribution to Science*, p. 240. Spittler graduated in chemistry at the Catholic University in Washington, DC, to later become a priest and teach for decades at John Carroll University, a Jesuit Catholic university in Ohio.
[73] Eduardo Vitoria, *La ciencia química y la vida social. Conferencias de vulgarización científica dadas en el paraninfo de la Universidad Literaria y en el Centro Escolar y Mercantil de Valencia del 23 al 28 de novembre de 1915, por el _____ Dr. en Ciencias. Director del Laboratorio Químico del Ebro* (Barcelona: Tipografía Católica Pontificia, 1916), p. 39.
[74] Ibid., p. 55.   [75] Ibid., p. 99.

Figure 4.5 Eduardo Vitoria in the laboratory, aged 92, together with Manuel Sanz at the *Instituto Químico de Sarriá* in 1956.
Reproduced with permission of the *Institut Químic de Sarrià*, Barcelona.

[...] God has rejoiced in showing his Wisdom and Power, depositing a flow of energy and reactivity in a such an original yet simple molecule, an outpouring of energy and reactive aptitudes that cause admiration among those who study it [...] being so easy in our land to obtain a good calcium carbide, and therefore acetylene [...] if we do not produce it in Spain, we will have to import it from abroad.[76]

Similarly, in 1944, in a public address at the Academy of Sciences in Barcelona, Vitoria also expressed the way in which chemistry and priest–chemists mediated between God and nature through natural products, a significant branch of organic chemistry:

In fact, the chemist has studied [...] natural products, which are many other heralds of the Omnipotence and Wisdom of the Supreme Maker, many other open books before his eyes. [...] in many of them [the chemist] discovers the presence of ethylene groups, and functions such as aldehydes, ketones, acids,

---

[76] Eduardo Vitoria, 'El acetileno: Sus polimerizaciones. Sus condensaciones', *Ión*, 17(2) (1942), 889–6, p. 896. See also: Eduardo Vitoria, *El pan y el vino eucarísticos: Estudio químico-litúrgico* (Bilbao: El Mensajero del Corazón de Jesús. La Editorial Vizcaína, 1944).

esters, amides. [...] all depending on double bonds and carbonyl groups, now isolated, now associated to oxydriles, amines, halogens [...][77]

Although the application of Jeffrey Herf's concept of 'reactionary modernism' to the Jesuit case has recently been debated,[78] there are good reasons to use it for the sake of a deeper analysis of the way in which the Jesuit and Opus Dei chemists reconciled modern chemistry and theology.[79] Herf explained the Nazi success in gaining power through the reconciliation between traditional, anti-modernist, romantic, 'irrational' ideas and modern technology.[80] Although Nazism held an areligious–atheistic worldview, the substitution of the romantic, 'irrational' ideas with theological, religious factors paves the way to applying this kind of analysis to the strategy of Opus Dei and the Society of Jesus, and, in particular, to the roles they played in Francoist Spain.[81] In fact, the IQS was in tune with the Society of Jesus's project of creating new scientific centres 'that would be the cradle of future scholars capable of achieving eminence',[82] which was also in tune with the Opus Dei project of personal perfection. Vitoria conceived the curriculum for the training of first-class industrial chemists and considered that man's ingenuity in the study of molecules came straight from God's wisdom.[83] In practical terms,

---

[77] Eduardo Vitoria, 'Discurso de contestación', *Memorias de la Real Academia de Ciencias y Artes de Barcelona*, 563 (1944), 95–119, pp. 113–14.
[78] See: Néstor Herrán, '"Science to the Glory of God." The Popular Science Magazine *Ibérica* and Its Coverage of Radioactivity, 1914–1936', *Science & Education*, 21(3) (2012), 335–53; Lino Camprubí, *Engineers and the Making of the Francoist Regime* (Cambridge, MA: MIT Press, 2014).
[79] Jeffrey Herf, *Reactionary Modernism: Technology, Culture and Politics in Weimar and the Third Reich* (Cambridge: Cambridge University Press, 1984).
[80] 'Reactionary modernism is an ideal typical construct [...] converting technology from a component of alien, Western civilization into an organic part of German *Kultur*. They combined political reaction with technological advance'. Herf, *Reactionary Modernism*, p. 1.
[81] Néstor Herrán, 'Science to the Glory of God'. 'The reactionary modernists were modernists in two ways: First [...] they wanted Germany to be more rather than less industrialized, to have more rather than fewer radios, trains, highways, cars, and planes. They viewed themselves as liberators of technology's slumbering powers, which were being repressed and misused by a capitalist economy linked to parliamentary democracy. Second, they articulated themes associated with the modernist vanguard [...] the triumph of spirit and will over reason. [...] When modernists turned to politics, they sought engagement, commitment and authenticity, experiences the Fascists and Nazis promised to provide'. Herf, *Reactionary Modernism*, p. 6. On reactionary and alternative modernism, see: Roger Griffin, 'Modernity, Modernism and Fascism: A "Mazeway Resynthesis"', *Modernism/Modernity*, 3 (1996), 1–22; Roger Griffin, *Modernism and Fascism: The Sense of a Beginning under Mussolini and Hitler* (Basingstoke: Palgrave/Macmillan, 2007).
[82] Valera, 'Eduardo Vitoria S.J.', p. 162.
[83] Vitoria, *La ciencia química y la vida social*, p. 104.

Christian rhetoric for the reconciliation of science and religion soon bridged academic and industrial contexts. Similarly, the so-called *Christian Concept of Autarky* (1940) – referring to the book published by Jesuit José Agustín Pérez del Pulgar (1875–1939) – materialised into a science-based, totalitarian, charitable political economy. It set industrialists, scientists and priests to work in a common patriotic endeavour within a framework into which the Jesuit and Opus Dei chemists' ideologies fitted very well.[84]

***

Totalitarian ambitions after the war materialised on different fronts. Under the banner of hostile statements against Republican science, the new Francoist chemists set up an ambitious programme of useful, applied chemistry that bridged their religious values, their alliances in power and their scientific ambitions. An 'applied chemistry' rhetoric, a right-wing Catholic framework that provided chemists with useful values issued from several readings of natural theology, a passion for research often perceived as a passion for God's nature and the country's progress and the fascist-like fascination with organisation, control and totalitarianism presented the 'managers' of raw materials as crucial actors in the construction of the new Spain.

Despite the inevitable disagreements among different factions – from fascist, Falangist control over the chemical industry to Opus Dei's values that inspired Albareda in his research policies – the totalitarian atmosphere of the 1940s and the early 1950s shaped a new ethos of chemistry for internal consumption, used international references as a sign of prestige and modernity – in reactionary terms – and built up new political, religious and economic alliances. Chemists eased the construction of a continuum between the government, the CSIC and the PJC, Opus Dei, the Society of Jesus and the *Sindicato*. In addition, as chemist–historians, they contributed to the building of historical accounts that legitimated the status of the dictatorship in relation to the old JAE times.

The new Francoist chemistry was therefore constructed in strong opposition to the supposedly overly theoretical, elitist, and 'ideological' research of the JAE in order to now promote a 'real' applied chemistry with industrial links across the whole territory. The totalitarian dream of the new regime reified in scientific terms in the CSIC, led by soil chemist Albareda, in close alliance with Lora-Tamayo, Rius Miró and Pascual, and in a new chemistry that had to serve an ambitious reorganisation of the economy of the country, supposedly beyond 'ideological' conflicts.

---

[84] Camprubí, *Engineers and the Making of the Francoist Regime*, p. 43.

Raw materials, industrial progress under strict centralised control and new applied chemistry research lines reinforced the new regime.

As a result, chemistry once again became a flexible wild card that, behind its image of objectivity, neutrality and industrial progress, was able to embed religious values – of right-wing Catholic organisations acquiring great power after their explicit support for Franco during the war, and totalitarian, fascist ambitions 'a la Mussolini' – into the ways in which raw materials and chemical products had to be produced and organised. The obsession with applied chemistry counterbalanced the potential of the 'liberal', 'pure' research of the INFQ and the scientific culture of the Republic. In Albareda's CSIC, chemistry was placed within the PJC and *Patronato Alfonso X el Sabio*, both representing Matter – as one of the three main branches of the *Arbor Scientiae* – together with Life and Spirit. Chemistry was therefore crucial for the making of the Francoist National Catholicism.

5 Autarchic Ambiguities

In 1938, Manuel Lora-Tamayo described how, during the Civil War, his chair in organic chemistry in Seville worked on several regional chemistry projects in the interest of self-sufficiency.[1] In fact, in a public address that same year, he referred to 'guided research' for the nation's benefit and looked to the French national research centres, the German *Kaiser-Wilhelm-Gesellschaft* and the Italian *Consiglio Nazionale delle Ricerche* (CNR) as examples.[2] Right at the end of the Civil War, in his opening address for the 1939–40 academic year, Lora-Tamayo once again insisted on the need for autarchic chemical nationalism[3] and launched the idea of copying the structure of the Italian fascist CNR to develop the new research policies in Franco's Spain. He originally conceived the whole project in the academic context of the university, but with industrial applications in fields such as mineralogy and metallurgy, liquid combustibles, cellulose, synthetic rubber and pharmaceutical products.[4] That seminal idea progressively grew and materialised in several institutes of the PJC.

The PJC established the *Instituto del Carbón* (coal) in the mining region of Asturias; the *Instituto de la Grasa* (fats) in Seville, close to the olive groves of Andalusia; the *Instituto de Tecnología Alimentaria* (food technology) in Valencia, next to orange orchards; the *Instituto de Industrias Pesqueras* (fishing) on the coasts of Barcelona, Cadiz and Vigo; and the *Instituto de Investigaciones Textiles* (textile research) in Catalonia. Construction and cement, iron and steel, non-ferrous metals, plastics, medical chemistry, fermentations and edaphology all had their own

---

[1] Manuel Lora-Tamayo, *La investigación química española* (Madrid: Alhambra, 1981), p. 194.
[2] Roberto Maiocchi, *Scienza e fascismo* (Rome: Carocci, 2004). The CNR had already obtained some tangible results since the 1920s in Mussolini's fascist regime.
[3] Manuel Lora-Tamayo, 'Recuerdos del CSIC en su 50ª aniversario', *Arbor*, 529 (1990), 99–115.
[4] Antonio Canales, 'La política científica de posguerra', in Amparo Gómez, Antonio Canales (eds.), *Ciencia y fascismo. La ciencia española de postguerra* (Barcelona: Laertes, 2009), pp. 105–36, 119.

research institutes in Madrid.[5] Similarly, it is in that context that the creation of the *Instituto Nacional de Industria* (INI) – inspired by the Italian *Istituto per la reconstruzione industriale* (IRI) – under the presidency of engineer, ultranationalist and close friend of Franco, José Antonio Suanzes, should be understood.[6]

In the 1930s, the dream of economic self-sufficiency was also in the public sphere in liberal regimes. In 1935, American engineer Frank A. Howard, from Standard Oil, published a paper on the role of chemistry in 'autarky', in economic terms. In the United States, the Wall Street Crash of 1929 had already motivated the protection of home industries as a safety net against future economic turbulence, and the chemical industry played a key role in avoiding dependence on imports. Howard stressed how the 'outstanding success of synthetic chemistry' – dyes, fertilisers, explosives, artificial fibres, synthetic rubber and gasoline – substituted for demand for imported natural products and meant industrial independence from Germany.[7]

In non-liberal contexts, however, the whole endeavour had other, hidden political purposes. In their study of the co-evolution of scientific research and the political economy in fascist regimes, historians Tiago Saraiva and Norton Wise made a useful distinction between autarky (referring to economic self-sufficiency) and autarchy (referring to autocratic political rule). While autarky emphasises economic policies in the making of the fascist state, autarchy focuses on top-down social control, in which scientists (chemists) appear as active creators of the regime.[8] In fact, beyond its economic postulates, the concept of autarchy has a strong political bias. Aggressive programmes of radical nationalism,

---

[5] The Committees of the *Patronato Juan de la Cierva Codorniu* at work in 1946 were the following: '*combustibles*', '*mineralogía y metalurgia*', '*aceite*', '*racionalización del trabajo*', '*fertilizantes*', '*aprovechamiento industrial de los productos del campo*', '*química forestal*', '*cemento*', '*azúcar*', '*agrios*' and '*comisiones generales de física y química*'. See: *Memoria de las actividades desarrolladas por el Patronato 'Juan de la Cierva Codorniu'* (1946), p. 8. http://simurg.bibliotecas.csic.es/viewer/image/CSIC000121685_Num01/7/#topDocAnchor (last accessed 29 January 2017). See also: 'Patronato Juan de la Cierva', 1 January 1954, VI04830, Realización y guión: Santos Núñez. Hermic Films. www.cienciatk.csic.es/Videos/PATRONATO+JUAN+DE+LA+CIERVA_2954.html (last accessed 19 December 2017).

[6] On Suanzes and the INI, see: Alfonso Ballestero, *Juan Antonio Suanzes 1891–1977. La política industrial de la postguerra* (León: LID Editorial Empresarial, 1993); Pablo Martín-Aceña, Francisco Comín, *INI, 50 años de industrialización en España* (Madrid: Espasa-Calpe, 1991); Elena San Román, *Ejército e industria: El nacimiento del INI* (Barcelona: Crítica, 1999).

[7] Frank A. Howard, 'Autarky in 1935. A Chemical Appraisal of the Self-Sufficient State', *Industrial and Engineering Chemistry*, July (1935), 770–3.

[8] Tiago Saraiva, Norton Wise, 'Autarky/Autarchy: Genetics, Food Production, and the Building of Fascism', *Historical Studies in the Natural Sciences*, 40(4) (2010), 419–28, p. 425.

anti-liberalism (including a strong criticism of liberal economies), a fascist-like administration and totalitarian strategies of sociopolitical control and repression of the civil population lay behind utopian projects of economic self-sufficiency and the obsession with national raw materials.[9] In the context of the international isolation of Francoist Spain in the early 1940s, autarkic projects of economic and industrial self-sufficiency gained prominence and guided the autarchic policies of the new regime, which also benefitted from earlier Civil War efforts (see Chapter 3). Raw materials, their geographical distribution and their efficient exploitation became state decisions to which chemists contributed by reframing academic disciplines and industrial endeavours. Historians of Francoism have traditionally considered that the regime's autarchy rested on three main pillars: an anti-liberal rationale for the economy; the full trust in national resources for economic growth; and foreign suspicion.[10]

Although early Francoist chemists offered their expertise for the efficient exploitation of local raw materials and reshaped applied chemistry training at the university level, paradoxically, they also contributed to diplomatic efforts to overcome international isolation. Since Spanish frontiers were, in practice, quite permeable to raw materials, chemical expertise and international conferences and publications, it is through the lens of chemistry that the standard account of the autarchic ideals of Francoism can be revisited in this chapter. For that purpose, I shall first analyse how, under the influence of fascist Italy and Nazi Germany, chemicals became such important agents for the materialisation of the autarchic projects of early Francoism. Second, I will discuss how industrial chemistry training evolved from traditional technical chemistry books and subjects to a more 'modern' approach to chemical engineering, which, in spite of the totalitarian context, soon included American concepts of 'unit operations'. Finally, chemical diplomacy and the chemists' active role as 'ambassadors' of the regime provide convincing reasons to revisit the concept of autarky/autarchy through chemistry, as well as highlighting some lines of continuity with the scientific culture of the pre-Civil War period.

'Our' Chemicals

In 1939, at the end of the Civil War, chemist Luis Bermejo, who had enjoyed a prominent position in the JAE, became enthusiastically in

---

[9] Rafael García Pérez, *Franquismo y Tercer Reich. Las relaciones económicas hispano-alemanas durante la Segunda Guerra Mundial* (Madrid: CEC, 1994), pp. 301–2.

[10] Ángel Viñas, *Guerra, dinero, dictadura. Ayuda fascista y autarquía en el España de Franco* (Barcelona: Crítica, 1984), p. 210.

favour of the Francoist cause (see Chapter 4). In one of the lectures at the University of Barcelona, invited by the new rector, Emilio Jimeno, Bermejo discussed the problem of combustibles in Spain.[11] In the technological competition between coal and petroleum, and very much influenced by German firms, Bermejo defended the Fischer–Tropsch process for producing gasoline from the hydrogenation of coal tar.[12] He referred to the 3rd International Conference on Coal, which took place in 1937 in Rome, fascist Italy, and also to German methods of low-temperature distillation of lignite. In Spain, Bermejo examined coal mines and ores for potential raw materials and also for hydrogenation processes.[13] Merging soils, mining, chemistry and autarchy – in political terms – Bermejo defined his plans as follows:

We must study all the different possibilities offered by Spanish soil and by the raw plant materials capable of producing fuels. All this must be done using united criteria, under single and effective management and with the most competent technicians [...] all this must be done with a view to the greater interest of the Homeland [...] Spain, which has been so mined by a political cancer, that before the war absolutely nothing was done.[14]

The autarchic dreams of Franco's chemists had much earlier roots in Italy. At the end of the Civil War, Lora-Tamayo and Jimeno had also stressed the need to imitate the Italian pattern of chemistry in the new state. In fact, in the 1920s, Mussolini's autarchic project had already had a notable impact on the production of chemicals such as combustibles (benzene, liquid fuel), textiles, cellulose, metals and other strategic products also closely linked to the war effort.[15] After the foundation of the CNR in 1923 and the early presidency of physicist Vito Volterra (1860–1940) – who later refused to support Mussolini's regime – in 1927 the famous inventor and Nobel Prize winner for physics in 1909, Guglielmo Marconi (1874–1937), became president. Chemist Nicola Parravano (1883–1938), a key figure in chemical autarchy, was made vice-president.[16] In tune with fascist policies, Parravano established strong links with chemical companies and became the director of the

---

[11] Luis Bermejo, 'El problema de los carburantes en España', in *Universidad de Barcelona. Aspectos y problemas de la nueva organización de España. Ciclo de conferencias organizado por la Universidad de Barcelona* (Barcelona: Universdad de Barcelona, 1939), pp. 231–53.
[12] Luis Bermejo, 'El problema de los carburantes', p. 231.
[13] In Asturias, Alicante, Teruel, Castellón, Utrillas, Mequinenza and Puertollano.
[14] Bermejo, 'El problema de los carburantes', pp. 251–3.
[15] Roberto Maiocchi, *Gli scieziati del Duce. Il ruolo dei ricercatori e del CNR nella politica autarchica del fascismo* (Rome: Carocci, 2003).
[16] Ibid., p. 24; Nicola Parravano, 'La industria de la leucita desde el punto de vista de la química física', *Anales* 259 (1929), 341–67 ('Conferencia en las Fiestas de las Bodas de Plata de la Real Sociedad Española de Física y Química, Abril 1929').

Istituto di Chimica in Rome. Similarly, chemist Giovanni Battista Bonino used his academic prestige as a theoretical chemist to defend Mussolini's autarchic policies, and became a key figure in fascist public propaganda. In the late 1920s, many Italian scientists perceived autarchy as a chance to legitimize science and technology and place it for the first time at the core of society in their cultural battle against the hegemony of Italian idealism and humanism.[17] Therefore, chemists such as Parravano and Bonino, but also Giuseppe Bruni (1873–1946) in Milan working on fertilisers and fuels,[18] Giulio Natta (1903–79) in Montecatini, and many others became obeying executors, organisers, propagandists and defenders of the intervention of the state,[19] strengthening the political power of fascism in the factories.[20]

The entire Italian fascist strategy had major resonance in Spain in terms of research policies and projects. This was the case of the gas generator (gasogen), for example, which was already developed in Italy in the 1930s, when the war with Ethiopia hastened the production of 'national' chemicals and the search for self-sufficiency in alcohol, natural gas (methane) and combustibles (Figure 5.1). Gas generators benefitted from the exothermic reaction of the partial combustion of coal to yield carbon monoxide, and its later treatment with water produced hydrogen. Gasogen was also very popular in Franco's Spain.[21] In 1942, industrial engineer José Antonio Artigas (1887–1977) provided full details of the different ways to produce gas fuel from solid combustibles in vehicles to avoid technological dependence on large chemical factories and to seek alternatives to car fuel scarcity.[22] Artigas placed gas generators within a project of 'chemical autarchy', which, in his view, responded to emergency, patriotic needs. After presenting the pros and cons of all combustibles, from wood to vegetable and mineral coals, he described several

---

[17] Roberto Maiocchi, 'Scieziati italiani e scienza nazionale (1919–1939)' in Simonetta Saldani, Gabriele Turi (eds.), *Fare gli italiani. Scuola e cultura nell'Italia contemporanea. II. Una società di massa* (Bologna: Il Mulino, 1993), pp. 41–86, 80–1.

[18] 'Congreso Italiano de Química Industrial, Milán, 12–17 abril 1924', *Anales*, 25 (1924), 111.

[19] The First World War had already proved the shortage of fertilisers, dyestuffs, combustibles and cellulose in the Italian market. In 1926, work on methyl alcohol became a paradigm of a new 'typically Italian' technology and an ideal solution for countries with no coal or oil. In addition, Bruni's work at the *Politecnico de Milano* on vulcanisation, in collaboration with Pirelli, was another landmark in the ideal contribution of chemistry (and, obviously, of chemists) to national defence. Maiocchi, 'Scieziati italiani e scienza nazionale (1919–1939)'.

[20] Maiocchi, *Gli scieziati del Duce*, pp. 150, 307.

[21] See, for example: F. de Zubeldia, 'El manejo de los gasógenos modernos', *Ión*, 14(2) (1942), 692–5.

[22] José Antonio Artigas, *El presente español de los vehículos con gasógeno* (Madrid: Asociación Nacional de Ingenieros Industriales, 1942).

Figure 5.1 The gasogen 'solution', as described in the early 1940s as a fuel for cars and tractors.
Fernando Martín Panizo, 'Posibilidades actuales y futuras de la tracción por gasógeno en España', *Ión*, 7(2) (1942), 140–4, p. 141. Reproduced with permission of the *Biblioteca de Catalunya*, Barcelona.

kinds of wood gas boilers and their assemblage on vehicles, all under the monopoly of CAMPSA, a national firm that controlled the distribution of all fuels. Fernando Martín Panizo, a pupil of Lora-Tamayo, also discussed the most efficient Spanish coal for the production of gasogen and its pros and cons when compared to methods of the hydrogenation of coal, gasoline from bituminous slates, and acetylene.[23]

In a similar vein, Luis Blas's experience with chemical weapons in service of the Francoist side during the war now shifted to support autarchic plans. Blas presented acetylene – a key molecule for Vitoria's programme of natural theology – as the main raw material for organic synthesis in countries that do not have natural oil resources.[24] Its synthesis from calcium carbide relied mainly on Spanish coal.[25] Equally, Blas discussed the 'autarchic' importance of bromine in Spain and its applications to anti-knock agents and fireproof compounds, as well as to

[23] Fernando Martín Panizo, 'Posibilidades actuales y futures de la tracción por gasógeno en España', *Ión*, 7(2) (1942), 140–4.
[24] Luis Blas, *Química de los plásticos* (Madrid: Aguilar, 1948), p. 65.
[25] Luis Blas, 'Presente y futuro de las industrias químicas derivadas del acetileno', *Ión*, 78 (1948), 3–6. Acetylene became a useful reagent for the synthesis of organochloride solvents, acetic acid and acetic anhydride, ketone, synthetic rubber (neoprene), styrene, vinyl plastics, benzene, oxalic acid and ethane. See also: Luis Blas, *Agenda del químico. Colección de tablas, constantes, datos matemáticos, físicos, físico-químicos, analíticos y técnicos* (Madrid: M. Aguilar, 1942).

photography.[26] Similarly, chemist and Minister for Public Works, José María Fernández-Ladreda – another of the heroes of the Civil War – stressed the need to carefully study Spanish sources of coal (lignite, in particular), which he considered superior to the German ores.[27]

Hitler's military support for Franco had already begun in July 1936, a few days after the war broke out. Spanish raw materials, in particular mines (Riotinto) and wolfram ores, were of particular interest to the Nazi army. During the Civil War, more than 20 German firms had economic interests in Spain – I. G. Farben, among them – and they kept close contact with the emerging Francoist authorities, which had established their headquarters in Burgos.[28] Bituminous slates, lignite distillation and hydrocarbon refining obtained public funding and Nazi German collaboration.[29] In spite of the dramatic consequences of the Second World War and the large-scale reorganisation experienced by the firms under Allied supervision, the German chemical industry still retained elements of its glorious past. Walter Reppe's (1892–1969) method of production of acetylene from coal at *Badische Anilin und Soda Fabrik* (BASF) and Fischer–Tropsch's reaction for the production of liquid hydrocarbons were two powerful strategies of autarchic-like forms of resistance from the old German science-based industry tradition in the search for alternatives to oil before the new petrochemical era.[30]

In 1941, it was precisely for that purpose that, as part of the first INI general plan for industrialisation, the Minister for Trade and Industry, Demetrio Carceller (1894–1968), travelled to Berlin to ask for German collaboration in fuel, synthetic rubber and artificial fibre production. Carceller had been trained as a textile engineer and was one of the founding members of the *Falange Española*. He acquired broad industrial experience in the firms of the Francoist banker Juan March (1880–1962) and become a key figure in the *Sindicatos Verticales* (Francoist trade unions) and a provincial chief of the *Movimiento* (the single party of the

---

[26] Luis Blas, 'El problema del bromo en España', *Ión*, 2(7) (1942), 117–19.
[27] In particular, lignites from Mequinenza, Fayón, Aliaga and Puente de García Rodríguez. Robert Wizinger, *Carbón, aire y agua. Revisado, puesto al día y adaptado al caso de España por el Dr. J.M. Fernández-Ladreda y Menéndez-Valdés, Coronel de Artillería. Catedrático de la Universidad de Sevilla* (Madrid: Ediciones A. Aguado, 1942), p. 46.
[28] García Pérez, *Franquismo y Tercer Reich*.
[29] Christian Leitz, *Economic Relations between Nazi Germany and Franco's Spain, 1936–1945* (Oxford: Clarendon Press, 1996), pp. 1–7. See also: Viñas, *Guerra, dinero, dictadura*; Christian Leitz, 'Nazi Germany and Francoist Spain, 1936–1945', in Sebastian Balfour, Paul Preston (eds.), *Spain and the Great Powers in the Twentieth Century* (London: Routledge, 1999), pp. 127–50.
[30] Raymond G. Stokes, *Opting for Oil: The Political Economy of Technological Change in the West German Chemical Industry, 1945–1961* (Cambridge: Cambridge University Press, 1994), pp. 1–9.

regime). Carceller planned to secure Nazi German collaboration in coal, lubricants, gasoline, nitrogen fertilisers and other synthetic products. As a result of his diplomatic efforts, in 1942 the INI established a permanent office in Berlin.[31]

In the 1940s, autarchic applied chemistry officially became of national interest.[32] Early projects of the PJC had a strong chemical bias and focused on the production of coke from low-quality coal, scrap metal from low-quality iron with the Krupp–Renn process of direct reduction of ores, chemicals from agriculture waste, lubricants from oil shale and metals from pyrites.[33] In 1942, among the INI projects, many required specific chemical expertise: synthetic gasoline, rubber, mining and iron and steel. The *Instituto del Combustible*, for instance, worked on the artificial production of hydrocarbons, mainly through the use of the Fischer–Tropsch and hydrogenation methods on local raw materials such as olive oil, colophony and beeswax, with varying degrees of success.[34]

Albareda's school of soil chemistry represented another excellent opportunity for the efficient study and exploitation of the country's raw materials nationwide.[35] In the 1940s, he built up an influential research school and launched new journals such as the *Anales de Edafología, Ecología y Fisiología Vegetal*.[36] In his book *El Suelo* (the soil), with a preface by his mentor Rocasolano, Albareda appropriated physical chemistry (and colloid chemistry, in particular) for autarchic purposes.[37] In the 1950s, agricultural chemistry and edaphology centres spread throughout

---

[31] Ignasi Riera, *Els catalans de Franco* (Barcelona: Plaza y Janés, 1998).
[32] Through the state promotion of fertilisers, mining and metallurgy, textile cellulose, artificial fibres, plastics, iron and steel, penicillin and antibiotics. José Manuel Sánchez Ron, *Miguel Catalán. Su obra y su mundo* (Madrid: CSIC, 1994), p. 378. See also: García Pérez, *Franquismo y Tercer Reich*, pp. 346, 550.
[33] Santiago López, 'El Patronato Juan de la Cierva (1939–1960), I. Parte: Las instituciones precedentes', *Arbor*, 619 (1997), 201–38, p. 204. See also: Santiago López, 'El Patronato Juan de la Cierva (1939–1960), II. Parte: La organización y la financiación', *Arbor*, 625 (1998), 1–44; Santiago López, 'El Patronato Juan de la Cierva (1939–1960), III. Parte: La investigación científica y tecnológica', *Arbor*, 637 (1999), 1–32.
[34] V. Gómez Aranda, 'Una contribución española a la producción artificial de hidrocarburos', *Ión*, 8(2) (1942), 197–205.
[35] José María Albareda, Enrique Gutiérrez Ríos, 'Equilibrio de los cationes de cambio en perfiles de suelos especiales en relación a sus condiciones de formación', *Anales de Física y Química*, 40 (1944), 365–78.
[36] Important names in Albareda's soil chemistry school were Enrique Gutiérrez Ríos, Ángel Santos, Ángel Hoyos, Mariano Tomeo, Vicente Aleixandre and Fernando Burriel, among many others. Lora-Tamayo, *La investigación química*, pp. 299–304.
[37] José María Albareda, *El Suelo. Estudio físico-químico y biológico de su formación y constitución*, prólogo del Dr Antonio de G. Rocasolano (Madrid: Sociedad Anónima Española de Traductores y Autores, 1940). The project also counted on the experimental stations of Aula Dei (Zaragoza) and El Zaidín (Granada).

the country.[38] In 1955, in terms of permanent positions, the *Instituto de Edafología* led the CSIC, followed by the *Instituto Alonso Barba de Química*, under Lora-Tamayo's direction, and in third place was the 'Rocasolano' Institute of Physical Chemistry.[39] Equally, other agrochemical and food technology centres appeared in the fertile regions of Murcia and Valencia, again as paradigms of applied science to serve regional and local needs.[40] The *Instituto de la Grasa* in Seville focused its applied chemistry research on olive oil, as its journal *Grasas y Aceites* described in detail.[41]

In the late 1940s, a working group on plastics began its project in the chemistry section of the *Instituto Alonso Barba*. Chemist Juan Luis de la Ynfiesta, another of Lora-Tamayo's pupils, devoted his early efforts to the application of plastics to industrial needs, but the group quickly expanded.[42] In 1953, the *Centro Español de Plásticos* opened its doors in Barcelona, aiming to popularise the use of the new materials and train industrialists and organise conferences, at a time when Catalonia held 70 per cent of the plastics market.[43] In spite of the autarchic context, the journal *Revista de Plásticos* contributed to provide updates on international novelties and to promote local research on several polymers, as well as providing a useful space for Spanish chemical firms to advertise their products. Nevertheless, the shortage of raw material for the industrial synthesis of plastics continued to be a bottleneck for successful results. In 1955, Ynfiesta praised the Department of Plastics for its capacity to exploit new national raw materials in order to overcome the domestic scarcity of reagents for the synthesis of polymers.[44]

[38] In Galicia, Salamanca, Seville, Almería, Murcia, Barcelona and the Canary Islands.
[39] María Jesús Santesmases, 'El legado de Cajal frente a Albareda. Las ciencias biológicas en los primeros años del CSIC y los orígenes del CIB', *Arbor*, 631–2 (1998), 305–32, p. 327.
[40] Lora-Tamayo, *La investigación química*, p. 303.
[41] Ibid., pp. 313–15. It is also worth noting the case of the rector the University of Salamanca and professor of analytical chemistry, Felipe Lucena, who created a Centre for Edaphology. Siro Arribas, *Introducción a la historia de la química analítica en España* (Oviedo: Universidad de Oviedo, 1985), p. 51; Isabel Ramos, *Profesores, alumnos y saberes en la Universidad de Salamanca en el Rectorado de D. Antonio Tovar (1951–1956)* (Salamanca: Ediciones Universidad de Salamanca, 2009); Fernando Burriel, Felipe Lucena, Siro Arribas, *Química analítica cualitativa: Teoría y semimicrométodos* (Madrid: Paraninfo, 1959).
[42] Juan Luis de la Ynfiesta, José Alemán Vega, 'Derivados del pineno de interés en la industria de plásticos', *Anales de la Asociación Española para el Progreso de las Ciencias (AAEPC)*, 20 (1955), 779–90.
[43] *Centro Español de Plásticos. 50 Aniversario, 1953–2003* (Barcelona: Centro Español de Plásticos, 2003).
[44] Juan Luis de la Ynfiesta, José Alemán Vega, 'Derivados del pineno', *AAEPC*, 20(2) (1955), 779.

Other academic chemists soon became handmaidens of the autarchic 'applied research' programme. In 1951, Carlos Nogareda, at that time a professor of physical chemistry at the University of Salamanca, organised a summer school in the town of Béjar, which was again devoted to chemical applications to industry. It is quite striking, however, to see a man such as Nogareda, who had prestigious pre-war training – first with Moles, later in Cambridge at the Colloid Science Department, working with physical chemist Eric K. Rideal (1890–1974) on high vacuum pressures, and who had also enjoyed the personal contact of Nobel Prize winners there[45] – shifting to local, regional applications and publishing only amateur history of chemistry and educational materials.[46]

Paradoxically, although Jesuits were clearly opposed to the leftist Republican government, after the Civil War, the IQS did not join the PJC, and neither did it enjoy a stable relationship with the CSIC. Nevertheless, the IQS set up a strong alliance with the chemical industry through a new board, the *Patronato*, for the funding of applied chemistry research. Founding members of the *Patronato* were José Pellicer – director of the *Fábrica Nacional de Colorantes y Explosivos* (FNCE) and a key actor in the German investment in Francoist Spain; José Lipperheide – the owner of a powerful metallurgical firm from northern Spain; Demetrio Carceller – the aforementioned prominent Falangist, then working at the *Compañía Española de Petróleos S.A.* (CEPSA); José Valls Taberner – a reputable chemical entrepreneur at the Cros chemical company; and José Guitart Solé – the founder of the association of alumni of the IQS.[47] In the early 1950s, the *Patronato* acted as a useful mediator between chemical industries, political power and academic chemistry.[48] Presidents of the *Patronato* would coincide with the presidents of the *Diputación*, such as Joaquín Buxó and Juan Antonio Samaranch, the latter being another prominent Francoist and some years later the well-known president of the International Olympic Committee. Chemical companies such as Unicolor, Carburos Metálicos, Titan and SAFA, and some international

---

[45] Joseph John Thomson, Ernst Rutherford, Francis Aston, James Chadwick and Patrick Blackett, among others. See also: Carlos Nogareda, *Rayos cósmicos: Discurso inaugural del curso académico 1950–51* (Salamanca: Universidad de Salamanca, 1950).
[46] Fernando Galán, 'Sobre la callada desaparición de un viejo y excelente maestro de la Universidad de Salamanca: El profesor Carlos Nogareda Domènech. In Memoriam', *Salamanca. Revista de Estudios*, 33/34 (1994), 349–58. I am indebted to José Nogareda (Carlos Nogareda's son) for providing me with data and personal memories of his father.
[47] Núria Puig, Santiago López, *Ciencia e industria en España: El Instituto Químico de Sarriá, 1916–1992* (Barcelona: Instituto Químico de Sarriá, 1992).
[48] In 1955, its president was Joaquín Buxó, and he was also president of the '*Diputación de Barcelona*' – a layer of government responsible for the administration of the city councils – and a prominent Francoist with a close personal relationship with the *Caudillo*.

146   Autarchic Ambiguities

firms such as Ciba and Nestlé, gradually joined the *Patronato* and benefitted from the IQS's technical assessment services.[49]

Far from being in isolation, in October 1955, the 18th International Conference of Industrial Chemistry was held in Madrid, with close to 1,000 delegates attending from 28 countries and more than 300 papers presented.[50] Side by side with prestigious academic papers, the *Sindicato Nacional de Industrias Químicas* presented several economic articles on the state of the art of the Spanish chemical industry.[51] In the closing session, the conference honoured the most distinguished Spanish representatives: José Agell, Lora-Tamayo, industrialist Juan Abelló (1895–1983) and engineer Carlos Abollado (1902–90).[52] The latter three were active agents in autarchic projects that held applied chemistry goals as intrinsic parts of their involvement in politics. Abollado was a professor of physical chemistry at the Industrial Engineering School in Madrid, but he also became a key figure of the Falangist *Sindicato*. Lora-Tamayo's distinguished career as an organic chemist placed him in a strong position at the PJC and the CSIC, and some years later he would become Minister for Education (see Chapter 6). Trained as a chemist and pharmacist, Abelló became a powerful pharmaceutical industrialist, the president of the Madrid Board of Trade (*Cámara de Comercio*) and a very influential political figure, as well as a personal friend of Franco.[53] All three were MPs in the Francoist *Cortes Españolas*. The fourth awardee was José Agell,

---

[49] Nuria Puig, Santiago López, 'Chemists, Engineers and Entrepreneurs. The Chemical Institute of Sarrià's Impact on Spanish Industry (1956–1992)', *History and Technology*, 11 (1994), 345–59.

[50] *La Vanguardia Española*, 23 October 1955, p. 12.

[51] *La Vanguardia Española*, 24 October 1955, p.4.

[52] In the 1920s, industrial engineer Carlos Abollado worked at the Electrochemical Department of Siemens and acquired extensive industrial experience. Back in Spain, he worked at *Pavimentos Asfálticos*, a local subsidiary of Shell. In 1940, Abollado became the general secretary of the Falangist *Sindicato de Industrias Químicas* and later joined the Ministry of Industry. He took an active part in several INI autarchic projects, such as on bituminous slates in Puertollano and alcohol dehydration. Abollado became a professor of physical chemistry at the *Escuela de Ingenieros Industriales de Madrid* (1947–72) and an MP (*procurador*) of the Francoist parliament (1943–6) by direct appointment of the *Caudillo*. Pablo Valbuena Vázquez, *Historia de la Escuela Técnica Superior de Ingenieros Industriales de Madrid desde 1901 hasta 1972* (Madrid: E.T.S.I. Industriales (UPM), 1996). http://oa.upm.es/5331/1/PFC_PABLO_VALBUENA_VA ZQUEZ.pdf (last accessed 5 December 2016). See also: Carlos Abollado, 'La industria química y la química industrial', *Ión*, 45 (1945), 248, cited by Ángel Toca, 'Dos profesiones para un solo cometido. La introducción de la ingeniería química en España durante el primer franquismo', *Dynamis*, 26 (2006), 253–85, p. 282.

[53] In the 1910s, Juan Abelló (1895–1983) obtained BAs and a PhD in Science and in Pharmacy in Madrid. Starting out as a secondary school teacher of physics and chemistry and running a small pharmacy, he made good profits through the commercialisation of drugs. He later went on to have a very successful entrepreneurial career with his *Laboratorios Abelló*. In the 1940s, he became president of the Madrid Board of Trade

who had led reforms in chemical training in the 1910s and 1920s (see Chapter 1), had survived professionally through the Civil War and during early Francoism and directed the chemical company of synthetic fibres, SAFA.

In the closing session of the conference, Joaquín Planell (1891–1969), the Minister for Industry, spoke to the delegates in a very optimistic, propagandist tone on the future of the Spanish chemical industry. In the 1930s, Planell had been a military attaché in Washington, and he joined the Francoist side during the war. He became vice-president of the INI and the founding president of the *Empresa Nacional Calvo Sotelo* (ENCS) and the *Instituto del Combustible* at the PJC. Obviously, raw materials were once again the central element of the speech. He talked about marine salt, pyrites, vegetable extracts, mineral coal, iron ores and the efforts made to search for oil fields on Spanish land. In Puertollano, the ENCS had established a plant for the distillation of bituminous slates in which, in 1955, the catalytic hydrogenation of 150 million tonnes per year of crude oil provided lubricants and paraffin, as well as useful fuels such as kerosene and diesel.[54] In fact, the synthetic fuel programme had already begun during the Second World War and with the Francoist alliance with Nazi Germany, but it continued in the 1950s and the 1960s with new deals with BASF and other chemical companies as a milestone of the new petrochemical industry (see Chapter 6).[55]

A journey through the *Catálogo del Libro Español de Química (1920–1955)*[56] (catalogue of Spanish chemistry books) issued for the 1955 Exhibition of Chemical Industry provided impressive figures to

---

(*Cámara de Comercio*) and, in 1948, a member of the Francoist parliament. Gerardo Monge, *Figuras de la industria química española* (Barcelona, 1955), pp. 532–3.

[54] *La Vanguardia Española*, 28 October 1955, p. 6.

[55] 'Spain [...] developed a synthetic fuel program at Puertollano following a 1944 deal between the pro-Axis government and Germany. In 1950 Spain's government signed new deals for technology with BASF and others; production started in 1956 and lasted until 1966'. David Edgerton, 'Not Counting Chemistry: How We Misread the History of Twentieth-Century Science and Technology', *Distillations*, Spring 2008. www.sciencehistory.org/distillations/magazine/not-counting-chemistry-how-we-misread-the-history-of-20th-century-science-and (last accessed 29 September 2018).

[56] *Catálogo del Libro Español de Química (1920–1955), con motivo del homenaje al P. E. Vitoria S.J*, *Afinidad (n° extraordinario, fuera de serie). Mayo de 1955*. 'At the request of Science Service, through its Director, Watson Davis, who has been cooperating with the Division of Cultural Relations of the U.S. State Department in the sponsorship of the translation and publications of textbooks in Spanish and Portuguese for the aid of the educational work in the Latin American countries, a committee was appointed by the American Chemical Society in 1947 to submit recommendations as to titles of books in the field of Chemistry and Chemical Technology which should be translated'. Wallace R. Brode, 'Bibliography of Chemistry, and Chemical Technology Textbooks in the Spanish and Portuguese Languages', *Journal of Chemical Education*, 26(10) (1949), 553.

Figure 5.2 Front cover of the *Catálogo del libro Español de Química (1920–1955)*, a special issue of the journal *Afinidad*, May 1955. Private collection of the author.

disprove the notion of scientific isolation (Figure 5.2). Thirty-four publishing houses produced chemistry books for the Spanish-speaking market.[57] Concerning authors, of 700 names, close to 400 were Spanish and had produced original texts, whereas 300 were foreign

---

[57] *Catálogo del Libro Español de Química (1920–1955)*, p. 283.

and had been translated into Spanish. In total, the exhibit displayed 1,200 books. The catalogue borrowed information from the IQS library, the library of the *Seminario de Química* at the University of Barcelona and the *Instituto Nacional del Libro Español*. All of the books were classified into 22 categories, and the list of journals with chemical content numbered 41. Beyond the more traditional academic fields (inorganic, organic, analytical, technical, engineering), they obviously included all industrial, applied subjects (steel, leather, oil, concrete, paints, coal, pharmaceuticals, plastics, textiles) that had been promoted since the early 1940s by the CSIC, the PJC and its research institutes.[58] These figures, in fact, confirm that 'our chemicals' were a sign of the early vitality of chemical companies with strong Italian and German connections, the organisation of prestigious international conferences and considerable production of chemical literature, which included translations of foreign authors.

'Technical' Chemistry

Although the actual professionalisation of chemical engineering was not seen in Spain until some decades later,[59] the boundaries of industrial chemical training shifted considerably during the autarchic period.[60] The 'applied chemistry' programme of the new regime aimed to update chemists' training for more efficient integration with industry, and it reformed the chairs in 'technical chemistry'. The project oscillated from the former tradition of industrial chemistry and the new chemical engineering, which, under a strong American influence, progressively gained ground, particularly after the German defeat in the Second World War, and even before the agreements between the United States and Franco's Spain of 1953. The results were not even, but industrial training for chemists changed considerably in the 1940s and 1950s. As in the case of the emergence of chemical engineering in Britain in the early twentieth century, the reasons for the appearance of a new discipline go beyond

---

[58] *Catálogo del Libro Español de Química (1920–1955)*, pp. 281–2. For an earlier chemical bibliography, see: Maria Serrallach, *Bibliografía Química. Documentación científico-industrial. Con un prólogo del Dr. D. José Pascual Vila* (Barcelona: Imprenta Clarasó, 1946). In 1946, Serrallach obtained a fellowship at the University of Notre Dame for her advanced training in the chemical bibliography and the use of the *Chemical Abstracts*.

[59] Ángel Toca highlighted some years ago that, in autarchic Spain, mainly in the 1940s and 1950s, routes of learning and expertise under the label of 'technical chemistry' were varied and complex. Toca, 'Dos profesiones para un solo cometido'.

[60] 'Editoral: Técnicos para la industria química', *Ión*, 38 (1944), 591.

a simple response to industrial demands, and often have to do with a variety of interest groups – our chemists in this case.[61]

Antonio Rius Miró soon took the lead in the establishment of the new Francoist technical chemistry. After earning his PhD in electrochemistry, he enjoyed JAE grants for his training from 1917 to 1920 at the University of Basel and the Technical School of Dresden. Back in Spain, in 1922 he taught electrochemistry and physical chemistry at the Industrial Engineering School in Zaragoza and contacted Rocasolano's group at the Faculty of Science. As part of joint projects between the university and the engineering school, Rius Miró supervised several PhD students, Albareda among them.[62] In 1930, he became Professor of Industrial Organic Chemistry and Industrial Chemical Analysis at the *Escuela Industrial* in Madrid and Abelló's main advisor, who brought him close to Franco.[63] At the end of the Civil War, in 1940, Rius Miró was awarded the chair in technical chemistry at the University of Madrid and, under the protection of Albareda, he reached the peak of his political power in the new regime. Benefitting from the dismantling of Moles's school of physical chemistry, he entered technical chemistry to introduce aspects of the emerging field of chemical engineering, and he strongly promoted industrial links with the pharmaceutical industry and research on alkaloids.[64] Rius Miró was Albareda's personal advisor and a member of the permanent committee of the CSIC. He also directed the Physical Chemistry Section of the *Instituto Alonso Barba* and became president of the RSEFQ (1940–9). In 1944, he created a doctorate programme in 'industrial chemistry', which was followed in the 1950s by unsuccessful

---

[61] James F. Donnelly, 'Chemical Engineering in England, 1880–1922', *Annals of Science*, 45(6) (1988), 555–90. See also: James F. Donnelly, 'Defining the Industrial Chemist in the United Kingdom, 1850–1921', *Journal of Social History*, 29 (1996), 779–96.

[62] The collaboration between the *Escuela Industrial* and the Science Faculty in Zaragoza provided a chair in electrochemistry for Rius Miró and eased the development of his own research school, which included names such as Albareda, Vilas, García Marquina, Buscarons, Arnal-Yarza, García de la Puerta, Gay, Labetas, Alvira and Medrano. Antonio Rius Miró, *Real Academia de Ciencias Exactas, Físicas y Naturales. Discurso leido en el acto de su recepción por el Excmo. Sr. _____ y Contestación del Excmo. Sr. D. Francisco Navarro Borrás, el día 21 de noviembre de 1945* (Madrid: Real Academia de Ciencias Exactas, Físicas y Naturales, 1945). Other names from Rius Miró's research school who signed joint papers at *Anales* were J. A. Knecht, J. Royo Iranzo, J. Llopis and I. Calleja. See, for instance, the papers appearing in *Anales de Física y Química*, 392 (1945).

[63] In the early 1940s, Rius Miró worked at the *Laboratorios de Investigación de la Fábrica de Productos Químicos y Farmacéuticos Abelló*. Antonio Rius Miró, 'Sobre el método, para la determinación cuantitativa de la morfina en opio', *Anales*, 36 (1940), 51–6.

[64] Martínez Moreno, Otero de la Gándara, A. Crespí and Joaquín Ocón. Lora-Tamayo, *La investigación química*, pp. 290–1. See also: Antonio Rius Miró. *Introducción a la Ingeniería Química* (Madrid: Alfa, 1944).

attempts to found a chemical engineering faculty. To that end, he also counted on the support of officer and chemist, Fernández-Ladreda.[65]

Rius Miró's *Introducción a la Ingeniería Química* (1944) made a clear distinction between 'chemical engineering' – based mainly on standard operations, quantitative measures and abstract laws on large-scale production – and 'industrial chemistry', which he perceived as a descriptive set of raw materials and industrial methods of production from a more pragmatic perspective. He also considered that the name of his chair 'technical chemistry' lacked theoretical ambition, which foreign Anglo-American trends of 'unit operations' could successfully fulfil. In the United States, instead of describing every industrial process in detail – as was the case in the nineteenth-century industrial chemistry tradition – in the late 1910s and early 1920s, the old empirical data were grouped under fluid flow, heat transfer, mass transfer and thermodynamic and mechanical processes, which included standard processes such as transportation, evaporation, filtration, distillation, extraction, absorption, liquefaction and refrigeration.[66] Ideas from American industrial chemists such as Arthur Dehon Little, William Walker, Warren Lewis and William McAdams – the 'founding fathers' of chemical engineering[67] – were appropriated in Francoist Spain. It that context, autarchic policies paradoxically integrated the latest foreign innovations such as the American unit operations framework, but in a totalitarian atmosphere.[68]

In the 1940s and 1950s, more than 20 chemical engineering textbooks (foreign translations and original Spanish texts) circulated widely in engineering and technical schools and science faculties. This was, for example, the case for Ángel Vián's textbook, *Elementos de ingeniería química: Operaciones básicas*, which appeared in 1952. That same year, Vián was finally able to obtain a chair in industrial chemistry in Salamanca. Facing marginalisation due to his link with the Moles's school

---

[65] José María Fernández-Ladreda, *El Doctorado en Química Industrial y la formación de los químicos para la industria. Discurso correspondiente a la apertura del curso académico 1949–1950* (Madrid: Artes Gráficas Estades, 1949).

[66] On the history of chemical engineering, see: T. S. Reynolds, 'Defining Professional Boundaries: Chemical Engineering in the Early Twentieth Century', *Technology and Culture*, 27 (1986), 694–716; William F. Furter (ed.), *History of Chemical Engineering* (Washington, DC: American Chemical Society, 1980); Clive Cohen, 'The Early History of Chemical Engineering: A Reassessment', *The British Journal for the History of Science*, 29 (1996), 171–94; Colin Divall, 'Education for Design and Production: Professional Organization, Employers, and the Study of Chemical Engineering in British Universities, 1922–1976', *Technology and Culture*, 35 (1994), 258–88.

[67] Furter, *History of Chemical Engineering*.

[68] For the concept of reactionary modernism in Germany, see: Jeffrey Herf, *Reactionary Modernism: Technology, Culture and Politics in Weimar and the Third Reich* (Cambridge: Cambridge University Press, 1984).

Figure 5.3 Barnett F. Dodge (1896–1972), Professor of Chemical Engineering at Yale, in a cartoon by Manuel del Arco in *La Vanguardia Española*, 16 February 1954, p. 12.
Courtesy of Xavier Castellnou del Arco.

during the Civil War (see Chapter 3), Vián worked outside the INFQ, at the *Fábrica de Pólvoras de Murcia*, where he acquired industrial skills and experience in his later career.[69] Other university and industrial chemists wrote some chapters, and Fernández-Ladreda contributed with a preface, in which he emphasised the pragmatic function of universities for the training of professionals to serve the interests of the nation and disregarded what, in his view, were 'abstract', 'useless', academic debates.[70]

In 1948, during the harsh years of international isolation, American chemical engineers Donald Othmer and Robert S. Aries were already weighing up several industrial projects in Spain.[71] In 1953, once the regime had achieved the first stages of its international recognition, the US Department of State asked Barnett F. Dodge (1896–1972), Professor

---

[69] Vián co-authored the book with Joaquin Ocón, both being professors of technical chemistry in Salamanca and Santiago, respectively. Ángel Vián, Joaquín Ocón, *Elementos de Ingeniería Química (Operaciones Básicas). Prólogo de J.M. Fernández-Ladreda* (Madrid: Aguilar, 1952). On Ángel Vián, see: Ángel Vián, *Publicaciones científicas (1936–1984)* (Madrid: Técnicas Reunidas, 1985).

[70] Vián, Ocón, *Elementos de Ingeniería Química*, p. ix. See also: Ángel Vián, 'La química técnica como enseñanza universitària', *Química e Industria*, 15 (1968), 73.

[71] *Ión*, 80 (1948), 185.

and Head of the Department of Chemical Engineering at Yale, to travel to Spain to lecture for two months on chemical engineering (Figure 5.3).[72] Dodge corresponded with Fernando Calvet, one of the luminaries of the Silver Age, who had finally obtained a chair in technical chemistry at the University of Barcelona – unfortunately, far from his early biochemical background (see Chapter 3) – and was probably one of the few chemists with enough cultural affinity to be able to properly receive Dodge in Spain.

Dodge travelled to Spain in January 1954 to give an entire course of 23 lectures in two months (three lectures per week) on chemical engineering, and he was 'the first American professor going to Spain under the new State Department policy'.[73] Calvet had informed Dodge about the kind of audience he would encounter: chemistry students, academic staff, practicing chemists and engineers from industries in and around Barcelona. The first lecture, on the profession of chemical engineering, took place at the main university lecture hall with more than 300 attendees. Held in smaller rooms, the other lectures constantly achieved high attendee figures of 150–200 people.

Dodge taught about the education of a chemical engineer (1 lecture), basic concepts, principles and methods of calculation (2), material balances (2), the application of thermodynamics to chemical engineering (4), fluid flow (4), heat transfer (5) and unit operations based on phase distribution (5).[74] However, he was disappointed by the lack of feedback at the end of his talks: 'apparently there is not the practice in Spain to ask the Professor lots of questions'.[75] After lecturing, he visited the Barcelona Engineering School and the university labs, which he considered 'meagre and antiquated'.[76] Although he appreciated Calvet's efforts to introduce his students to foreign chemical engineering textbooks and praised Vián's textbook, the lack of specialised laboratories for chemical engineering also disappointed him. Dodge got a slightly better impression during his visit to the IQS, but he lamented its poor library and laboratory facilities, as well as the fact that the institute was supported by tuition fees alone. He was well aware of the political context he found himself in, and of the dictatorial censorship imposed on any kind of publication or public address. He found it very

---

[72] Barnett J. Dodge, 'A Chemical Engineer in Spain', *Chemical Engineering Progress*, 50(9) (1954), 36, 62, 64, 66, 68–9. See also: Miguel Masriera, 'El curso del Profesor Dodge', *La Vanguardia Española*, 12 March 1954, p. 5. Masriera stressed the need for a stronger intersection between chemical training in science faculties and engineering schools, and he praised Calvet for having organised such a prestigious course from his technical chemistry chair at the University of Barcelona (see Chapter 7).
[73] Dodge, 'A Chemical Engineer in Spain', p. 36.   [74] Ibid., p. 62.   [75] Ibid., p. 64.
[76] Ibid., p. 67.

hard to reach foreign newspapers and he noticed the silence of his local chemist colleagues: 'none of our friends were inclined to discuss political questions'. Dodge described his meeting with a local scientist – perhaps Masriera – 'who had formerly been a University Professor, but who had been somewhat active on the "wrong" side in the Civil War and has never been able to obtain another University or teaching post'.[77]

Dodge's involvement in the atomic bomb project at Oak Ridge[78] sparked the interest of local journalists, but local authorities censored part of the conversation on this topic in an interview with the newspaper *La Vanguardia Española*, and it was never published. After Dodge's visit to two chemical plants near Barcelona, one for phosphate fertilisers and the other devoted to organic products (dyes, pharmaceuticals and insecticides), he concluded: 'Little research was done. Most of the equipment and the "know-how" for these plants were imported from Germany'. Dodge bluntly stated, 'Industrially and mechanically Spain is a backward country [...] The standard of living in Spain is very low [...] Fuel is scarce, and everything combustible, even nut shells, is carefully saved for fuel purposes'.[79]

That apparent isolation and backwardness of the country did not hinder attempts to retain some aspects of the old German science-based industry, nor to introduce American unit operations into the Spanish academic system. The autarchic context of the early decades of the dictatorship provided ample academic room to appropriate from abroad and develop new training programmes for the new experts of the Francoist chemical industry.[80] All of these industrial chemistry plans – closely associated with the autarchic rhetoric of isolation and self-sufficiency – in practice never closed the borders to foreign raw materials, foreign innovations and publications. The Falangist *Sindicato* and its strong anti-liberal agenda did not hinder the publication of foreign book reviews and translations of foreign papers and novelties. Equally, foreign chemical engineering – particularly American textbooks and projects after the 1953 US agreement with Franco's Spain – progressively leaked into the academic training plans of major figures such as Rius Miró, and were in tune with Dodge's visit to Barcelona. Reforms in industrial chemistry

---

[77] Ibid., p. 68.
[78] 'Barnett F. Dodge, a former dean of the Yale University School of Engineering who worked on the Manhattan Project to develop the atom bomb, died yesterday at Yale-New Haven Hospital after surgery. He was 76 years old and lived in New Haven'. *The New York Times*, 17 March 1972.
[79] Ibid., p. 69.
[80] Ángel Toca, 'Ingeniería Química en España: los orígenes (1850–1936) (I)', *Anales de Química*, 103(2) (2007), 47–53; Ángel Toca, 'Ingeniería Química en España: los orígenes (1850–1936) (II)', *Anales de Química*, 103(3) (2007), 59–66.

reinforced the Francoist modernising programme of 'applied chemistry', as the CSIC *Revista de Ciencia Aplicada* clearly epitomised.

Chemical Diplomacy

As economic historian Fernando Guirao has shown in recent years, autarchy had its own contradictions and paradoxes. Guirao explained the survival of the regime, within the new logic of the Cold War following the Allies' victory in 1945, in terms of the weak international opposition to the repression of the dictatorship. Francoist Spain soon became an island of political stability, an anti-communist vanguard and a key player – albeit a fairly marginal one – in the economic recovery of Europe.[81] In Guirao's words: '[...] the ostracism Spain suffered after 1945 was not as pronounced as had been claimed and [...] trade relations offered an avenue of escape'.[82] In spite of the international isolation of the regime, French diplomacy publicly admitted, for example, that Francoist Spain was right in its pioneering warnings of the danger of communism, and in 1948 had already reopened the Spanish border on quite generous terms.[83]

Once the liberal Allies defeated Nazi Germany and fascist Italy, chemistry became a useful tool for the establishment of new foreign relations and further international legitimation. The United Nations (UN) condemnation of Franco's regime in 1946 and the ban by the British Labour Government on allowing Spain to enter the Marshall Plan in 1947 were two worrying signs for the new elites.[84] It was in that context that chemist Lora-Tamayo described the role of science in the 1940s and 1950s in the following terms:

[...] during that period, with no foreign diplomatic relations, convalescing from recent war, Spanish professors [...] and *pensionados* [...] constituted the most robust link abroad, recreating former bonds, creating new ones, and, at such a time of isolation enabled foreign scientists to visit Spain and give their courses and lectures, contrasting with their own, impartial view of Spain's reality the very different negative images that were being spread around. Right from the start this nascent research team offered an invaluable service to Spanish politics, but one which due to ignorance of its existence has not been fully appreciated.[85]

---

[81] Fernando Guirao, *Spain and the Reconstruction of Western Europe, 1945–1957. Challenge and Response* (Basingstoke: Macmillan, 1998).
[82] Ibid., p. 4.
[83] Borja de Riquer, *Historia de España (IX): La dictadura de Franco* (Barcelona/Madrid: Crítica/Marcial Pons, 2010), p. 119.
[84] Denis Smyth, 'Franco and the Allies in the Second World War', in Balfour, Preston (eds.), *Spain and the Great Powers*, pp. 185–209.
[85] Lora-Tamayo, *La investigación científica*, p. 8.

Just as the prominent physicist Werner Heisenberg acted as a goodwill ambassador for the Nazi regime,[86] so many Francoist chemists, with their 'scientific', 'apolitical' lectures abroad, also became fellow travellers and ambassadors of the new totalitarian policies. In 1950, Lora-Tamayo and Rius Miró were the chemists among a group of scholars who offered a collection of books and journals to the *Caudillo* at the central premises of the CSIC, many of them far below international standard and not peer reviewed.[87] Nevertheless, in contrast with this endogenic tradition of the publication policy at the CSIC, Spanish chemical research appeared in a good number of international journals and books.[88]

From Lora-Tamayo's personal account, we know that in the autarchic years the country held international conferences on pure and applied chemistry, industrial chemistry, food science and technology, etc. Tacitly repeating some pre-war practices, chemists attended international meetings, sent students abroad and progressively created their own research schools.[89] Paradoxically, in the early 1940s, and still benefitting from Nazi German contacts, Albareda, one of the leaders of the crusade against the JAE, used his newly founded Edaphology Institute as a sort of seminal foreign office. Albareda's book, *El Suelo*, had appeared in 1940 and constituted a synthesis of his early international training in soil chemistry and the academic network he successfully managed to build.[90] In April 1941, Professor Hans Pallmann from the ETH Zurich, one of Albareda's mentors of the Silver Age, came to Madrid and Barcelona to give a course on soil chemistry and geobotanics.[91] From his prominent position at the CSIC, after the Second World War, Albareda soon established new contacts with foreign national research centres and universities in France, Britain and Germany, especially the early CNRS, the *Deutscher Akademischer Austauschdienst* (DAAD; German Academic Exchange Service) and the British Council, as well

---

[86] Mark Walker, 'Physics and Propaganda', in *Nazi Science: Myth, Truth and the German Atomic Bomb* (Cambridge, MA: Perseus, 1995), pp. 123–51.

[87] The full list of names was Mariano Bassols, Santiago Alcobé, José María Otero Navascués, Tomás Carreras Artau, Juan Solé Sabarís, José María Millàs Vall·llicrosa and Pedro Laín Entralgo. Antoni Malet, 'Las primeras décadas del CSIC: Investigación y ciencia para el franquismo', in Ana Romero and María Jesús Santesmases (eds.), *Cien años de política científica en España* (Madrid: Fundación BBVA, 2008), pp. 211–56, 250.

[88] From the late 1940s onwards, Albareda's, Lora-Tamayo's and Pascual's schools published their research in prestigious international journals.

[89] Manuel Lora-Tamayo, *Lo que yo he conocido (Recuerdos de un viejo catedrático que fue ministro)* (Puerto Real: Federico Joly y Cia- Ingrasa, 2002), pp. 324–8.

[90] Enrique Gutiérrez Ríos, *José María Albareda, Una época de la cultura española* (Madrid: CSIC, 1970), pp. 216–20; José María Albareda, *El Suelo: Estudio físico-químico y biológico de su formación y constitución* (Madrid: Sociedad Anónima Española de Traductores y Autores, 1940).

[91] Later followed by John E. Russell, Fritz Scheffer and Walter Kubiena.

as the ETH in Zurich and the Universities of Bonn, Mainz, Utrecht, Leuven and Geneva.[92]

In 1946, Lora-Tamayo and Pascual undertook a long European journey, aiming to re-establish some of their earlier international contacts.[93] The two had met at the University of Seville before the Civil War. Both enjoyed *pensionados* for their training and shared common views regarding organic chemistry research, as well as religious and political views that placed them in influential positions. In 1940, Pascual had obtained the chair in organic chemistry that had been vacant during the war after García Banús's exile. Pascual and Lora-Tamayo started the trip by visiting Kurt Meyer in Geneva, an expert on macromolecules, who accepted later an invitation to lecture in Spain. They then moved to the prestigious ETH in Zurich, which had been a target institution for the chemical training of the pre-Civil War generation (see Chapter 1). There, they contacted Leopold Ruzicka (1887–1976), the director of the ETH and winner of the Nobel Prize for Chemistry in 1939 for his work on steroid hormones, and Paul Karrer, a Nobel Prize winner in 1937 for his work on carotenoids and vitamin A. Ruzicka introduced Lora and Pascual to Vladimir Prelog – Ruzicka's successor and a great expert on reaction mechanisms, who later won a Nobel Prize in 1975.

In London, they met Christopher Ingold at UCL and established the basis for a future collaboration, which subsequently allowed Lora-Tamayo's pupil, Rafael Pérez Álvarez-Osorio, to do his PhD in Ingold's group. At Cambridge, they contacted Alexander Todd, the future Nobel Prize winner for Chemistry in 1957 for his synthesis of nucleotides, and at that time president of the IUPAC. They were keen to establish a future collaboration with Todd's group and, in particular, to send graduate students to his laboratory.[94] At Oxford, they unsuccessfully tried to meet Robert Robinson, the founding editor of the chemistry journal *Tetrahedron Letters*, which some years later became a publishing target for the new generation of Spanish chemists. With a rhetoric of political neutrality, Lora-Tamayo emphasised that '[...] *the world of science lies beyond political divergences among nations* [...] In the political isolation that we experienced [...] without even trying, the journey involved a pleasant resonance of understanding'.[95]

In 1947, after the trip, Lora-Tamayo invited Ingold to give three lectures at the *Instituto Alonso Barba* in Madrid. Ingold lectured on the

---

[92] Gutiérrez Ríos, *José María Albareda*, p. 244.
[93] Lora-Tamayo, *Lo que yo he conocido*, pp. 324–32.
[94] Among them, the case of Antonio González, the future director of the *Instituto de Productos Naturales orgánicos* at La Laguna (Tenerife), is worth mentioning.
[95] Lora-Tamayo, *Lo que yo he conocido*, p. 330 (emphasis added).

electronic interpretation of organic reactions, on substitution reactions of saturated compounds and on the modern ideas on steric hindrance. All of the lectures were published in *Anales* after a translation into Spanish by Álvarez-Osorio, who, some years later, after staying at UCL with Ingold, published a reference text in Spanish on the mechanisms of organic reactions.[96] During the whole process, and in spite of the official claims of autarky/autarchy, Lora-Tamayo's school closed the circle of scientific permeability abroad through an acquaintance with such a British luminary on chemistry as was Ingold. Also in 1947, Pascual was formally invited to the conference to celebrate the centenary of the Chemical Society.[97]

In West Germany, a certain continuity in science policies after the Second World War and their shared conservative and anti-communist values played an important role in German–Spanish relations during the early period of autarchy.[98] West German ideological affinity was central to the establishment of new scientific collaborations. Lora-Tamayo's contact with Kurt Alder – an expert in the organic synthesis of dienes – in Kiel led to his invitation to Madrid in 1949 for the celebration of the tenth anniversary of the foundation of the CSIC, which was perceived as a 'great world scientific embassy'.[99] In 1954, Albareda gave a formal lecture in Düsseldorf before a good number of German scientists. He presented the organisation of the CSIC and the figures of international exchanges and fellowships over recent decades, and now, unlike in his domestic addresses, he tacitly recognised the contribution of the JAE before the war.[100] That same year, the analytical chemist and at that time rector of the University of Barcelona, Francisco Buscarons (1906–89) (see Chapter 7), visited the *Max-Planck Gesellschaft* (MPG).[101]

---

[96] Javier Pérez-Castells, 'In Memoriam Rafael Pérez Álvarez-Ossorio', *Anales de Química*, 113(2) (2017), 139–40; Rafael Pérez Álvarez-Ossorio, *Mecanismo de las reacciones orgánicas* (Madrid: Alhambra, 1976–7; 1st Edition, 1958).

[97] Universidad de Navarra. Archivo General. Fondo José María Albareda. Correspondence (1 July 1947), 014/003.

[98] Karl Ziegler won a Nobel Prize in 1963 for the invention of a new method of catalysis of the process of polymerisation, together with the Italian chemist Giulio Natta. Ziegler was a member of the SS and received awards during the Nazi period. Albert Presas, 'La inmediata posguerra y la relación científica y técnica con Alemania', in Romero, Santesmases (eds.), *Cien años de política científica en España*, pp. 173–209, 192.

[99] This was the description of Enrique Gutiérrez Ríos, one of the most distinguished of Albareda's pupils and one of his later biographers. Gutiérrez Ríos, *José María Albareda*, p. 250.

[100] Albareda's address appeared later in 1956 as 'Die Entwicklung der Forschung in Spanien'. See Albert Presas, 'Nota histórica: Una conferencia de José María Albareda ante las autoridades académicas alemanas', *Arbor*, 631–2 (1998), 343–57.

[101] On Francisco Buscarons, see: Joaquim Sales, *La química a la Universitat de Barcelona* (Barcelona: Publicacions i Edicions de la UB, 2011), pp. 167–70.

Chemical Diplomacy 159

Figure 5.4 'Foreign *Savants* as Guests of Spanish Science', under the auspices of the CSIC in Madrid, *La Vanguardia Española*, 16 April 1950. Pérez de Rozas. Courtesy of the *Arxiu Fotogràfic de Barcelona*.

After the Second World War, and particularly in the 1950s, professional links with Anglo-American chemists also gradually increased.[102] In 1950, even before the international recognition of the regime, the RSEFQ managed to invite 17 recognised foreign chemists and physicists (Figure 5.4). That same year, the *Sociedad* received a formal invitation to attend the International Conference of the American Chemical Society (ACS) for its 75th anniversary in New York. Also in 1950, the *Asociación Española de Químicos de España* (ANQUE), which had been founded in 1946 under the leadership of Rius Miró and Fernández-Ladreda,[103] organised the first 'Chemistry Week' in Santander. Again, invitations of foreign experts soon paved the way for an international collaboration in the field of catalysis.[104] This was later to be a crucial step in the creation in 1956 of a catalysis section at the *Instituto 'Rocasolano'*, which worked on theoretical topics such as the electronic effect in oxide catalysis, but also on industrial projects such as the production of ethylene from naphtha.[105]

In the early years of official autarchy, Emilio Jimeno had already become an honorary member of the Royal Society of Edinburgh; in

[102] Manuel Valera et al., 'La Guerra Civil española y la investigación científica en química. Estudio Preliminar', in Javier Echevarría, M. Sol de Mora (eds.), *Actas del III Congreso de la Sociedad Española de Historia de la Ciencia y de la Técnica. San Sebastián, 1–6 Octubre 1984* (San Sebastian: Editorial Guipuzcoana, 1984), Vol. III, pp. 395–407.
[103] 'Primera Asamblea Nacional de Químicos Españoles', *Ión*, 63 (1946), 520–5.
[104] W. E. Garner and D. A. Dowden, the latter from the catalysis laboratory of ICI. In particular, chemist Juan Francisco García de la Banda obtained a Ramsay Fellowship to work in Bristol in 1951–2.
[105] Joaquín Pérez-Pariente, Javier Soria Ruiz, Juan Francisco García de la Banda, 'The Origin of the Institute of Catalysis and Petroleum Chemistry of the Spanish National Research Council and Its Relationship with the Development of Catalysis in Spain', *Catalysis Today*, 259 (2015), 3–8.

1953 he was a corresponding member of the Royal Society Corrosion Committee, and in 1954 he was awarded an honorary doctorate from the *Hannover Technische Hochschule*.[106] Thanks to the re-established network of international connections and the fluid relations with the French government, quantum chemistry reached Spain in the 1950s. Chemist José Ignacio Fernández (1917–2012) stayed in Paris with Raymond Daudel (1920–2006) at the *Centre de chimie théorique de France*, and later at CALTECH with Linus Pauling (1901–94), and at the Mathematical Institute in Oxford with Charles Alfred Coulson (1910–74). His address in 1955 at the University of Valencia on the application of quantum theory to carcinogenic substances epitomised his achievements.[107] Equally, other research schools on spectroscopy, radiochemistry, calorimetry, catalysis and physical chemistry acquired international recognition.[108]

In 1953, the meeting celebrating the 50th anniversary of the RSEFQ attracted more than 600 delegates, among whom were more than 120 foreign scientists representing the most important chemical societies.[109] Analogous with the pre-Civil War cosmopolitan practices, four Nobel Prize winners (Kurt Alder, Otto Hahn, Alexander Todd and Arne Tiselius) were welcomed to Francoist Spain, and the number of papers published in English increased substantially. Also in 1953, Rius Miró (electrochemistry section) and Pascual (nomenclature) attended the IUPAC Conference in Stockholm. Masriera, now a science journalist and freelance chemist outside the new academic system, accompanied the Spanish delegation. Hermann Staudinger's keynote speech was a golden opportunity for Masriera to meet up again with his old ETH mentor. He even reported in the daily press on the lively controversy regarding isomerism and resonance between Pauling and the Russian chemist Yuri Zhdanov (1919–2006) during the conference. In fact, in 1951, Pauling's resonance theory and Ingold's reaction mechanisms in organic chemistry had been condemned by the Soviet Academy of Sciences for their metaphysical, idealist character that was supposedly

---

[106] Felipe Calvo, *Emilio Jimeno Gil. Semblanza de un maestro* (Santander: Amigos de la Cultura Científica, 1984), p. 59.

[107] José Ignacio Fernández, 'Hacia una teoría cuántica de las sustancias que muestran actividad cancerígena. Lección inaugural del curso 1955–56', *Anales de la Universidad de Valencia*, 29(I) (1955), 1–106.

[108] This was the case for Octavio Foz in Valencia (and later in Madrid) and his group working on molecular spectroscopy (Jesús Morcillo), radiochemistry (José Miguel Gamboa), calorimetry (Manuel Colomina), catalysis (Francisco García de la Banda), surface physical chemistry (Juan Llopis), kinetics (María Josefa Molera) and thermodynamics (Andrés Pérez). Lora-Tamayo, *La investigación química española*.

[109] The American Chemical Society, the Chemical Society (British), the *Deutsche Chemische Gesellschaft*, the *Società chimica italiana* and the *Société chimique de France*, among others.

Chemical Diplomacy 161

opposed to dialectical materialism.[110] Masriera has presented the case as a paradigm of the vitality of chemistry at the international level, an example of the growing US–Soviet rivalry and a sign of the supposed 'openness' of the Franco regime.[111]

The UNESCO international conference on science popularisation in Madrid of 1955 also played an important role in terms of foreign legitimation.[112] At the Industrial Engineering School in Madrid, scientific films, UNESCO publications, public lectures and exhibitions and academic papers provided a cosmopolitan atmosphere. This was in the same year that the UN had officially recognised Franco's regime. On 19 October, the inaugural session of the conference took place at the *Instituto de Investigaciones Astronómicas* in the presence of the Minister for Foreign Affairs and the President of the CSIC and former Minister for Education, José Ibáñez Martín.[113] Miguel Catalán, one of the big names of the Silver Age, having returned from the post-war purge, lectured on Elhuyar, the Spanish chemist who had discovered tungsten – again, history of chemistry for the making of professional identity – and was one the Spanish representatives among the delegates of the national associations of science writers.

***

In 1945, Abollado discussed in *Ión* the basis for a chemical autarchy in Spain. He highlighted that only a true training in chemistry, research and technical skills could provide a strong chemical industry as a secure path to autarchy. In his view, national expertise had to prevail over financial, economic leadership.[114] In fact, names such as Albareda, Lora-Tamayo, Rius Miró, Carceller, Abelló and others reshaped the boundaries of training in industrial chemistry and introduced foreign novelties within a context of totalitarian politics. Their obsession with national raw

---

[110] For the context of that controversy, see: Ethan Pollock, *Stalin and the Soviet Science Wars* (Princeton: Princeton University Press, 2006).

[111] Miguel Masriera, 'Glosario científico: Primeras impresiones del Congreso de Estocolmo', *La Vanguardia Española*, 4 August 1953. See also: Agustí Nieto-Galan, 'From Papers to Newspapers: Miguel Masriera (1901–1981) and the Role of Science Popularisation under the Franco Regime', *Science in Context*, 26(3) (2013), 527–49.

[112] Organización de las Naciones Unidad para la Educación, la Ciencia y la Cultura (UNESCO), 'Festival de Ciencias. Madrid. 19 a 22 octubre de 1955. Informe de la Secretaría', *Anales de la Asociación Española para el Progreso de las Ciencias*, 21 (1956), 212–23.

[113] Ibid., p. 214. The Spanish delegates from the *Asociación Española para el Progreso de las Ciencias* (AEPC) to the conference were Rafael Estrada, Manuel Bermejillo and José María Torroja, and those from the national associations of science writers were Julio Rey Pastor, José Antonio Artigas and Miguel Catalán.

[114] Carlos Abollado, 'La industria química y la química industrial (Conclusión)', *Ión*, 45(5) (1945), 245–51, p. 248.

materials did not hinder foreign collaborations with fascist Italy and Nazi Germany and later with the United States and the Allied European countries. Professional contacts abroad soon became powerful diplomatic tools that apparently were a good match with the quantitative and qualitative data on chemical reactions and industrial processes.

The circulation of chemicals and chemists in the Francoist autarchy provides new clues as to the ways in which foreign contacts were used for the legitimation of the regime and how international conferences, industrial projects and educational reforms coexisted with an official rhetoric of inner resistance, foreign suspicion and nationalistic pride. Although in the early 1940s, and especially during the Nazi regime, the concept of a 'new Europe' could have brought Spanish elites, including chemists, to approach the analysis of raw materials at a continental level, the fact is that, after 1945, the Allies' influence forced Franco's regime and its scientific experts to expand their geographical references in material, institutional and political terms.[115]

Although in years following the Second World War the United States established only very limited diplomatic relations with Francoist Spain, the strategic position of the country, its raw materials and military interests soon changed the whole picture.[116] In 1953, Spain and the United States signed an official agreement, but perhaps President Eisenhower's visit to Madrid in 1959 truly epitomised the signs of the new era, in which chemistry benefitted substantially from the new geopolitics and from the establishment of new plans for economic liberalisation that changed the chemical landscape of the country in the 1960s.[117]

---

[115] On the Nazi concept of 'New Europe' as an extension of national autarchy, see: David Brydan, 'Axis Internationalism: Spanish Health Experts and the Nazi "New Europe"', 1939–1945', *Contemporary European History*, 25(2) (2016), 291–311. I am indebted to Santiago Gorsotiza for this reference.

[116] Boris Liedtke, 'Spain and the US, 1945–1975', in Balfour, Preston (eds.), *Spain and the Great Powers*, pp. 229–44.

[117] María Jesús Santesmases, 'Orígenes internacionales de la política científica', in Ana Romero and María Jesús Santesmases (eds.), *Cien años de política científica en España* (Bilbao: Fundación BBVA, 2008), pp. 293–328, 318. See also: Lorenzo Delgado, 'Cooperación cultural y científica en clave política. Crear un clima de opinión favorable para las bases USA en España', in Lorenzo Delgado, María Dolores Elizalde (eds.), *España y Estados Unidos en el siglo XX* (Madrid: CSIC, 2005), 207–44; Antonio Cazorla-Sánchez, *Políticas de la victoria. La consolidación del Nuevo Estado franquista (1938–1953)* (Madrid: Marcial Pons, 2000).

6   Technocratic Progress

Since the First World War, ideas of technological efficiency had gained influence in the European public sphere. American Taylorism and Fordism emphasised the value of 'rationalisation', expertise, productivity and mechanisation as metaphors for efficient organisation, but in practice acted as ways to escape ideological conflicts and class confrontation.[1] Although the shock of the 1929 Wall Street Crash discredited many promises of industrial efficiency, the belief that more science and technology would provide more public good soon gained ground again and remained alive on both continents even after the Second World War. The idea that scientific information provides an objective basis for resolving political disputes has influenced expert decisions in democracies and dictatorships throughout the twentieth century.[2] In the 1970s, political scientist Robert Putnam defined technocrats as 'those who exercise power by virtue of their technical knowledge'.[3] In his view, in many periods and places, the idea that technicians must replace politicians has spread, and hostility against politicians and political institutions has become common in public debates against the openness and equality of democracy. Social and political conflicts have often been misguided or contrived, whereas pragmatism has too often been considered to come before moralism, and technological progress and material productivity before social justice.[4]

We know, for example, that research institutes under Nazi Germany were not islands of freedom in a sea of propaganda, since all benefitted

---

[1] Charles S. Maier, 'Between Taylorism and Technocracy: European Ideologies and the Vision of Industrial Productivity in the 1920s', *Journal of Contemporary History*, 5(2) (1970), 27–61. See also: P. Meiksins, 'The Myth of Technocracy: The Social Philosophy of American Engineers in the 1930s', *History of Political Thought*, 21(3) (2000), 501–23.
[2] Richard Olson, *Scientism and Technocracy in the Twentieth Century: The Legacy of Scientific Management* (London: Lexington Books, 2016), p. xvi.
[3] Robert Putman, 'Elite Transformation in Advanced International Societies', *Comparative Political Studies*, 10 (1977), 383–412.
[4] Olson, *Scientism and Technocracy*, p. xi.

from the regime in financial terms. Many chemists offered their services for their own professional benefit, but formally declared few 'political' interferences and responsibilities.[5] As in other totalitarian contexts, they remained formally 'apolitical', but effectively served the interests of the regime.[6] Nazi experts such as Fritz Todt (1891–1942) – the father of the autobahn system – and Albert Speer (1905–81) – a key organiser of the regime – perceived themselves as non-politicians. In Latin dictatorships such as Mussolini's, Franco's and Salazar's, experts played a crucial role in the stability of the non-democratic political contexts in which these dictatorial regimes could only hold, consolidate and legitimise their power through a changing network of administrators or 'technocrats' that ran the economic modernisation along capitalist lines and totalitarian values. Those technocratic elites also played a key role in the making of corporate, capitalist, regulatory states.[7]

In the United States, at the end of the Second World War, Vannevar Bush's *Science: The Endless Frontier* (1945) emphasised some of the technocratic aspects of science and technology that had a great impact on many other countries.[8] In Bush's words:

In 1939 millions of people were employed in industries which did not even exist at the close of the last war – radio, air conditioning, *rayon and other synthetic fibres, and plastics* are examples of the products of these industries. But these things do not mark the end of progress – they are but the beginning if we make full use of our

---

[5] As historian Mitchell Ash discussed two decades ago: '[...] chemists, too, willingly offered their services to business and the state under Nazism, as they had in the past; they assumed, usually correctly, that they would be able to work with minimal interference if their results were judged to be promising for the war effort'. Mitchell Ash, 'Essay Review: Science, Technology and Higher Education under Nazism', *Isis*, 85 (1995), 458–62, p. 462.

[6] As stated by theoretical sociologist Niklas Luhmann: '[T]hough attempts at political influence on science have plainly been made, these are registered only as temporary "irritations", that tend to disappear sooner or later'. Quoted in: Mitchell Ash, 'Scientific Changes in Germany, 1933, 1945, 1990: Towards a Comparison', *Minerva*, 37 (1999), 329–54, p. 329. Rienhard Siegmund-Schultze, 'The Problem of Antifascist Resistance of "Apolitical" German Scholars', in Monika Renneberg, Mark Walker (eds.), *Science, Technology and National Socialism* (Cambridge: Cambridge University Press. 1994), pp. 312–23. 'Far from political commitment, a common self-perception of individual scientists working under dictatorships was paradoxically that their work had nothing to do with politics'. Amparo Gómez, Brian Palmer, Antonio Canales, 'Science Policy under Democracy and Dictatorship: An Introductory Essay', in Amparo Gómez, Brian Palmer, Antonio Canales (eds.), *Science Policies in Twentieth-Century Dictatorships: Spain, Italy and Argentina* (Farnham: Ashgate, 2015), p. 19.

[7] Paul H. Lewis, *Latin Fascist Elites: The Mussolini, Franco and Salazar Regimes* (London: Preager, 2002). See also: Paul H. Lewis, 'The Spanish Ministerial Elite, 1938–1969', *Comparative Politics*, October (1972), 83–106.

[8] Vannevar Bush, *Science: The Endless Frontier. A Report to the President on a Program for Postwar Scientific Research. July 1945* (Washington, DC: National Science Foundation, reprinted July 1960).

scientific resources. New manufacturing industries can be started and many older industries greatly strengthened and expanded if we continue to study nature's laws and apply new knowledge to practical purposes.[9]

Under a rhetoric of basic, pure science, Bush stressed the potential applications of science for security, health, jobs, standard of living and cultural progress. In fact, in the Cold War context, the production of knowledge often appeared in public as being detached from its moral, political implications. Science supposedly held objectivity and reason, whereas politics was trapped in subjectivity and sometimes even irrationality; therefore, the scientist could be separated from the citizen, the former creating new knowledge and the latter applying it, often wrongly.[10] As Kurt Vonnegut denounced in the case of Irving Langmuir, separation between means and ends has also become a standard feature of professionalism in many modern societies.

Historians agree that, in the 1960s, Franco's cabinets worked intensively in favour of what Tony Judt called 'the depolitisation of the culture'.[11] It was an ambitious project of depolitisation of the masses through radical Spanish nationalism, kitsch populism, strict Catholic morality and hard-line censorship,[12] which was often complemented by elite claims of 'technocratic neutrality'. At the heart of the Cold War logic, in a context of forceful internal stability, Franco's Spain started a pattern of economic modernisation and growth in which a skilled but politically docile workforce was obviously more than welcome by the elites, which implemented several educational reforms in accordance with a growing American influence and increasing internationalisation.[13]

In 1959, the same year in which the Valley of the Fallen (*Valle de los Caídos*), the symbol of Franco's victory over the Republic, was completed, the government launched a new stabilisation plan aiming to reduce inflation and open the country up to trade and foreign investment. Soon afterwards came admission to the International Monetary Fund (IMF), the General Agreement on Tariffs and Trade (GATT) and the Organisation for Economic Co-operation and Development (OECD). Around 1965, Spain was not 'a developing nation' anymore in terms of the UN criteria. Tourism, emigrants' remittances and a steadily growing

---

[9] Ibid., p. 10 (emphasis added).
[10] On the concept of the citizen scientist, see: Mark Walker, *German National Socialism and the Quest for Nuclear Power, 1939–1949* (Cambridge: Cambridge University Press, 1989), p. 4.
[11] Tony Judt, *Postwar: A History of Europe since 1945* (London: Penguin Books, 2005), p. 517.
[12] Borja de Riquer, *Historia de España (IX): La dictadura de Franco* (Barcelona/Madrid: Crítica/Marcial Pons, 2010), pp. 330–44.
[13] Carolyn P. Boyd, *Historia Patria: Politics, History and National Identity in Spain, 1875–1975* (Princeton: Princeton University Press, 1997), p. 273.

GDP did the rest, together with major industrial growth, which obviously included the chemical industry. The technocratic educational reforms of the 1960s focused on the economic growth and industrial needs of 'developmentism'.[14] This was an ideal terrain to provide fresh technical expertise for the new chemical corporations.

In the mind of Laureano López Rodó (1920–2000) – Minister for Economic Growth and member of Opus Dei – the technocratic state of the Development Plans provided the basis for economic growth and further monarchic restoration as an ideal transition after Franco's death. Although López Rodó's plan was far from any real move towards a liberal democracy, he strengthened the determinism between economic growth and political change.[15] López Rodó's thesis reinforced in practice modernisation theories that often consider economic growth as a teleological precondition for democracy. In that framework, the 1960s technocracy and *'desarrollismo'* have often been presented as necessary precedents for further democratisation of Spain in the 1970s after Franco's death, and as subtle ways by which to describe the Second Spanish Republic as a 'failure' in economic (and political) terms, so Franco's coup d'état became a sort of necessary evil.[16] Rather than stressing the regime's cruel repression, its human suffering and, most importantly, its totalitarian nature, standard accounts of Franco's dictatorship (this could be extrapolated in part to other non-liberal periods of the century) tend to prioritise the political instability that supposedly caused the Civil War and the 1960s technocratic economic growth as further preconditions for the establishment of a Western-like liberal democracy in the Cold War context.[17]

It is therefore within that framework that the chemical industry and the chemists who ran that technological system spread an 'apolitical', 'technocratic' discourse in public addresses, research projects, publications and personal contacts. In the context of the so-called 'technocratic' cabinets of that time, several chemists undertook major academic reforms and searched for new international support, all in a strong alliance with the new corporate power of the chemical industry that was disseminated throughout the country. In fact, in the 1960s, chemistry and technocracy can be approached through the prominent political role played by chemist Manuel Lora-Tamayo as Franco's minister, expert and educational

---

[14] Carlos París, 'La pretensión de una universidad tecnocrática', in Juan José Carreras, Miguel Ángel Ruiz Carnicer (eds.), *La Universidad española bajo el régimen de Franco (1939–1975)* (Zaragoza: Instituto Fernando El Católico, 1991), pp. 437–55.
[15] Laureano López Rodó, *Memorias* (Barcelona: Plaza y Janés, 1990).
[16] Carles Sirera, 'Neglecting the Nineteenth Century: Democracy, the Consensus Trap and the Modernization Theory in Spain', *History of the Human Sciences*, 28(3) (2015), 51–67.
[17] Michael Richards, *A Time of Silence (1936–45): Civil War and the Culture of Repression in Franco's Spain* (Cambridge: Cambridge University Press, 1998).

reformer. It is therefore mainly through Lora-Tamayo and some of the strategies of his closest co-workers that I shall trace back the technocratic nature of chemistry in the following sections of this chapter.

'Neutral' Expertise

Although several German chemical companies were responsible for mass murder and slave labour during the Nazi period, they managed to ensure the continuity of their industries after the war, emphasising their 'non-ideological', 'managerial' and technocratic cooperation in the 1960s.[18] We cannot detach the construction of the I. G. Farben Buma IV plant in Auschwitz for the production of synthetic rubber, the Leuna works, Agfa film production in Wolfen and plants in the Halle-Merseburg region, among many others, from Nazi mass exterminations in concentration camps.[19] Under a public rhetoric of professional independence, unique leadership and powerful action, many German chemists ended up consciously or unconsciously supporting Nazi policies.[20] However, once the nightmare of the war faded away, the golden years of the German chemistry-based industry, with the massive production of artificial dyes and drugs, appeared again as an ideal stage for bypassing the uncomfortable war years. In fact, after the Second World War, numerous chemical companies substituted the old I. G. Farben intelligentsia with young chemists and managers. They constructed their own biased corporate histories to legitimate their status in the Cold War[21] and strongly influenced Franco's Spain in the 1960s.

Technocracy and apolitical statements were at the core of the rhetoric of the regime's academic reforms.[22] In the 1950s, neutral statements by chemist Lora-Tamayo had become useful tools for the establishment of

---

[18] Georg Wagner-Kyora, 'Continuities in the Identity Construction of Industrial Chemists, 1949–1970', *German Studies Review*, 29(3) (2006), 611–19.

[19] Bernd C. Wagner, *IG Auschwitz. Zwangsarbeit und Vernichtung von Häftlingen des Lagers Monowitz 1941–1945* (Munich: K. G. Saur Verlag, 2000).

[20] Mark Walker, *Nazi Science: Myth, Truth and the German Atomic Bomb* (Cambridge, MA: Perseus, 1995), pp. 269–71.

[21] As Georg Wagner-Kyora emphasizes, 'Identity construction [...] is a virtual political aspect of everyday acting and cultural self-imagining [...] it defines a kind of communicative social discourse, which derives not only from images but also from individual and collective memories'. Wagner-Kyora, 'Continuities', p. 613.

[22] The Nazi Ministry for Education and Science described the German scholarly tradition as objective and conforming to the scholarly ethos to serve the subject, not political interest: science as a neutral, objective and (to some extent) international activity. This trend continued after 1945. For many German professors after 1945 – we should not forget the long-standing influence of the German academic world on the Spanish system after the Second World War – 'to be proficient in their field of study remained above the petty squabbles of politics'. Notker Hammerstein, 'National Socialism and the German Universities', *History of Universities*, 18(1) (2003), 170–88, pp. 175, 177.

international relations, mainly with anti-communist West Germany, and later with France and the United States (see Chapter 5). Lora-Tamayo strongly criticised the negative influence of Soviet ideology on the development of Soviet science for its lack of freedom of scientific thought. However, he did not point out any contradictions in the adaptation of scientific research, which was considered 'neutral', in Franco's dictatorship.[23] In the late 1960s, referring to Albert Wohlsteller, one of the well-known experts on Cold War defence policies and theories of deterrence in the United States, Lora-Tamayo considered that, 'Since they are only interested in the clarification of truth, scientists are free from dishonesty and from the ambiguous motives of traditional actors in the political arena'.[24] That kind of technocratic account, often combined with scientific cosmopolitanism, certainly contributed to the international legitimisation of the regime.

In 1962, Lora-Tamayo entered Franco's government as Minister for Education and Science (Figure 6.1). The new cabinet was the result of a reshuffle after public controversy arose at a conference held in Munich in June 1962. What the Francoist media described as the *Contubernio de Múnich* (Munich conspiracy) was an attempt by political opposition members to force the regime to adopt democratic reforms.[25] The response was a programme of economic modernisation 'without democracy', which was soon labelled as 'technocratic'. In a context of Cold War escalation – with the Cuban Missile Crisis and the building of the Berlin Wall – the anti-communist flag of Franco's regime efficiently combined the modernisation of the Spanish economy with political repression.

The 1962 technocratic cabinet into which Lora-Tamayo entered had the mark of a younger generation of Opus Dei members – Albareda died in 1966 – which has attracted huge historical interest. It is obviously hard to explain how a fundamentalist, right-wing conservative organisation, reactionary to many aspects of the modern world, could lead the modernisation of Spain in the 1960s, a period that economic historians consider to be the point at which the country's industrialisation really took off.[26] As discussed in Chapter 4, a certain Weberian Protestant-like ethic was certainly involved. Opus Dei doctrine emphasised the importance of the professional 'calling' and success in professional life. Moreover, the

---

[23] Manuel Lora-Tamayo, *Moral profesional del investigador* (Madrid: CSIC, 1953).
[24] Manuel Lora-Tamayo, 'La investigación al servicio de la defensa', *Arbor*, 281 (1969), 5–19, p. 18.
[25] Juan Carlos Laviana (ed.), *1962: Del Contubernio de Múnich a la huelga minera* (Madrid: Unidad Editorial, 2006).
[26] José V. Casanova, 'The Opus Dei Ethic, the Technocrats and the Modernization of Spain', *Social Science Information*, 22(1) (1983), 27–50; Joan Estruch, *Saints and Schemers: Opus Dei and Its Paradoxes* (Oxford: Oxford University Press, 1995).

'Neutral' Expertise 169

Figure 6.1 Manuel Lora-Tamayo as new Minister of Education and Science at the *Palacio del Pardo* in Madrid, 11 July 1962.
Reproduced with permission of the *Archivo José F. Demaría 'Campúa'*.

recovery of the Spanish golden age of civilisation and its early-modern Catholic Empire had to be achieved pragmatically through Europeanisation. But beyond that grandiloquent rhetoric, economic progress left the totalitarian structures of the regime intact, in which chemistry and the chemical industry became an ideal example of technocracy. The 'modernising' elite included Opus Dei ministers, but also industrialists and the financial powers. It aimed to avoid the contradiction between economic growth and the 'traditional' social and political structure of the country. Both the totalitarian ambitions of the 1940s and early 1950s and the new technocracy of the 1960s rejected any liberal separation between the state and civil society. Modernity had to be confined to the economy and, why not, to the 'neutral' use of chemistry.[27] That was the context in

[27] Casanova, 'The Opus Dei Ethic', pp. 37–8.

which Lora-Tamayo worked hand in hand with those in Franco's cabinet that Tony Judt called the 'modernising crypto monks' of Opus Dei.[28] In Lora-Tamayo's view, the 'political neutrality' of chemists supposedly freed them from ethical commitments in their research work and further applications.[29]

In 1963, Franco's cabinet authorised the executions of the communist leader Julián Grimau (1911–63) and anarchists Joaquin Delgado and Francisco Granados, and created the *Tribunal de Orden Público* (TOP), a repressive special court for public order. In 1964, the campaign *'25 años de paz'* (25 years of peace) commemorated the 25th anniversary of Franco's victory in the Civil War.[30] Perhaps not by chance, that year also coincided with the celebration of the 200th anniversary of the Barcelona Royal Academy of Natural Sciences and Arts (RACAB), which was founded in 1764 by a group of local natural philosophers.[31] Now, José Pascual, a prominent organic chemist and close friend of Minister Lora-Tamayo, was keen to commemorate the event in conjunction with official political propaganda. In practice, it was 'neutral' chemistry that therefore had to serve Pascual's ambitions at the local level.[32] His close contacts with Barcelona mayor José María de Porcioles (1904–93) and with the local political and economic elites became crucial for the building of a new CSIC research institute on the periphery of the city, where the new university campus grew spectacularly.[33]

Since the 1940s, Pascual had developed an ambitious applied chemistry programme for technical, industrial assessments through the *Instituto de investigaciones técnicas* (IIT), which hosted sections focusing on electrotechnics, industrial chemistry, metallurgy and metallography, cement and concrete, industrial physics, tanning and cellulose. He was also in charge of an organic chemistry section at the *Instituto Alonso Barba* at

---

[28] Judt, *Postwar*, p. 518.
[29] In the same vein he used to express his criticism of the ideological content of *Arbor*, the official journal of the CSIC, which, in his view, seldom left room for 'real, objective' scientific papers – by contrast, for instance, with the more 'objective' journal *Revista de Ciencia aplicada*. Lora-Tamayo, *Lo que yo he conocido*, p. 267.
[30] Manuel Fraga Iribarne (1922–2012), Lora-Tamayo's colleague in cabinet and Minister for Information and Tourism, was in charge of the public celebrations and further legitimation of the dictatorship.
[31] Agustí Nieto-Galan, Antoni Roca-Rosell (eds.), *La Reial Acadèmia de Ciències i Arts de Barcelona. Història, Ciència i Societat* (Barcelona: RACAB/IEC, 2000).
[32] *Reial Acadèmia de Ciències i Arts de Barcelona*. Archive. José Pascual's Correspondence. Barcelona.
[33] *Universidad de Navarra. Archivo General.* Fondo José María Albareda. Correspondence Albareda-Pascual. See also: Lluís Calvo (ed.), *El CSIC en Cataluña (1942–2012): Siete décadas de investigación científica* (Barcelona: CSIC, 2012).

the CSIC, which, in 1952, evolved into a department of applied organic chemistry.[34] In the 1960s, Pascual took active part in the 32nd International Conference of Industrial Chemistry,[35] in a week devoted to the promotion of textile technology, as well as in a campaign for the support of national industry in 1964. In 1967, a new research centre, the *Centro de Investigación y Desarrollo* opened its doors in Barcelona as a result of Pascual's long-standing determination to create it. It included chemical applications to textiles, pharmacology and tanning.[36] In 1969, Pascual also joined the team of Banco Urquijo to assess the fuel needs of Catalan industry, and they reported their results to the Ministry for Industry. In that context, Pascual's expertise aimed to renew the old IIT, which he associated with a long-term local tradition of technical education that originated in the 1910s, followed by engineer Esteban Terradas and Minister Joaquín Planell's projects in the autarchic years. Pascual believed that, in the regime's technocratic context of applied chemistry, industry and research centres had to benefit each other.[37]

In a context of increasing international recognition and considerable industrial and demographic growth, the 1964 celebration at the RACAB also coincided with the 25th anniversary of the foundation of the CSIC. For the latter event, a crowd of 100 invited foreign experts and six Nobel Prize winners came together in Madrid and were welcomed by Franco in person.[38] Among them were names such as Peter Debye (1884–1966), winner of the Nobel Prize for Chemistry in 1936 for his work on the diffraction of X-rays and electrons in gases,[39] Severo Ochoa (1905–93), an exiled Spanish biochemist working in the United States who won the Nobel Prize for Physiology or Medicine in 1959 for the discovery of the mechanisms of ribonucleic acid and deoxyribonucleic acid,[40] and Kurt Alder (1902–58), the Nobel Prize winner for Chemistry in 1950 for his discovery of diene synthesis, with whom Lora-Tamayo had established a close scientific collaboration.

---

[34] Agustí Nieto-Galan, Joaquim Sales, 'Josep Pascual Vila (1895–1979): Una aproximación biográfica', *Anales de Química*, 113(1) (2017), 10–47.
[35] *Diario del XXXII Congreso Internacional de Química Industrial* (Barcelona, 1960).
[36] Nieto-Galan, Sales, 'Josep Pascual Vila', pp. 40–6.
[37] *Universidad de Navarra. Archivo General*. Fondo José María Albareda. Correspondence Albareda-Pascual.
[38] Lora-Tamayo, *Lo que yo he conocido*, p. 265.
[39] Controversy about Debye's Nazi past has been discussed in Jurrie Reiding, 'Peter Debye: Nazi Collaborator or Secret Opponent', *Ambix*, 57(3) (2010), 275–300.
[40] María Jesús Santesmases, 'Severo Ochoa and the Biomedical Sciences in Spain under Franco, 1959–1975', *Isis*, 91 (2000), 706–34.

In 1965, during the discussion of a new education bill at the Spanish *Cortes*, Lora-Tamayo once again emphasised the neutral and objective nature of his project: 'In the University as an institution, as in the Church, the Army or the Judiciary, there cannot be groups or political factions [...] everyone should leave their personal ideals at the door [...]'. In fact, the entire reform of the universities hinged on a rhetoric of apoliticism.[41] Although Lora-Tamayo accepted a certain level of freedom of expression from every chair – a controversial issue in Spanish universities since the last decades of the nineteenth century[42] – he insisted on the exclusion of any political agitation or proselytism and described himself as a useful reformer, belonging neither to the *Falange* nor Opus Dei, nor indeed to any other political factions of the regime.[43] Lora-Tamayo brought his reputation for being a prestigious scientist – well connected internationally, but now acting as a chemist-reformer – to being an agent of technocratic modernisation from inside the regime.[44]

Spanish economic and demographic growth contributed to the rise in social demand for better education from the emerging middle classes, and this had a great impact on the universities. Student university figures almost doubled from 35,000 at the end of the 1940s to more than 60,000 by the early 1960s, and soared to 200,000 students by the early 1970s.[45] In response, Lora-Tamayo's university reforms aimed to create new teaching staff, new faculties in the provinces, more public funds for research, closer collaboration with the CSIC and new private universities under the control of the Catholic Church.[46] However, the reforms were met with some resistance. University full professors (*catedráticos*) opposed the official recognition of private Catholic universities, whereas other

---

[41] Manuel Lora-Tamayo, 'Estructura de las facultades universitarias y su profesorado', *Revista de Educación*, 174 (1965), 17–26, pp. 21–2. See also: Robert Proctor, *Value-Free Science? Purity and Power in Modern Knowledge* (Cambridge, MA: Harvard University Press, 1991).

[42] See, for instance: Agustí Nieto-Galan, 'A Republican Natural History in Spain around 1900: Odón de Buen (1863–1945) and His Audiences', *Historical Studies in the Natural Sciences*, 42(3) (2012), 159–89.

[43] But in numerous public events, Lora-Tamayo praised the *Caudillo*. Ángel Bayod (ed.), *Franco visto por sus ministros* (Barcelona: Planeta, 1983), p. 127.

[44] International publications from Lora-Tamayo's school were remarkable. See, for instance: Manuel Lora-Tamayo, Ramón Madroñero, *1,4-Cycloaddition Reactions: Imino Compounds as Dienophiles* (New York: Academic Pres, 1965); Manuel Lora-Tamayo, Ramón Madroñero, *Dienes Containing Two Nitrogen Atoms in the Conjugated Systems* (New York: Academic Press, 1966).

[45] Agustí Nieto-Galan, 'Reform and Repression: Manuel Lora-Tamayo and the Spanish University in the 1960s', in Ana Simões, Kostas Gavroglu, Maria Paula Diogo (eds.), *Sciences in the Universities of Europe, 19th and 20th Centuries: Academic Landscapes* (Dordrecht: Springer, 2015), pp. 159–74.

[46] In the opening session of the 1962–63 academic year at the University of Madrid.

Catholic groups perceived the reforms more positively.[47] In spite of that controversy, Lora-Tamayo attempted to reduce the power of the *catedráticos*, reorganised faculties and created new departments and the new category of *profesor agregado* (a sort of senior lecturer at an intermediate level between *catedráticos* and *ayudantes* [assistant lecturers] and *adjuntos* [pre-tenured lecturers]). He also promoted full-time devotion to university tasks, which was still an exception in the Spanish academic system.

After decades of CSIC hegemony, the reforms also provided stable funds for scientific research in universities, issued in part from the budget of the Development Plan (1964–7). The project also included official recognition of the Opus Dei *Universidad de Navarra*, with Albareda becoming its rector in 1962, and renewed the earlier recognition of the *Universidad Pontificia de Salamanca* (1940) and the Jesuit *Colegio de Estudios Superiores de Deusto* (1953). Lora-Tamayo's ideological affinity with right-wing Catholic organisations and his close friendship with Albareda eased the final decision.[48] In a similar line, in 1965, the IQS and its degree in chemical engineering achieved official state recognition, a tangible result of Jesuits' long-standing attempts to gain more influence in the educational policies of the regime.[49]

Lora-Tamayo strikingly perceived all of these controversial, tough decisions from the cabinet as technocratic, a sort of logical extrapolation of his chemistry skills as applied to educational matters under a supposed banner of objectivity and neutrality. Many years later, already well into the democratic period after Franco's death, Lora-Tamayo was asked about his feelings during his time as Minister for Education (1962–8), and he answered straight away:

> Bad. I had a very bad time. I am not a politician. I was Minister because it was at that time a moral obligation, but I also resigned when I considered it was morally necessary [...] When I was asked, I first thought of not accepting the position – I had many reasons not to. [...] *I am completely out of politics. I have only been a researcher.*[50]

---

[47] In collaboration with his pupil, the chemist Rafael Pérez Alvárez-Osorio, he opened a period of debate among *catedráticos*, the result being quite disappointing because of the reluctance to accept any major change to the totalitarian system of the 1943 Act. Lora-Tamayo, *Lo que yo he conocido*, p. 184.

[48] Lora-Tamayo, *Lo que yo he conocido*, p. 202.

[49] The IQS survived the stronger hegemony of Opus Dei in the CSIC and in the universities. Not by chance, IQS recognition by Lora-Tamayo occurred some years later, when in 1981 he received an honorary doctorate from the Institute. Lluís Victori, *L'Institut Químic de Sarrià, 1905–2005* (Barcelona: IQS, 2005); Lora-Tamayo, *Lo que yo he conocido*, p. 164.

[50] *Ya*, 4 March 1996 (emphasis added). Also quoted in Nieto-Galan, 'Reform and Repression', pp. 159–74, 170.

These words endorsed the idea of a 'neutral' chemistry, very much focused on applied projects, institutional growth and stronger collaboration with industry, but detached from real political commitments. Paradoxically, Pascual's influence on Barcelona's local power and Lora-Tamayo's lofty status as Minister for Education and Science appeared formally to stand apart from the practice of chemistry and the technocratic values they associated with it. Similarly, in 1965, Ángel Vián, from his chair in industrial chemistry in Madrid, discussed the weaknesses of the Spanish chemical industry and the need to improve technical training as a national goal in the context of the Development Plan. In his view, outmoded science faculties, corporatism in engineering schools and foreign know-how and patents had to be replaced by the technocratic plans of the chemist–minister Lora-Tamayo. Bypassing economic and obviously political factors, Vián suggested the 'efficient', 'scientific', 'technical' location of new chemical companies.[51]

Albareda's death in 1966 coincided with large university student protests as signs of liberal dissent (see Chapter 7) and also with the scandal of the Palomares atomic bombs, which jeopardised diplomatic relations with the Unites States and challenged the ability of the censorship apparatus of the regime to control public opinion.[52] In spite of those early symptoms of political instability, public homage to Albareda, to his work at the CSIC and to the international reputation of his research school on soil and agricultural chemistry was given from all over the country. When social agitation and student turmoil forced Lora-Tamayo's resignation from the ministry in 1968, his organic chemistry research school was presented as being more robust than ever.[53] In 1969,

---

[51] Ángel Vián, 'La industria química y el desarrollo económico', *Química e Industria*, 12(2) (1965), 37–47; Ángel Vián, 'Tendencia de la industria química mundial y su reflejo en la española', *Química e Industria*, 12(4) (1965), 103–4.

[52] On January 1966, two US military planes crashed in the sky over the town of Palomares (Almería). Three plutonium nuclear bombs fell on land near the coast and leaked their contents and a fourth fell into the sea. Clara Florensa, 'James Bond, Pepsi-Cola y el accidente nuclear de Palomares (1966)', in Lino Camprubí, Xavier Roqué and Francisco Sáez de Adana (eds.), *De la Guerra Fría al calentamiento global. Estados Unidos, España y el nuevo orden científico mundial*. (Madrid: La Catarata, 2018), pp. 17–38.

[53] Since the late 1940s, it had published important papers in international journals such as *Chemische Berichte*, *Nature*, *Biochemistry Journal*, *Biochimica et Biophysica Acta*, *Journal of the Chemical Society*, *Tetrahedron Letters*, *Chemistry and Industry*, *Journal of Medical Chemistry*, *Journal of Heterocyclic Chemistry* and *Bulletin de la Société chimique de France*. Lora-Tamayo's research school included the following names: Antonio Alemany, Gregorio Alonso, Gonzalo Baluja, Manuel Bernabé, José Calderón, Carlos Corral, Manuel Dabrio, Carlos Elorriaga, Francisco Fariña, Edilberto Fernández Álvarez, José A. Fuentes, Mercedes Fuertes, Guillermo García Muñoz, Vicente Gómez, Antonio González, Ramón Madroñero, Ángel Martín Municio, Fernando Martín Panizo, María Victoria Martín Ramos, Roberto Martínez Utrillo, M. Ángel Murado, Ofelia Nieto López, Rafael Pérez

Lora-Tamayo presented Spanish science as the result of spectacular progress, again through the shift towards applied science and the creation of a huge number of research institutes.[54] All of this occurred in combination with the growing presence of Spanish scientists in international scientific organisations, publications in prestigious international journals, frequent exchanges of researchers with foreign groups and the CSIC's growing international joint ventures.

Cold War Allies

The events of 1953 led to a triple US–Spain official agreement, plus another secret one, with substantial military and economic aid. In exchange for the establishment of US military bases in its territory, Spain received $600 million in military aid, $600 million in economic aid, access to $1 billion in private US credits in the next ten years, plus other foreign investments.[55] The US agreement also had a scientific dimension. Late-1940s' scientific contact with the United States and the formal agreement of 1953 created a bilateral collaboration that soon reached Spanish chemists, with a rhetoric of basic, pure research from one side and applied science, often as masked military projects, on the other. Political support for the regime brought together training programmes and funds for grants and research projects. Vannevar Bush had clearly stated that the promotion of basic research had supposedly brought liberal values to a devastated post-war Europe,[56] but in Francoist Spain, the anti-communist issue halted any attempt at political liberalisation.[57]

Álvarez-Osorio, Mariano Pinar, Joaquín del Río, Benjamín Rodríguez, Alberto Sánchez, Mamfredo Stud Schulter, Serafín Valverde and Salvador Vega. Manuel Lora-Tamayo, *Sucinta historia de un instituto de investigación. Discurso leído en el acto de su recepción como académico de honor por _____ y contestación de Antonio Cortés Lladó, el día 24 de mayo de 1973* (Seville: Real Academia de Medicina de Sevilla, 1973).

[54] In his view, the late 1960s represented the robust state of the art of chemical fields such as catalysis, colour, enzyme chemistry, biochemistry, spectroscopy, chemical engineering, macromolecules and plastics, organic reaction mechanisms, metallurgy, radiochemistry and medical chemistry, thermochemistry and soil chemistry. Lora-Tamayo, *Un clima para la ciencia* (Madrid: Gredos, 1969).

[55] Boris Liedtke, 'Spain and the US, 1945–1975', in Sebastian Balfour and Paul Preson (eds.), *Spain and the Great Powers in the Twentieth Century* (London: Routledge, 1999), pp. 229–44; Ángel Viñas, *Los pactos secretos de Franco con Estados Unidos: Bases, ayuda económica, recortes de soberanía* (Barcelona: Grijalbo, 1981).

[56] Vannevar Bush, *Science the Endless Frontier*; John Krige, *American Hegemony and the Postwar Reconstruction of Science in Europe* (Cambridge, MA: MIT Press, 2006), pp. 9–11.

[57] In the early 1950s, Pascual had already used in a public address several British and American examples to discuss the nature of scientific research. José Pascual, *La investigación científica y sus diferentes clases* (Barcelona: CSIC. Delegación de Barcelona, 1953), p. 23.

In 1956, John T. Reid, the cultural attaché of the American Embassy in Madrid, contacted Miguel Masriera to organise a series of lectures[58] that would cover American theatre and poetry and foreign American policy, but also the modern use of nuclear power– in tune with the 'Atoms for Peace' campaign.[59] In April 1956, the programme continued though the *Coloquios Íntimos de Estudios Norteamericanos* and brought together several chemists: Moles's old pupil, Tomás Batuecas,[60] who had managed to remain in the academic system after the Civil War; the eminent quantum chemist José Ignacio Fernández; physical chemist José Ibarz, one of Jimeno's pupils who at that time occupied a chair at the Faculty of Science of the University of Barcelona; and Masriera, again acting as a science journalist.[61]

In the early 1960s, American funds supported new chemistry research and progressively included Spain in the 'instrumental revolution' – in other words, in the profound changes that chemistry labs and practices were undergoing at that time with the introduction of infrared and ultraviolet spectroscopy, nuclear magnetic resonance, mass spectrometry and X-ray diffraction (Figure 6.2).[62] Members of Pascual's school of organic chemistry enjoyed new opportunities to purchase scientific instruments and laboratory equipment at the university and to improve their training abroad.[63] Funds came from the CSIC (where Pascual had an influential position), the Juan March Foundation, private corporations such as Sandoz, the US government and the so-called *Comité Conjunto Hispano Americano*. There were other cases of collaboration with the US Army on surface chemistry[64] and thermochemistry.[65] Josep Font, one of

---

[58] Although processes of coproduction between the United States and Europe during the Cold War period have been widely studied, the role of peripheral countries still needs further investigation. John Krige's highly regarded book *American Hegemony*, for instance, barely mentions other European countries beyond the big three. John Krige, *American Hegemony*.

[59] John Krige, 'Atoms for Peace, Scientific Internationalism, and Scientific Intelligence', *Osiris*, 21 (2006), 161–81.

[60] On the biography of Tomás Batuecas, see: J. I. Fernández Alonso, 'Vida universitaria: Biografía del profesor T. Batuecas'. *Nova Acta Científica Compostelana. Bioloxía*, 26 (1964), 163–83.

[61] Miguel Masriera Archive. Box 14, R.7, Correspondence, 23 March 1956. *Institut d'Estudis Catalans*. Barcelona.

[62] Peter Morris (ed.), *From Classical to Modern Chemistry: The Instrumental Revolution* (London: RSC/Science Museum/CHF, 2002).

[63] Josep Castells in Manchester, Manuel Ballester in Harvard and Félix Serratosa at MIT in Boston. On Félix Serratosa, see: Pere Bonnín, *Fèlix Serratosa* (Barcelona: Fundació Catalana per a la Recerca, 1995); Félix Serratosa, '¿Qué investiga Vd? Carta abierta a D. Pedro Rocamora, Director de Arbor', *Arbor*, 295–6 (1970), 5–19.

[64] In the 1960s, trained in Cambridge and at King's College London in surface chemistry, Juan Llopis, a pupil of Rius Miró, obtained research funds from the US Air Force. Lora-Tamayo, *La investigación química*, p. 230

[65] After his training at the National Bureau of Standards in Washington, DC – a reference centre for Spanish chemists such as Emilio Jimeno – and his later work on the

Cold War Allies 177

Figure 6.2 The US ambassador, Robert F. Woodward (third from the left), visits the *Instituto Químico de Sarriá* in Barcelona to supervise the use of American funds for the acquisition of new chemical instruments, *La Vanguardia Española*, 20 November 1963.
Pérez de Rozas. Courtesy of the *Arxiu Fotogràfic de Barcelona*.

Pascual's pupils, stated in a recent paper that, 'in spite of all the ideological constraints of the dictatorial system', the US connection worked very well.[66] In 1960, American aid and the Juan March Foundation funded a new industrial lab at the IQS, and the International Agency for Development allocated $100,000 for the acquisition of new scientific instruments.[67] For that special occasion, on 16 May 1960, Franco visited the IQS in person.[68] Cheered by students and by the political, religious and academic authorities, the *Caudillo* visited the new pilot chemical plant. The director of the IQS, Pedro Ferrer Pi, offered some words to

---

thermochemistry of alkylbenzoic acids, Manuel Colomina obtained a research contract with the US Army in 1959, in collaboration with Pérez-Osorio, one of the most distinguished members of Lora-Tamayo's research school. Ibid., p. 241.

[66] Josep Font, 'Química 1950–2000', *Anales de Química*, 111(1) (2015), 21–4, p. 21. Pascual's school played a significant role in the organization of the International Conference of Industrial Chemistry from 23 to 30 October that same year in Barcelona.

[67] Nuria Puig, Santiago López, *Ciencia e Industria en España. El Instituto Químico de Sarriá, 1916–1992* (Barcelona: IQS, 1992), pp. 132–4.

[68] *ABC*, 17 May 1960, pp. 47–8.

the audience, and Franco thanked the Society of Jesus for its 'patriotic' contribution to the progress of chemistry and its strong alliance with chemical companies.[69] Franco's visit seemed to have borne fruit for Jesuit chemistry, as a gas chromatography apparatus arrived in 1961. In the years to come, from 1964 to 1968, the budget for the *Planes de Desarrollo* allocated new research funds to the IQS, plus an IBM 1130 computer.[70]

In 1967, the official opening of the PJC research institute in Barcelona, under Lora-Tamayo's ministry, also received American funding for several chemical projects.[71] In the period of 1966–9, Pascual developed his research in terpenic alcohols (methols) through a contract with the US Department of Agriculture, which produced nine PhDs and 12 papers.[72] Thanks to these foreign funds, and in contrast with the endogenic tradition of the publication policy at the CSIC,[73] Pascual and his group had papers published in prestigious international journals,[74] members of Lora-Tamayo's school belonged to reputable international scientific boards and other Spanish chemists had a voice in international organisations.[75]

In addition, foreign relations with France and de Gaulle's cabinet also had their chemical side. On 16 November 1962, Lora-Tamayo was awarded an honorary doctorate from the Sorbonne in Paris. In the context of the Algerian War, and despite a critical reception by French students, distinguished Gaullist figures, such as Louis Jacquinot, Minister of State, and Gaston Monnerville, the president of the Senate, chaired the ceremony at the amphitheatre of the Sorbonne. The opposition of the rector, the historian Jean Sarrailh, was counterbalanced by

[69] Ibid.
[70] 1964: 960,000 pesetas, 1965: 3,300,000 pesetas, 1966: 2,500,000 pesetas, 1967: 4,500,000 pesetas, 1968: 3,100,000 pesetas. Puig, López, *Ciencia e Industria en España*, pp. 132–4.
[71] *ABC*, 30 September 1967.
[72] Josep Castells, Félix Serratosa (eds.), *Obra científica del Profesor Dr. José Pascual Vila (1895–1979). Publicación subvencionada por la Fundación 'Juan March'* (Barcelona: EUNIBAR, 1982), p. xiii.
[73] Antoni Malet, 'Las primeras décadas del CSIC: Investigación y ciencia para el franquismo', in Ana Romero and María Jesús Santesmases (eds.), *Cien años de política científica en España* (Madrid: Fundación BBVA, 2008), pp. 211–56.
[74] *Nature* (1948, 1954), *Chemische Berichte* (1952), *Journal of Chemical Society* (1958), *Bulletin de la Société chimique de France* (1960), *Tetrahedron Letters* (1962) and *Journal of Organic Chemistry* (1966).
[75] In fields such as soil chemistry, colour chemistry, electrochemistry, analytical chemistry, food chemistry, polymer chemistry, etc. Lora-Tamayo, *Un clima para la ciencia*, p. 138; Lorenzo Delgado, 'Cooperación cultural y científica en clave política. Crear un clima de opinión favorable para las bases USA en España', in Lorenzo Delgado, María Dolores Elizalde (eds.), *España y Estados Unidos en el siglo XX* (Madrid: CSIC, 2005), pp. 207–44.

some of Lora-Tamayo's 'friends' in the government body,[76] particularly biochemist Jean Roche (1917–2006), who succeeded Sarrailh as rector.[77]

In 1963, the invitation of Adolf Butenandt (1903–95) to Spain became a well-publicised public event.[78] Butenandt had won the Nobel Prize for Chemistry (1939), and he was a prestigious figure in steroid chemistry (sex hormones), as well as the president of the Max-Planck Society. The Spanish considered him an ideal candidate for further scientific collaboration and political propaganda. Butenandt received an honorary doctorate from the University of Madrid and honorary membership of the CSIC. During the so-called 'Butenandt Festival', he visited Madrid and Seville, and the Spanish authorities were particularly keen to arrange an interview with Franco (Figure 6.3). That same day, the *Caudillo* hosted former German Chancellor Franz von Papen (1879–1969), who, in the 1930s, had played an important role in endorsing Hitler's rise to power. In a formal session chaired by Lora-Tamayo, Butenandt lectured at the CSIC on the German research policies of the Max-Planck Society and on potential collaboration with Spain. He later travelled with Albareda to Seville to visit the university, the *Instituto de la Grasa* – a key PJC research institution for the 'applied chemistry' project – as well as the soil chemistry station *El Cuarto*.[79]

Nevertheless, if perceived from outside Franco's dictatorship, Butenandt was an uncomfortable guest. He had joined the Nazi Party in 1936 and directed the Kaiser Wilhelm Institute (KWI) for Biochemistry in Berlin-Dahlem during the Nazi period, as well as making a considerable contribution to several military projects. He also endorsed Nazi policies between 1937 and 1945 and supported the mobilisation of the institute during the war.[80] Butenandt's visit to Spain in 1963 also had its counterpart in the MPG invitation to Lora-Tamayo to visit Germany in June 1969. As president of the CSIC at that time, Lora-Tamayo visited 12 of the 50 institutes of the MPG, including Butenandt's Biochemistry Institute in Munich.[81] At

---

[76] Jean Courtois, a professor of biochemistry, and Raymond Delaby, a professor of chemistry.
[77] This is Lora-Tamayo's version of events, at least, in his autobiography. Lora-Tamayo, *Lo que yo he conocido*, p. 340.
[78] See the official propaganda newsreel: No-DO, 23 December 1963 – No. 1094B – RTVE. es. 'El profesor alemán Adolfo Butenandt en Madrid. Investido "doctor honoris causa" de la Universidad Central'. www.rtve.es/filmoteca/no-do/not-1094/1472865 (last accessed 24 September 2018); Albert Presas, 'La inmediata posguerra'.
[79] 'Hoy llega a Sevilla el Profesor Butenandt', *ABC*, 14 December 1963, p. 49.
[80] Achim Trunk, 'Biochemistry in Wartime: The Life and Lessons of Adolf Butenandt, 1936–1946', *Minerva*, 44 (2006), 285–306.
[81] History and experimental medicine (Göttingen), physical chemistry, law and nuclear physics (Heidelberg), plasma physics (Garching), aeronautics (Lindau), agriculture (Bad Kreuznach), documentation (Frankfurt) and behavioural physiology (Seewiesen).

Figure 6.3 (From left to right) Manuel Lora-Tamayo, Helmut Allardt (the ambassador of the German Federal Republic) and Adolf Butenandt in an interview with Franco in the *Palacio del Pardo*, Madrid, 11 December 1963.
Reproduced with permission of the *Archivo José F. Demaría 'Campúa'*.

the official farewell dinner, he shared the high table with the rector of the university, the president of the Bayern Academy of Science and several directors of Max-Planck Institutes. All of this led to an increasing exchange of senior and junior researchers between the Max-Planck Society and the CSIC.[82]

Other chemists such as Erich Pietsch (1902–79) played a central role in bridging German and Spanish interests in the 1960s. Pietsch was the editor of the *Gmelin Handbuch der Anorganischen Chemie*, which contributed to improving accessibility to the results of the international chemistry community.[83] Trained at the Physical Chemistry Institute of the University of Berlin, in 1936 (following the dismissal of its Jewish former director, Richard Meyer) Pietsch became head of the Gmelin Institute for Inorganic Chemistry and held that position throughout the Nazi

[82] 'Estancia del profesor Lora-Tamayo en Alemania a invitación de la Sociedad Max-Planck', *Arbor*, 283–4 (1969), 319–20.
[83] W. Boyd Rayward, Mary Ellen Bowden (eds.), *The History and Heritage of Scientific and Technological Information Systems* (Philadelphia: Chemical Heritage Foundation, 2002).

period and the war, right up to 1967. He worked in research at the War Office from 1939 and was made president of the *Deutsche Gesellschaft für Dokumentation* (1955–61). In 1959, Pietsch became an honorary member of the CSIC and helped to better equip the *Instituto de Química Alonso Barba*. In 1962, negotiations regarding German technical support to Spain ended with Butenandt's visit in the following year and with Pietsch's honorary doctorate at the University of Madrid in 1967. However, after Butenandt's visit, some voices at the MPG perceived the CSIC more as an ideological agent of the regime than as a reliable research institution, but despite this, Pietsch managed again to turn the discussion towards the technocratic arena, focusing on specific research projects, trying to leave 'politics' aside and strengthening the basis of future exchanges between Franco's Spain and the German Federal Republic.[84]

Under the rhetoric of an apparently pure, apolitical, technocratic science, chemistry and leading chemists such as Lora-Tamayo and Pascual, but also their foreign partners Butenandt and Pietsch, built up international networks of professional affinities, diplomatic agreements and pompous public events that contributed to establishing chemistry as a crucial element in the logic of the Cold War. From the perspective of the European periphery, in a dictatorial country that was strategically significant but excluded from the Marshall Plan, Spanish chemists were desperate to find new funds for their research projects and professional ambitions, always under the banner of technocracy.

Corporate Chemistry

In November 1961, Suanzes, the powerful president of the INI, lectured on 'Franco and the Economy' to celebrate the 25th anniversary of the *Caudillo*'s coup d'état.[85] In a biased and optimistic speech, Suanzes presented the progress of the country through impressive

---

[84] Presas mentions the names of some prestigious Spanish chemists who benefited from that exchange programme, especially after Butenandt's visit to Spain in 1963: Eduardo Primo, José María Gamboa, Manuel Ballester, J. M. Sistiaga, J. F. García de la Banda, etc. Albert Presas, 'La inmediata postguerra y la relación científica y técnica con Alemania', in Romero, Santesmases (eds.), *Cien años de política científica*, pp. 202–4. On Butenandt, see: Wolfgang Schieder, Achim Trunk (eds.), *Adolf Butenandt und die Kaiser-Wilhelm-Gesellschaft. Wissenschaft, Industrie und Politik im Dritten Reich* (Göttingen: Wallstein Verlag, 2004); Robert Proctor, *Adolf Butenandt (1903–1995). Nobelpreisträger, Nationalsozialist und MPG-Präsident. Ein esrter Blick in der Nachlass* (Berlin: Max-Planck-Gesellschaft zur Förderung der Wissenschaften, 2000).

[85] José Antonio Suanzes, *8 Discursos, 'Franco y la economía'. Conferencia pronunciada por Suanzes en Burgos, el 28 de noviembre de 1961, con motivo del XXV aniversario de la exaltación del Caudillo Francisco Franco a la Jefatura del Estado Español* (Madrid: Centro de Estudios Económicos y Sociales del Instituto Nacional de Industria 1963).

figures regarding the production of coal, iron, steel, nitrogen fertilisers, cement and sulphuric acid and the capacity for petroleum refining.[86] Beyond Suanzes' propaganda, 1961 represented the beginning of a period that economic business historians consider crucial for the launch and modernisation of Spanish industrialisation. After the stabilisation plan of 1959, the production of chemicals such as fertilisers, pesticides, artificial fibres, rubber and pharmaceuticals grew substantially, all of them linked to technological changes in the petrochemical industry.[87]

In 1956, the *Empresa Nacional Calvo Sotelo* (ENCS) had established a factory for the production of lubricant oils from the distillation of local bituminous slates in Puertollano. In the period of 1956–61, the ENCS produced more than 600,000 tonnes of distilled fractions. In addition, hydrogenation of coal provided several distilled fractions of lubricants, carburant, paraffin and solvents, distributed by *Compañía Arrendataria del Monopolio de Petróleos Sociedad Anónima* (CAMPSA). At the *Refinería de Petróleos de Escombreras* (REPESA), near Cartagena, imported oil from Saudi Arabia provided several distilled fractions for the internal consumption of the country. Throughout the 1960s, the opening of new concentrations of chemical plants known as *'polos químicos'* led to new oil refineries and the manufacturing of synthetic materials on an industrial scale. Oil was substituted for shale oil when ENCS and REPESA merged into *Empresa Nacional de Petróleo* (ENPETROL), which managed to successfully develop and commercialise oil-bearing processes in international markets. With the scientific support of the CSIC, ENPETROL produced solvents, benzene, cyclohexane, nonaromatic naphtha and querosenes, as well as desulphurised oil distillation products.[88] Similarly, the plastics industry signed major contracts with the CSIC[89] and worked on the synthesis and applications of polymers. In 1968, *Electroquímica de Flix* developed its own procedure to obtain tri-chlorethylene. In the 1970s, following its merger with the Cros chemical company, Flix developed the 'Deacon' process to obtain chlorine from hydrochloric acid and oxygen gas and for the de-hydrochlorination of

---

[86] Ibid., pp. 157–8.
[87] The main chemical industries in Spain in 1970 were Unión de Explosivos Riotinto, Empresa Nacional Calvo Sotelo, Cros, Dow Uniquesa, SNIACE, Fertilizantes Iberia, Calatrava Petroquímica, REPESA, Amoníaco español, Alcudia, Sociedad Española del Oxígeno, DuPont, ICI, Union Carbide, Hoechst, BASF, Montecatini-Edison, BAYER, AZKO N. N., Rhône-Poulenc and Monsanto. Emilio de Diego, *Historia de la industria en España. La química.* (Madrid: Escuela de Organización Industrial. Actas Editorial, 1996), pp. 103–49, 135.
[88] Lora-Tamayo, *La investigación química*, pp. 297–9.
[89] With Ynfiesta, Fontán, Guzmán, José María Gómez Fatou and José González Ramos.

chlorinated hydrocarbons in collaboration with the *Instituto de Catálisis y Petroleoquímica*.[90]

Nevertheless, all of these cases were thought of as exceptions in a general context of growing foreign technological dependence under the hegemony of the new corporate, international chemical industries. They reinforced the alliance between the economic lobby of chemical entrepreneurs that had supported Franco during the Civil War and the technocratic plans of the dictatorship. Corporations targeted Franco's Spain as a very useful country for expansion.[91] To the 1959 chemical firms – Cros, Solvay, Lipperheide, Urquijo, Ibys, Abelló, Foret, Carceller, Sandoz, Alday, Andreu and ENCS, among others – in the 1970s we have to add a good number of new international corporations: Bayer, Hoechst, BASF, Dow, Ciba Geigy, Monsanto, Lever, Shell (Figure 6.4), ICI, Henkel, Rhône-Poulenc, L'Air Liquide, Standard Oil, Houghton-Puteaux, Agfa, Kuhlmann, Avon, Gal, Procter and Gamble, DuPont, ICI, Union Carbide and Azko NN. Figures of more than 6,000 chemical firms and 134,000 workers in 1963 rose to close to 8,000 firms and almost 200,000 workers in 1972.[92] In 1969, in the same wave of self-legitimating optimism, Lora-Tamayo underlined the success of several applied science fields (coal, fats, metallurgy, fermentations, plastics and soils), now with new Cold War allies such as foreign chemical corporations, who were keen to establish their factories in Franco's Spain.[93]

Corporations progressively occupied the land that the government had provided at the *polos químicos*. Under the Development Plan, the introduction of new petrochemical plants aimed to counter the lack of hydrocarbons for combustibles, the lack of a chemical industry network and the weaknesses of industrial research. In the mid-1960s, the ENCS in Puertollano included prestigious corporations such as ICI, Phillips Petroleum, Cros, Foret and Montecatini for the production of ethylene, propylene, polyethylene, butadiene and glycols. In Tarragona, Shell,

---

[90] Lora-Tamayo, *La investigación química*, pp. 254–5.
[91] Nuria Puig, *Bayer, Cepsa, Repsol, Puig, Schering y La Seda. Constructores de la química española* (Madrid: LID, 2003); Puig, López, *Ciencia e industria en España*, pp. 53–9. Emilio de Diego provides a longer list of the major chemical industries established in Spain in 1970, which is worth reading: de Diego, *Historia de la industria en España*, p. 135. See also: Nuria Puig, *La nacionalización de la industria farmacéutica en España. El caso de las empresas alemanas, 1914–75* (Madrid: Fundación Empresa Pública, 2001); Eugenio Torres (ed.), *Los 100 empresarios españoles del siglo XX* (Madrid: Editorial Empresarial, 2000). See also: Josep Font, 'The Impact of Chemistry in Catalonia's Industrial Development during the Twentieth Century', *Contributions to Science*, 6(2) (2010), 243–8.
[92] *Anuario de la industria química española* (Madrid: Sindicato Vertical de Industrias Químicas, 1963); idem (1972).
[93] Lora-Tamayo, *Un clima para la ciencia*.

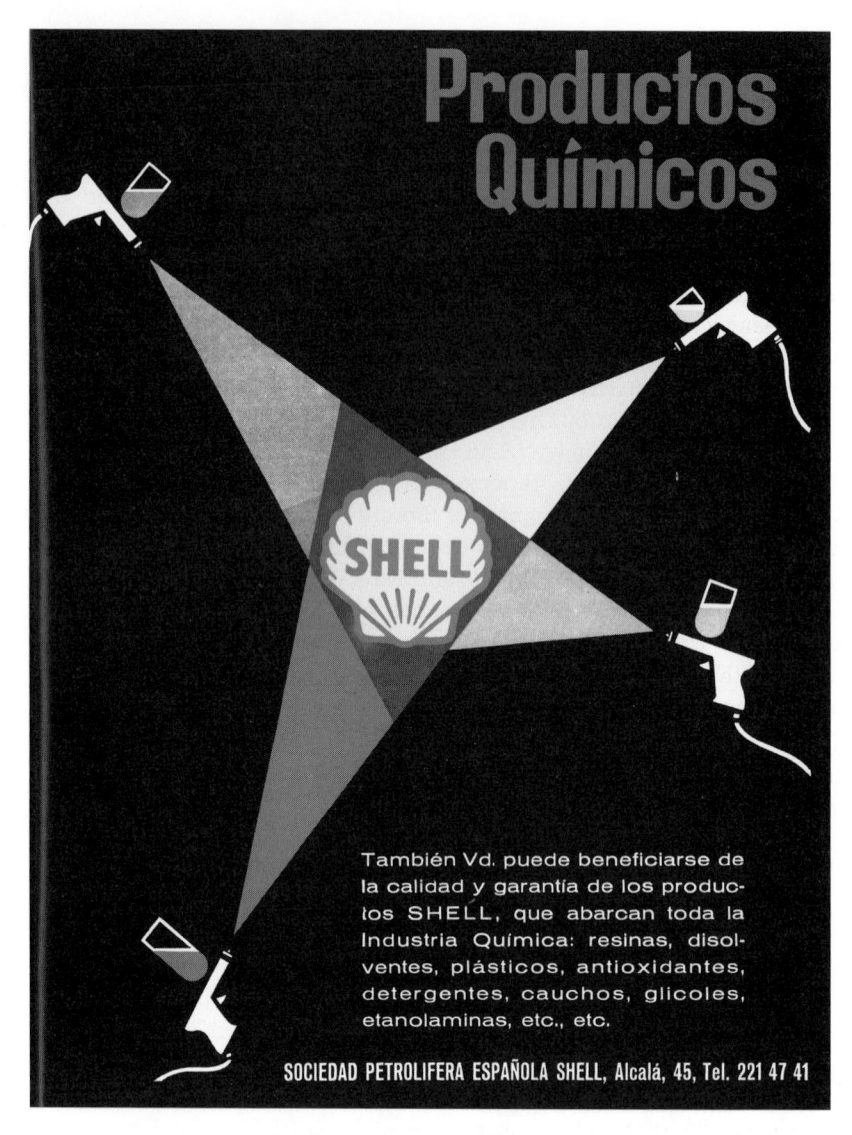

Figure 6.4 *Sociedad petrolífera española Shell*. Advertisement for petrochemicals in the mid-1960s: resins, solvents, plastics, antioxidants, detergents, rubber, glycols and ethanol-amines.
Química e Industria. Revista de la Asociación Nacional de Químicos de España, 13(5) (1966) (leaflet). Private collection of the author.

Hoechst, Dow, Monsanto and Catalana de Gas produced thousands of tonnes of ethylene derivatives, alcohols, glycols and vinyl chloride. In Huelva, *Minas de Riotinto* and Gulf Oil provided more than 100,000 tonnes per year of aromatic compounds. The petrochemical complexes increased the productivity and export rates of smaller chemical plants and opened the door to some industrial research.[94] Low wages and low levels of taxation provided an 'excellent' environment for the new corporations to establish their factories in Spain. Despite the public rhetoric of connecting research with industrial needs, these companies perceived the Spanish research system to be weak, not very reliable and, in their view, far from having a solid alliance with industry. They did not establish research departments and appointed local chemists as subaltern experts that often had to be trained at the central branches of the companies.[95]

In December 1965, the first international applied chemistry fair, '*Expoquimia*', took place in Barcelona. José Agell, a leading figure in the chemical industry, was responsible for organising the event. These were the years surrounding the celebration of the '25 years of peace', which were associated with the material progress brought about by the chemical industry, regardless of its national origin. Agell declared that in the period of 1953–64, national production of sulphuric acid had almost doubled, from 650,000 tonnes to more than 1 million tonnes. Production growth was also significant for nitric and hydrochloric acids, oil derivatives, plastics, artificial textiles, pharmaceuticals and cellulose. Agell acknowledged, however, that the continuous growth in the production of new chemicals would require further government efforts to support the chemical industry.[96]

In 1966, in a debate on the present and future of the Spanish chemical industry,[97] government officers again presented an optimistic view of the growing process of internationalisation since the early 1960s. They pointed out the progress of the 'macrochemical' industry in Santander, Tarragona, Castelló, Puertollano and Huelva, mainly focused on nitrogen fertilisers, cellulose and insecticides, but also on the main weaknesses

---

[94] F. González Sánchez, 'Panorama de la petroleoquímica en España', *Química e Industria. Revista de la Asociación Nacional de Químicos de España (ANQUE)*, 13(5) (1966), 168–73.

[95] Puig, *Bayer, Cepsa, Repsol, Puig, Schering y La Seda*, pp. 48–9; de Diego, *Historia de la industria en España*, p. 149. See also: Nuria Puig, 'La ayuda económica de Estados Unidos y la americanización de los empresarios españoles', in Delgado, Elizalde (eds.), *España y Estados Unidos en el siglo XX*, pp. 181–206.

[96] *La Vanguardia Española*, 11 May 1965, p. 27.

[97] *Afinidad*, XXIII(242) (1966): 'Entrevista con el Ilm. Sr. D. Mario Álvarez-Garcillán, Director General de Industrias Químicas', pp. 95–8; 'Ciclo de Conferencias "50 años de Química": Presente y futuro de la industria química española, por el Ilm. Sr. Mario Álvarez-Garcillán', pp. 99–116; 'Líneas de desarrollo de la industria química por el profesor Dr. Karl Winnacker', pp. 117–24.

of the whole endeavour: little quality control, unstable prices, poor research and marketing and the existence of too many small chemical industries spread through the country.[98] In spite of all the promising figures issued from the INI, the CSIC and the PJC research institutes, the report bluntly stated that 'the profession of researcher does not exist in our country'.[99] In the discussion, José Castells, one of the most distinguished members of Pascual's school of organic chemistry, once again raised the pure–applied issue and tried to counterbalance the official hegemony of 'applied chemistry', which had been shaped in the totalitarian years of the autarchy, and even in the new era of the international chemical corporations seemed not to leave much room for basic, fundamental, 'free' research done by Spanish chemists themselves.[100] Real expertise shifted towards the decision centres of the new foreign corporations, whereas arguments such as Castell's for the establishment of a scientific culture of pure, basic – perhaps too 'liberal' – chemistry remained rather marginal.[101]

In 1969, already well established in the *polo químico* in Tarragona, the German chemical company BASF frequently appeared in advertisements in the press and epitomised foreign dependency. BASF advertised its 5,000 products for the 'progress of humankind': plastics, resins, finishing products (glues, paper, textiles and packaging), raw materials for lacquers, varnishes, antifreeze, dyestuffs, fertilisers and plant protection products. BASF España was presented as a company with Spanish and German capital and with strong investment in research (240 million deutschmarks in 1967).[102] In the early 1970s, after a decade of corporate growth, chemistry represented 9.8 per cent of the industrial income of the country, with a growth rate of 15 per cent per year.[103] As in other countries, all of those changes were linked to the growth of the petrochemical industry with the introduction of new raw materials, new industrial processes and the inevitable reshaping and modernisation of the whole technological system. In 1971, there were 8,240 chemical firms in Spain, but more than 6,000 had fewer than 50 workers – at least, such

---

[98] Ibid., pp. 100–1.   [99] Ibid., p. 101.   [100] Ibid., p. 108.
[101] See also Paul Lucier, 'The Origins of Pure and Applied Science in Gilded Age America', *Isis, Focus: Applied Science*, 103 (2012), 527–36, p. 527. Lucier describes how in late nineteenth-century United States, '"pure" was the preference of scientists who wanted to emphasize their nonpecuniary motives and their distance from the marketplace. "Applied" was the choice of scientists who accepted patents and profits as other possible returns on their research. In general, the frequent conjoining of "pure" and "applied" bespoke the inseparable relations of science and capitalism in the Gilded Age'. See also: María Rubio, J. María Cavanillas, 'La industria química española', *Revista de Economía Política*, 45 (1967), 35–64.
[102] *La Vanguardia Española*, 17 May 1969.   [103] Ibid., 3 November 1972.

was the diagnosis of the leaders of the Spanish chemical industry at the 1972 annual meeting of the International Society for Industrial Chemistry. In their view, these small-scale companies, their geographical dispersion, their high production costs and the proliferation companies that were again too marginal explained the wide gap between European and Spanish chemical firms.[104]

Despite its inevitable foreign dependence, in terms of expertise and basic research, the international chemical industry of the 1960s contributed to industrialisation taking off in Spain and reinforced discourses of modernisation and progress. In their daily practice and public addresses, the elite chemists of the regime constructed a technocratic culture in the climate of 'developmentism', which paradoxically combined the modernisation of the country with university upheavals and political repression (see Chapter 7).

***

Rooted in early twentieth-century 'civil discourse' (see Chapter 1),[105] which had supposedly separated the scientific policies of the country from its political struggles, technocracy and its chemical dimension took a new form during Franco's dictatorship. Pascual's and Lora-Tamayo's scientific productivity was always presented as being detached from any political agenda. Nevertheless, Lora-Tamayo reached the highest political position as minister, and Pascual also played his political cards well in the local context of Barcelona, by presenting his molecules as cosmopolitan achievements that merited international publication. In fact, from the early totalitarian dreams to the later technocracy, Francoist chemists became useful tools for the establishment of international relations in the context of the Cold War. From the 1960s onwards, growth of the chemical industry and its multinational character reinforced the alliance between chemical entrepreneurs and the technocratic plans of the government.

Chemists developed a huge range of apolitical attitudes that benefitted their scientific careers. Although in public many often enforced the 'neutral' status of chemistry in terms of political commitments that allowed them to develop their research subjects and receive research funds without any ethical or deontological debate, they had no qualms about publicly expressing their enthusiasm for the *Caudillo* if needed. As in the case of our chemists, Mark Walker sees scientists under the Nazi

---

[104] *Anuario de la industria química española* (Madrid: Sindicato Vertical de Industrias Químicas, 1972).
[105] Thomas Glick, *Einstein in Spain: Relativity and the Recovery of Science* (Princeton: Princeton University Press, 1988).

regime as 'fellow travellers' who neither resisted Hitler nor embraced National Socialism. They became 'apolitical' professional scientists doing 'good' science independently of their employer.[106] In Spain, progress in chemistry was in tune with economic progress and foreign corporations in a new liberal economy with no room for basic – liberal – chemistry or for liberal democracy. Apolitical discourse, uncritical support for industrial projects, changes in expertise (often accepting the scientific hegemony of foreign chemistry in the corporations of the *polos químicos*), little room for pure research and a tacit optimism regarding industrial growth shaped the technocratic spirit of chemistry in the 1960s up to the end of the dictatorship in 1975. Perhaps that entire framework could even be extrapolated to other assessments of the democratic status of expertise and the nature of a technocratically bounded rationality, in which political choices are always restricted to a reduced group of options that have been previously selected by experts.[107]

[106] Mark Walter, *Nazi Science: Myth, Truth and the German Atomic Bomb* (Cambridge, MA: Perseus, 1995), p. 4.
[107] Paul Erickson, Judy L. Klein, Lorraine Daston, Rebecca Lemov, Thomas Sturm, Michael D. Gordin, *How Reason Almost Lost Its Mind: The Strange Career of Cold War Rationality* (Chicago: University of Chicago Press, 2013).

7 Liberal Dissent

In 1958, on the eve of the early stages of economic liberalisation during Franco's regime, Isaiah Berlin (1909–97) lectured in Oxford on the 'Two Concepts of Liberty'. Berlin defended the value of pluralism in his fierce opposition to totalitarianism in all its guises: undemocratic, forced collectivism and the brutal, atavistic crimes of fascism. He warned of the dangers of any sort of monism, of any attempt to assign truth only to one set of values and to dismiss the rest. Berlin also stressed the need to fight imposed homogeneity and over-centralisation. Since liberalism – in terms of freedom of expression, the free press, freedom of religion and free markets – had been suppressed or defeated in many countries in the first half of the twentieth century, Berlin's reflections were particularly relevant to post–Second World War Western societies. He warned about the risks of overly utopian political thinking and described the present as a time when:

[...] so large a number of human beings, in both the East and the West, have had their notions, and indeed their lives, so deeply altered, and in some cases, violently upset by fanatically held social and political doctrines. Dangerous, because when ideas are neglected by those who ought to attend to them [...] they sometimes acquire an unchecked momentum and an irresistible power over multitudes of men that may grow too violent to be affected by rational criticism.[1]

Chemistry was obviously not isolated from debates on liberalism and political and economic freedom. We know, for instance, that the liberal values of the nineteenth-century German chemical community – in

---

[1] Isaiah Berlin, 'Two Concepts of Liberty', in Isaiah Berlin, *Four Essays on Liberty* (Oxford: Oxford University Press, 1969), p. 1. See also: Alan Ryan (ed.), *The Idea of Freedom: Essays in Honour of Isaiah Berlin* (Oxford: Oxford University Press, 1979); Mark Lilla, Ronald Dworkin, Robert Silvers (eds.), *The Legacy of Isaiah Berlin* (New York: New York Review of Books, 2001); David Weinstein, 'Nineteenth- and Twentieth-Century Liberalism', in George Klosko (ed.), *The Oxford Handbook of the History of Political Philosophy* (Oxford: Oxford University Press, 2011). For a short summary of Berlin's thought and his personal relation to Michael Ignatieff, see: Derrick O'Keele, *Michael Ignatieff: The Lesser Evil?* (London: Verso, 2011), p. 31.

particular, the cases of Hermann Emil Fischer (1852–1919) and August Wilhelm von Hofmann (1818–92) – were shaped in terms of their responses to war, national and international solidarity and the capacity for 'post-war reconciliation among scientists from previously contending nations'.[2] Similarly, Carl Bosch (1874–1940), the Nobel Prize winner for chemistry for his famous method of producing ammonia industrially from atmospheric air (with Fritz Haber), and Carl Krauch (1887–1968), the chemical industrialist and Nazi war criminal – both executives at I. G. Farben – held opposed conceptions of the German political economy.[3] Similarly, as Peter Hayes pointed out some years ago, the failure of liberalism in German academia tragically 'paved the way for National Socialism, the Second World War and the Holocaust'.[4]

In the Spanish case, the attempts to build up a liberal state in the nineteenth century have attracted the attention of historians for decades.[5] The 1812 liberal Constitution of Cádiz was followed by long periods of absolutist rule, which have often been associated with a lack of a strong civil society and an absence of a widely established industrial, bourgeois culture.[6] Even in the twentieth century, the construction of a modern liberal state seemed weak and unstable. Individual freedom, cultural and religious pluralism and efficient state protection against any kind of discrimination hardly applied in Spain for the whole period covered in this book. The monarchic restoration and Primo de Rivera's dictatorship did not provide a context for the effective defence of liberal civil rights, and the short Republican period was soon followed by the long Francoist dictatorship.[7]

The Civil War and the totalitarian ideology of Franco's regime made reconciliation between the liberal and anti-liberal sides too difficult and

---

[2] 'In the end, perhaps the most important test of liberal internationalism in science is its response to war, and the possibility of post-war reconciliation among scientists from previously contending nations [...]'. Jeffrey Allan Johnson, 'Dilemmas of Nineteenth-Century Liberalism among German Academic Chemists: Shaping National Science Policy from Hofmann to Fischer, 1865–1919', *Annals of Science*, 72(2) (2015), 224–41, p. 241.

[3] In Hayes's words: 'Bosch's understanding of his industry, his nation, and scientific progress led him to oppose the Nazis, but also to lay the basis for their recruitment of Krauch and the German chemical industry for their expansionist purposes'. Peter Hayes, 'Carl Bosch and Carl Krauch: Chemistry and the Political Economy of Germany, 1925–1945', *Journal of Economic History*, 47(2) (1987), 353–63, p. 353.

[4] Ibid.

[5] Josep Fontana, *La época del liberalismo*. Vol. 6. *Historia de España* (Barcelona/Madrid: Crítica/Marcial Pons, 2013).

[6] Owen Chadwick, 'On Liberalism', in *The Secularization of the European Mind in the Nineteenth Century* (Cambridge: Cambridge University Press, 1975), pp. 21–47.

[7] Francisco Romero, Angel Smith (eds.), *The Agony of Spanish Liberalism: From Revolution to Dictatorship, 1913–23* (Basingstoke: Palgrave Macmillan, 2010).

paved the way for the war wounds that brought irreversible division in the chemical community (see Chapter 3). Even in the obscurity of the 1940s, physician and historian of medicine, Pedro Laín Entralgo (1908–2001), dared to publish a book entitled *España como problema* (1948) (Spain as a problem), in which he called for a certain degree of reconciliation with the defeated and exiled after the Civil War. Nevertheless, Opus Dei member and editor of the journal *Arbor*, Rafael Calvo Serer (1916–88), immediately countered Laín Entralgo's book with a volume entitled *España sin problema* (1949) (Spain without a problem), which denied the existence of the 'other Spain' and defended the irreversibility of Franco's victory in 1939.[8]

It is therefore in that context that we have seen how many chemists built up their careers and co-evolved within a political regime that was fiercely hostile to liberalism, even signing up enthusiastically to totalitarian values. Nevertheless, others merged their chemical knowledge with more or less explicit liberal political agendas[9] and subtly refused uncritical adhesion to a specific regime and its political and economic elites. As happened with nineteenth-century Darwinism and thermodynamics,[10] chemistry became a polyhedral field under multiple appropriations for political and professional ambitions through the twentieth century.

Tracing back liberal chemistry discourses and practices, this chapter goes backwards chronologically, including some references to the Silver Age, and ends in the 1970s, in the final years of Franco's dictatorship. It provides further evidence of the 'liberal chemistry' that coexisted in opposition to the hegemonic anti-liberal culture of twentieth-century Spain. For that purpose, I will first analyse the ethos of the Spanish chemical community in exile after the Civil War and the ways in which chemists survived as political refugees in Latin America, in strong opposition to Franco's regime and with a passionate commitment to educational and social reforms that endeavoured to provide equal opportunities for citizens on the other side of the Atlantic, particularly in Mexico. I shall

---

[8] Carolyn P. Boyd, *Historia Patria: Politics, History and National Identity in Spain, 1875-1975* (Princeton: Princeton University Press, 1997), p. 239.
[9] Historian Michael Richards brilliantly described how the core of Franco's cultural repression rested on strong anti-liberal violence. Richards stated, 'Collective memory in Spain has been formed through an attempt by the dictatorship to extirpate the sense of history once possessed by those who became the defeated'. Michael Richards, *A Time of Silence: Civil War and the Culture of Repression in Franco's Spain, 1936–1945* (Cambridge: Cambridge University Press, 1998), p. 2.
[10] See, for instance: Thomas Glick, *Darwin en España* (Barcelona: Península, 1982); Stefan Pohl-Valero, 'Thermodynamics, Social Thinking and Biopolitics in Spain under the Restoration', *Universitas Humanistica*, 69 (2010), 35–60.

later approach how other chemists became 'internal refugees' and subtly defended liberal values in the chemical industry, in universities and in the public sphere. Some entrepreneurs sheltered chemists that had been expelled from academia, particularly in the early decades of Franco's dictatorship. Others also tried to oppose totalitarian rules and censorship and struggled to build up new shelters of freedom.

The examples that follow cover different periods of the time span of the book and reflect how the co-evolution between science and politics often goes beyond officially well-established regimes and creates intermediate spaces on the fringes of opposition and tacit acceptance of the status quo. The chemists' different forms of liberal dissent are also useful examples of the long-standing tension between a totalising, centralised state power and individual freedom and civic resistance.

Chemists in Exile

As a natural reaction to defeat in the Civil War, exiled chemists tried to preserve as much as they could of the liberal ethos of the pre-war years.[11] In their view, it was the 'rational' study of nature – and the scientific agenda underlying it – that could provide useful ideas for the improvement of society in terms of the old utopias of the French Third Republic.[12] In fact, the 1930s Republican elites, which were an amalgam of bourgeois democrats, socialists and liberal intellectuals, often in contact with socialist, communist and anarchist groups on the left, dramatically clashed with the political agenda of Franco's Spain.[13] Around 70 chemists (PhDs, pharmacists, engineers and technicians) fled to Mexico after the Civil War and established themselves in academia and in industry.[14] To these figures should be added a younger generation that had been educated in

[11] Francisco Giral, *Ciencia española en el exilio (1939–1989)* (Barcelona: CIERE/Anthropos, 1994).
[12] Ángel Duarte, Pere Gabriel, 'Una sola cultura política republicana ochocentista en España?', *Ayer. Revista de Historia Contemporánea*, 39 (2000), 11–35.
[13] F. Giral, for instance, perceived his training at the INFQ as the fruit of the liberal, secular, modern spirit of the JAE. Francisco Giral, 'Química orgànica (1932–1939)', in *50 años de investigación en física y química en el edificio Rockefeller de Madrid, 1932–1982. Noviembre de 1982. Recuerdos históricos* (Madrid: CSIC, 1982), pp. 39–42.
[14] José Cueli, 'Matemáticas, física y química', in *El exilio español en México 1932–1982* (Mexico City: Salvat/FCE, 1982), pp. 531–43. The most relevant names were Pedro Bosch Giral, Rafael Oliván, J. Viciana, Luis Fanjul, Enrique Gay, Antonio Talayero, Alfonso Boix Vallicrosa, Leone Abramson, Laureano Poza Juncal, Julio Clón, Juan Xirau Palau, Adela Barnés, José Vázquez Sánchez, César Roquero, José Giral, Antonio Madinaveitia, Modesto Bargalló, Francisco Carrera (Civil Governor of Madrid, 1936), César Pi-Sunyer Bayo, Manuel Castañeda, Francisco Méndez Domínguez, Eugenio Muñoz Mena, Francisco Giral and José Ignacio Bolívar. Santiago Capella, José Antonio Chamizo, Julián Garritz, Andoni Garritz, 'La huella en México de los

Mexico, plus new Mexican co-workers and collaborators. In Mexico, exiled chemists found an ideal context for fruitful co-production between their Republican values and President Lazaro Cárdenas's (1895–1970) leftist political programme, which included the nationalisation of oil companies in 1938, major agricultural reform – the old dream of the FNICER (see Chapter 2) – a new law of industrial boards, which facilitated academic–industrial links, and an ambitious plan for cultural and educational reforms.[15] From a documented sample of more than 300 scholars, chemists and pharmacists represented roughly 15 per cent of exiled scientists in Mexico.[16] They also upheld the utopian views of pacifism and internationalism of the 1936 Popular Front, which had also borrowed ideas from the French *'front populaire'* of the 1930s.[17]

José Giral epitomises the figure of the Republican chemist–politician.[18] He combined his work at the *Instituto Politécnico Nacional* (IPN) in fields such as food chemistry, natural products and organic analysis with his presidency of the Spanish Republican Government in exile in 1945, and his long-standing diplomatic campaign in defence of the legality of the democratically elected cabinet of 1936 and against the international legitimation of Franco's dictatorship.[19] J. Giral was also one of the key figures in the organisation of exiled Spanish academics abroad. The *Unión de Profesores Universitarios Españoles en el Extranjero* (UPUEE) published a *Boletín* – a sort of newsletter – to build stronger links among

químicos del exilio español de 1939', in Antonio Bolívar (ed.), *Científicos y humanistas del exilio español en México* (Mexico City: Academia Mexicana de Ciencias, 2006), pp. 155–72.

[15] Andoni Garritz, 'Breve historia de la educación química en México', *Boletín de la Sociedad Química Mexicana*, 1(2) (2007), 3–24.

[16] Medical doctors (43%), engineers (27%), pharmacists (9%), architects (7%), chemists (6%), mathematicians (5%) and naturalists (4%). María Magdalena Ordóñez Alonso, 'Los científicos del exilio español en México. Un perfil'. http://clio.rediris.es/clionet/articulos/cientificos.htm (last accessed 21 November 2016).

[17] For a discussion about the nature of the community of exiled Spanish scientists after the Civil War, see: José Luis Barona (ed.), *El exilio científico republicano* (Valencia: Publicacions de la Universitat de València, 2010). See also: Alfredo Baratas, 'El exilio científico español de 1939. Diáspora y reconstrucción de una comunidad científica', in Manuel Llusia, Alicia Alted (eds.), *La cultura del exilio republicano español de 1939. Actas del Congreso internacional celebrado en el marco del Congreso plural: Sesenta años después (Madrid-Alcalá-Toledo, Diciembre de 1999)* (Madrid: UNED, 2003), I, pp. 707–14. See also: Alfredo Baratas, M. Lucena Girado, 'La Society for the Protection of Science and Learning y el exilio republicano español', *Arbor*, 588 (1994), 25–48; José María López Sánchez, 'El Ateneo Español de México y el exilio intelectual republicano', *Arbor*, 735 (2009), 41–5.

[18] 'Biografía de José Giral Pereira', *Ciencia*, XXIV (1–2) (1965), 9–12. Humberto Estrada, 'Francisco Giral González', in *Profesores eméritos. Semblanzas* (Mexico City: UNAM, 1989), pp. 15–19.

[19] Francisco Giral, *Vida y obra de José Giral Pereira. Relatada y comentada por* _____ (México: UNAM, 2004); Hebe Vessuri, 'La ciencia académica en América Latina en el siglo XX', *Redes. Revista de Estudios Sociales de la Ciencia*, 1(2) (1994), 41–76.

the community and also to keep all members informed about the latest publications in all fields of knowledge.[20]

Exiled chemists held very critical positions against the new science policies of Franco's Spain.[21] Often with anticlerical overtones, they explicitly denounced censorship and authorship usurpation in scientific books and even mutilations of their texts, and condemned the fierce repression against intellectuals. This was the case, for instance, of the Spanish version of Wilhelm Schlenk's and Ernst Bergmann's textbook on organic chemistry,[22] which omitted the names of its translators, Pilar Barnés and F. Giral, who were in exile, but kept Richard Willstätter's preface. It is worth remembering here that, after leaving Germany in 1927 in disagreement with anti-Semitic policies, Willstätter himself was a chemist refugee. He had become one of the mentors of the Silver Age chemists with a very close relationship with Madinaveitia (see Chapter 2). So, in their struggle for international recognition, the exiled chemists perceived Willstätter's appropriation by the Francoist regime as an unacceptable humiliation.[23]

F. Giral's annotated translation of Carl August Rojahn's textbook for the preparation of chemical and pharmaceutical products provides interesting clues as to how he established a Mexican research school.[24] The text was a joint venture between F. Giral, his father, J. Giral, Manuel Castañeda and Adela Barnés from the IPN and a large group of industrial chemists.[25] From 1941 to 1945, F. Giral held the chair in organic chemistry at the *Escuela Nacional de Ciencias Biológicas* at the IPN, and the position was compatible with a huge range of industrial projects.[26] He

---

[20] In 1944, for instance, the UPUEE wrote to the British Prime Minister, Winston Churchill, bitterly complaining about tacit British support for Franco's regime. 'Los profesores universitarios se dirigen a Mr. Churchill', *BUPUEE*, 11–12 (1944), 1–2.

[21] F. Giral, *Ciencia española en el exilio*, p. 36. In the *BUPUEE*, it is worth mentioning articles such as 'Contra la nueva legislación universitaria española', 'La cultural en España perseguida', 'El Estado franquista editor pirata', 'Supresión de nombres de autores y traductores' and 'Las publicaciones de los emigrados prohibidas en España'.

[22] Wilhelm Schlenk, Ersnt Bergmann, *Tratado de química orgánica*, 2 vols. *Prólogo de Richard Willstätter* (Madrid: Javier Morata, 1940).

[23] 'El Estado franquista editor pirata (III)', *BIUPUEE*, 9 (1944), 1–3.

[24] Carl August Rojahn, *Preparación de productos químicos y químico-farmacéuticos, traducido del original alemán y considerablemente ampliado por el profesor Francisco Giral, Catedrático de Farmacia de la Facultad de Santiago (España), Jefe de los Departamentos de síntesis orgánica en los Laboratorios Hormona, México*. 2 vols. (Mexico City: Atlante, 1942).

[25] José Vázquez Sánchez (*Laboratory Hormona*), César Roquero (*Laboratory Bios*), Alfonso Boix (*Laboratory Beick y Félix*), César Pi-Sunyer Bayo (*Laboratory Laquisa*) and Coronel Agustín Ripoll Morell (*Fabril de Aceites*) – Ripoll was a chief of the chemical weapons service of the Spanish Republican Army. Ibid., pp. xiii–xix.

[26] *Laboratorios Hormona*, Triarsan, the national oil company PEMEX, the *Laboratorio de Antipalúdicos Sintéticos* – which led a public campaign against malaria – and *Industria Nacional Químico-farmacéutica de México* (INQUIFA). From 1949 to 1955, F. Giral was the technical director of the Mexican chemical branch of

worked for ten years at *Laboratorios Hormona* in the production of steroids from the barbasco plant as an alternative to the extraction of that kind of substance from animal glands.[27] The American chemist Rosell Marker (1902–95) was a pioneer in that new research field, but F. Giral and his research school struggled for genuine Mexican contributions in the face of US scientific hegemony. In tune with his leftist Republican values, F. Giral later joined the *Industria Nacional Química Farmacéutica* (*Farquinal*), a national firm that was particularly committed to breaking the monopoly of the steroids market of the new private firm, Syntex, and keeping prices for progesterone and cortisone low.[28]

The Girals were also active in launching the journal *Ciencia*, a Latin American periodical devoted to pure and applied sciences (*Revista hispanoamericana de ciencias puras y aplicadas*).[29] *Ciencia* was a natural sciences generalist journal, nothing compared to *Anales*, but chemists and pharmacists played a very influential role on the editorial board, with names such as José and Francisco Giral, Madinaveitia, Moles (during his exile in Paris), García Banús from Colombia, Modesto Bargalló (1894–1981) and Augusto Pérez Vitoria. *Ciencia* contained sections on modern science (as a generalist topic), short research notes, foreign news and biographies, applied science, miscellanea, book reviews and abstracts of foreign papers (*Revista de Revistas*). The journal aimed to build up a new Latin American *ethos*, one that was different from the way in which the new Francoist state perceived the former Spanish colonies. As stated in the *Boletín* in 1944, *Ciencia* aimed to bring together:

[...] the best and most brilliant of the pure and applied sciences in Spain, working in a joint effort, and fraternal collaboration with the most distinguished Latin American colleagues. That is what hurts in Spain, because, *in the face of the grotesque Falangist 'Hispanidad'*, it represents a new way of understanding

---

Schering. From 1950 to 1961, he also became the founder and director of the Laboratorio Central de Investigación de Industria Nacional Farmacéutica. F. Giral, *Ciencia española en el exilio.*

[27] *El exilio español en México 1932–1982* (Mexico City: Salvat/FCE, 1982), pp. 151–2.

[28] In a similar line, in the 1940s, *Industrias Químicas Farmacéuticas Americanas* (IQFA) launched more than 25 drugs on the market under the scientific leadership of José Puche, a distinguished physiologist of Juan Negrín's school and rector of the University of Valencia during the Civil War. Other exiled chemists joined other labs such as Kriya, Eromex, Sevet, Cor, Uffimex and *Laboratorios Internacionales*. *Laboratorios Labys* produced serums and vaccines and counted among its scientific staff the physicist Blas Cabrera, the first director of the INFQ. Equally, Cabrera and chemist José Ignacio Bolivar played a key role in the industrial production of serums and vaccines against tetanus, diphtheria and typhus at *Laboratorios Zapata*. Alfonso Maya, 'Actividades productivas e innovacions técnicas', in *El exilio español en México, 1932–1982* (Mexico City: Salvat/FCE, 1982), 148–52.

[29] *Ciencia. Revista hispano-americana de ciencias puras y aplicadas. Publicaciones de la Editorial Atlante S.A.* (Mexico City, 1940).

and practicing cultural relations among all Spanish- and Portuguese-speaking people.[30]

For obvious reasons, Francoist Spain banned *Ciencia*.[31] In 1940, the editorial board sent 500 copies of the third issue to Spain, but they never reached the intended addressees. Published by Editorial Atlante, its economic survival was very much dependent on private donations, mainly from the metallurgical entrepreneur Carlos Prieto (1898–1991), president of the *Compañía Fundidora de Hierro y Acero* in Monterrey.

Although Moles's contribution to *Ciencia* was rather marginal, his dismantled research school of physical chemistry enjoyed some transatlantic continuity. César Roquero (1911–85) escaped from Franco's repression to Oran (Algeria), where he was subjected to forced labour, and he later moved to Mexico, where he joined several pharmaceutical labs and chemical businesses.[32] After years of prison and torture, María Teresa Toral continued her career in Mexico (see Chapter 3). She taught several physics and chemistry courses, published textbooks and translations and supervised some PhDs at the *Univerisdad Nacional Autónoma de México* (UNAM) and the IPN. Adela Barnés became a professor of inorganic chemistry at the IPN, where she contributed to the teaching of chemistry through the periodic table. The Mexican version of Moles's school now shifted to industrial jobs, but also to notable innovations in the teaching of chemistry.

One of the most prolific writers of chemistry papers in *Ciencia* was Modesto Bargalló. Educator, chemistry teacher and committed socialist, Bargalló went into exile at the end of the Civil War and, once in Mexico, became a chemistry lecturer at the IPN.[33] Bargalló's *Tratado de Química Inorgánica* was an updated general chemistry textbook that circulated widely among the Latin American community.[34] The book appeared in 1962 and was produced as a result of a joint project by Mexican and Spanish exiled chemists. It also included Linus Pauling's recent work on the chemical bond. In Mexico, Bargalló combined his chemistry teaching with his brilliant work in the history of chemistry, particularly as an expert

---

[30] 'Las publicaciones de los emigrados, prohibidas en España. La revista *Ciencia*', *BIUPUEE*, 10 (1944), 2 (emphasis added); José M. Cobos et al., 'El Boletín Informativo de la Unión de Profesores Españoles Universitarios en el Extranjero', *Llull*, 27 (2004), 27–60.
[31] F. Giral regretted that, even on his return from exile in 1977 after the end of the dictatorship, a complete set of *Ciencia* was nowhere available in Spain.
[32] Giral, *Ciencia española en el exilio*, pp. 121–2.
[33] Manuel Segura, Alberto Gomis, José María Sánchez Jiménez, 'Modesto Bargalló Ardévol (1894–1981), Maestro de maestros e historiador de la ciencia', *Lull*, 34 (2011), 419–42.
[34] Modesto Bargalló, *Tratado de química inorgánica* (Mexico City: Editorial Porrúa, 1962).

on the Spanish colonial history of mining and metallurgy in Latin America from the fifteenth to the eighteenth centuries.[35] In 1977, Bargalló obtained the Dexter Award of the American Chemical Society for his contribution to the history of chemistry, particularly to the history of metallurgy in the New World during the Spanish colonial period – a crucial topic for the making of a new identity among the exiled chemical community.[36]

Initially funded by the *Colegio de México*, a key institution of the Spanish Republic in exile, the *Instituto de Química* (IQ) contributed notably to shaping the ethos of the chemistry community.[37] Madinaveitia, the prestigious leader of the organic chemistry section at the INFQ in Madrid, became the director and natural leader of the new institute, in collaboration with the Mexican chemist Fernando Orozco. After earning his PhD at the UNAM, Orozco was internationally trained in Marburg (Germany), whereas Madinaveitia's scientific reputation was already widely known and appreciated by Cardenas's cabinet. Since his arrival in Mexico, ideological affinities contributed to placing Madinaveitia in the Mexican academic system. He soon focused on the study of Mexican natural products and committed himself to improving the lack of experimental training and research and to strengthening industrial connections.[38] His study of tequesquite, a sodium sesqui-carbonate that covered the bottom of Mexican lakes, became one of his early achievements, well representing the spirit of the *Instituto* and that of its founders.[39] In the English abstract of the paper, Madinaveitia and Orozco described their research project in the following terms:

We have studied the underground deposit of brines located at the bottom of the [...] Valley of Mexico, compared it with other similar ones existing in the Republic and in other countries. This deposit is of great interest because it may be used as a raw material for the alkali industry of the country.[40]

The IQ admitted students from the *Escuela Nacional de Ciencias Químicas* (ENCQ) and sent them to the United States and United

---

[35] Some years later, F. Giral and other exiled chemists also acted as chemist–historians for the making of their new identity, writing and lecturing on colonial chemistry episodes.
[36] Modesto Bargalló, *La minería y la metalurgia en la América española durante la época colonial* (Mexico City: FCE, 1955).
[37] Alberto Enríquez (ed.), *Exilio español y ciencia mexicana: génesis del Instituto de Química y del Laboratorio de Estudios Médicos y Biológicos de la Universidad Nacional Autónoma de México 1939–1945* (Mexico City: Colegio de México, 2000).
[38] See the correspondence between Madinaveitia and the president of the *Colegio de México*, Alfredo Reyes (15 October 1939, 31 October 1939). Enríquez, *Exilio Español y ciencia mexicana*.
[39] Antonio Madinaveitia, Fernando Orozco, 'Estudio del Yacimiento de Salmueras alcalines del Valle de México', *Boletín del Instituto de Química*, 1(1) (1945), 6–25, p. 6.
[40] Ibid., p. 25.

Kingdom for further training to be later appointed by the IQ.[41] Research results appeared regularly in the *Boletin del Instituto de Química* (*BIQ*).[42] In a report on the latest scientific results of the IQ in December 1941,[43] Orozco and Madinaveitia described in great detail their work on the Mexican alkaline lakes, the chemistry of natural essences, guayule (a kind of natural rubber) and nopal gum. However, beyond the immediate application of their research, both chemists wanted to study more theoretical, general aspects of catalytic hydrogenations, the links between colour and molecular structures and new analytical reagents. They were again dreaming of a 'liberal', pure, basic chemistry, close to the 1930 chemical culture of the INFQ that allowed them more freedom for basic research, but now compatible with industrial projects, very much in line with F. Giral's professional profile.

From 1942, the Rockefeller Foundation had supported the IQ with donations of scientific instruments and chemicals. Funds also came from the *Banco de México* and the *Comisión Impulsora y Coordinadora de la Investigación Científica*, while the *Comité Iberoamericano de Publicaciones Científicas* in New York completed the library with new international journals and books.[44] In 1951, the joint venture of the IQ with the *Laboratorios de Investigación Syntex* led to important contributions to the synthesis of steroid molecules such as cortisone and sex hormones, in collaboration with Carl Djerassi (1923–2015), who had become Associate Director of Research at Syntex in 1949. The synthesis of cortisone in 1951 had the support of Djerassi, George Rosenkranz, Jesús Romo and J. Pataki, and it placed Mexican organic chemistry very much at the centre of innovation. The *BIQ* became a prestigious journal with translations of top chemistry papers from the *Journal of the American Chemical Society*, *Journal of Organic Chemistry*, *Tetrahedron Letters* and *Chemistry and Industry*, but also original papers

---

[41] The main members of the team were José Iriarte, Octavio Mancera, Jesús Romo, Humberto Estrada, Alberto Sandoval and José F. Herrán. Andoni Garritz, 'Breve historia de la educación química en México', *Boletin de la Sociedad Química Mexicana*, 1(2) (2007), 3–24. See also: Capella, Chamizo, Garritz, Garritz, 'La huella en México de los químicos del exilio español'. On Madinaveitia, see: Manoel Toural, 'Antonio Madinaveitia, un científico republicano', in Pedro Bosch Giral et al., *Protagonistas de la química en España: Los orígenes de la catàlisis* (Madrid: CSIC, 2010), pp. 223–85.

[42] The whole research programme of the IQ focused on hydrogenation of quinones, polymerisation of anthracene, '*pulque*', the role of quicksilver in organic compounds and *papilonaceas silvestres*, as well as a long-term collaboration with the *Sosa Texcoco* chemical company. Maya, 'Actividades productivas e innovacions técnicas', pp. 148–9.

[43] Letter of 15 December 1941. Enríquez, *Exilio español y ciencia mexicana*, p. 121.

[44] Alberto Sandoval, 'Looking at the Origin of the *Boletín* after 20 Years of Uninterrupted Publication', *BIQ*, 22 (1970), 10–15.

leading to the discovery and successful application of the contraceptive pill.[45]

In 1970, going back to the liberal culture of his mentors after 20 years as the director of the *BIQ*, Alberto Sandoval (1918–2002) emphasised the need for and usefulness of promoting and funding 'pure' research in organic chemistry as a fundamental precondition for further large-scale, industrial collaboration. Through the seminal ethos that Madinaveitia sowed in the early 1940s, the IQ became a centre of innovation in steroid chemistry in the 1950s. Sandoval's argument for basic research in the 1970s seemed to be a long-standing position in favour of the liberal spirit he had acquired from Orozco and Madinaveitia, when basic research in chemistry had just been set up efficiently for the first time in Mexico.

Other young chemist refugees who arrived from Spain with their families at the end of the Civil War went on to enjoy distinguished careers with basic training in Mexico and periods abroad, mainly in the United States. Take, for instance, the case of José Ignacio Bolívar (1924–82), who received his scientific training in Boston, later became a professor at the Faculty of Chemistry of the UNAM, and excelled in the field of colloid chemistry. Equally brilliant was Pilar Rius de la Pola's training in quantum chemistry at Johns Hopkins University, and she later become another important figure in theoretical chemistry at the UNAM.[46] In 1943, Eugenio Muñoz, a former pupil of Miguel Catalán in Madrid, published *Introducción al estudio de la química*,[47] a widely read textbook published by Atlante, the same publisher that undertook the launching of *Ciencia*. Muñoz taught at the IPN and worked in several pharmaceutical labs.[48] Lidia Rodríguez-Hahn – one of the thousands of children that the Republican government shipped to the Soviet Union to protect them from bombing of civil population during the Civil War – graduated in chemistry at the UNAM, and in 1960 published her research results in the *Journal of the Chemical Society* with F. Giral.[49] She had done her PhD in London with Derek Barton (1918–98), Nobel Prize winner for chemistry in 1969, and worked on the

---

[45] J. Romo, J. Pataki, G. Rosenkranz, C. Djerassi, 'Síntesis de la cortisona', *BIQ*, 3 (1951), 5–12.
[46] Her sister, Magdalena Rius de la Pola, did her PhD in Germany and later became a professor of chemistry also at the UNAM, working on semiconductors. F. Giral, *Ciencia española en el exilio*, pp. 120–28.
[47] Eugenio Muñoz, *Introducción al estudio de la química* (Mexico City: Atlante, 1943).
[48] María Teresa Urbieta, José Llombart, 'Las actividades científicas del químico Eugenio Muñoz Mena en el País Vasco durante los años anteriores a la guerra civil', in *Los científicos del exilio español en México* (Mexico City: Universidad Michoacana de San Nicolás de Hidalgo, SMHCT, SEHCYT, 2001), pp. 235–58.
[49] J. R. Turvey et al., 'Notes', *Journal of the Chemical Society* (1960), 2366–79.

chemistry of terpenes.[50] Enrique García-Galiano worked on food chemistry at the chemistry faculty of the UNAM and played an important role in the Food and Agriculture Organization at the United Nations (FAO). Artur Bladé also obtained a chemistry degree in Mexico at the INP and counted among his teachers prestigious exiled chemists such as Toral, Bargalló, Muñoz, Adela Barnés and, as in many other cases, F. Giral, with whom he had a close relationship.[51]

Another striking example is the case of César Pi-Sunyer Bayo (1905–97), son of the highly reputed leader of the Barcelona School of Physiology, August Pi Sunyer (1879–1965). Pi-Sunyer Bayo earned degrees in pharmacy and chemistry with JAE fellowships at the *Pharmazeutisches Institut* and the *KWI für Biochemie*, both in Dahlem (Berlin) in 1929, and he became very committed to the Republican values of the 1930s. The academic and political autonomy of the University of Barcelona during the Republic and the experimental style of the *Institut the Fisiologia* (IF) made their mark on his liberal dissent. During the Civil War, Pi-Sunyer Bayo was in charge of the laboratory of the Military Hospital in Barcelona, and he later went into exile in Mexico, where he worked at several chemical companies,[52] and became president of the *Institut Català de Cultura* (1975–78) and the *Orfeó Català de Mèxic* (1971–74). He also supported several Catalan journals such as *Quaderns de l'Exili* and *Nova Revista*. Like F. Giral, he only returned to Spain after Franco's death, but in spite of the awards he received from the restored Catalan government, he returned to Mexico and died there in 1997.

In 1938, President Eduardo Santos invited Antonio García Banús to establish a new *Escuela de Química* in Colombia, with the result that he became a sort of founding father of chemical studies there (Figure 7.1).[53] García Banús designed the project for a new science faculty, becoming a professor of general chemistry and organic chemistry, and later the dean of the faculty. Using materials from the old laboratories of medicine, engineering and pharmacy, he organised a new chemistry department with a new teaching curriculum that included experimental syllabuses.

---

[50] F. Giral, *Ciencia española en el exilio*, p. 131.
[51] Artur Bladé. Personal interview, Barcelona (3 December 2013).
[52] Such as Syntogran and *Laboratorios de Física y Química S. A.*
[53] María Eugenia Martínez Garroño, 'Antonio García Banús en Colombia: Una aportación del exilio español de capital importancia para la química colombiana', in José A. Armillas (ed.), *VII Congreso Internacional de Historia de América*, 3 vols. (Zaragoza: Departamento de Educación y Cultura, 1998), pp. 491–505; Ramiro Osorio Osma, *Historia de la Química en Colombia* (Bogotá: Instituto Colombiano de Cultura Hispánica, 1985); Agustí Nieto-Galan, 'Free Radicals in the European Periphery: "Translating" Organic Chemistry from Zurich to Barcelona in the Early Twentieth Century', *The British Journal for the History of Science*, 37 (2004), 167–91.

Figure 7.1 Antonio García Banús in Caracas in 1955.
A. Nieto-Galan, 'Free Radicals in the European Periphery: "Translating" Organic Chemistry from Zurich to Barcelona in the Early Twentieth Century', *The British Journal for the History of Science*, 37 (2004), 167–91 © 2004 British Society for the History of Science, published by Cambridge University Press. Courtesy of Nuria Alomar.

He introduced new academic chairs in chemical engineering and organic chemistry.[54] In the late 1940s, that fruitful period turned into a much nastier experience when some exiled Republicans suffered persecution and harassment by the Colombian right-wing government. In fact, despite the aforementioned cases of collaboration between academic chemists and the chemical industry in the context of Cárdenas's government, F. Giral also reported hostility in Mexico from conservative groups, who sometimes perceived the chemist refugees as dangerous, revolutionary Marxists.

In 1947, after a tribute to him at the *Universidad Nacional*, García Banús moved to Venezuela, first to the *Universidad de Mérida* and later to Caracas, where he held other chairs and academic responsibilities. He then went to the *Universidad de Los Andes* in Mérida, where he was awarded a chemistry chair at the science faculty and where he founded and directed the *Laboratorio de Investigaciones de Química Orgánica*. In 1953, García Banús became professor of biological chemistry at the Faculty of Medicine at the *Universidad Central de Venezuela*, where he remained until his death in Caracas in 1955. For García Banús, teaching, popularising and institution-

[54] In 1947, in public recognition of his years of commitment to the *Universidad Nacional*, Banús was awarded an emeritus professorship.

building became his main tasks, a far cry from the top research papers he had published in his youth during his time as a JAE *pensionado* at the ETH (see Chapter 1).

The exiled chemists also became exiled chemist–historians, combining the universal past of the great European luminaries of chemistry from the alchemical times with the history of chemistry in Latin America – mining and metallurgy, in particular. They constructed this identity through the expert '*ensayadores*' of the mythical colonial times of mining and metallurgy, which Bargalló had described in detail, but also through eighteenth-century Spanish chemists such as Fausto de Elhuyar and Juan José Elhuyar as discoverers of tungsten in 1783, as well as Andrés del Río, the discoverer of vanadium. They ended up as contemporary generations of chemists: those escaping as refugees from the Spanish Civil War and adapting their skills to the Latin American constraints, but also a second generation of Spanish origin by family but already trained in Latin America and in close contact with the Anglo-American academic and industrial system.

The community also had its own core literature and publications: the generalist journal *Ciencia*, the more specialised *BIQ* and through the active role of Atlante both as a publisher and translator of foreign chemistry textbooks, fiercely opposing contemporary Francoist publications. Obituaries and academic celebrations frequently paid homage to the founding heroes of the new community such as José Giral and Blas Cabrera (for physics). New institutions for teaching and research as well as numerous chemical companies constituted another homeland in which top-level basic research often became subsidiary to more pragmatic local needs in terms of chemistry training and institutional growth. Despite that, many exiled chemists – F. Giral probably being the most paradigmatic example – struggled to achieve external recognition in the new countries where they had settled, as well as international prestige. Ideologically, and particularly under Cárdenas's government in Mexico, many contributed to strengthening public chemical companies, searching for local raw materials to be efficiently exploited, maintaining the old chemical cosmopolitanism and pacifism that they had inherited from the Second Republic and, in the case of J. Giral, even fiercely defending the legitimacy of the Republic in exile before the post-1945 international community.

From the other side of the Atlantic, the exiled chemists constituted a group representing strong liberal dissent against Franco's regime, as well as a line of continuity abroad with the values of the Silver Age. Although, in practice, they were in a minority position when compared with figures in academia and industry in Spain, and although they suffered from geographical isolation and fragmentation, their publications, institutions, labs and rituals became important shelters for those who had escaped from fascism.

Internal Refugees

As the historiography of the Spanish exile during and after the Civil War has clearly shown, the escape from war, dictatorship and repression took varied forms and strategies.[55] Beyond dissidents in exile, internal opposition strategies ranged from soft cultural resistance to more ideologically committed positions, which included several of our chemists particularly opposing Franco's dictatorship. They looked for liberal shelters in industry, but also in some cases in the universities, and even in the media, making their living through science popularisation, for instance.

In the early decades of the century, some liberal companies (in economic terms) defended free trade and expressed doubts about the efficiency of the Spanish state in the promotion of private industry. That feeling was in tune with other claims from the Catalan chemical industry against a centralised, authoritarian state. In 1924, for instance, at the beginning of Primo de Rivera's dictatorship, José Uhthoff had emphasised how private and regional chemistry schools could have their own traditions with no interference from the state.[56] Similarly, often feeling removed from official, centralised policies and regulations, Agell's *Institut de Química Aplicada* (IQA) also promoted a liberal chemistry in terms of its flexible alliances with private entrepreneurs, especially in Catalonia, but the constraints of the dictatorship seriously weakened his project. Even in the pro-industrialist context of the late 1920s, Agell's original organisation of chemistry training was perceived by the central authorities in Madrid as 'too liberal' and too close to political Catalanism.

In fact, liberal professions, among them chemists, contributed to the building of a Catalan national image that was sympathetic to the critics of the inefficiency of the Spanish liberal state and to an ideal of modernity very much focused on northern Europe and that praised Barcelona as the metropolis that could counterbalance Madrid as a powerful centre of the Catalan culture and economy.[57] The industrial dream that Agell and Uhthoff had taken the first steps towards early in the century

---

[55] Borja de Riquer, *La dictadura de Franco* (Barcelona/Madrid: Crítica/Marcial Pons, 2010), p. 298.

[56] José Uhthoff, 'El químico en nuestro país. Lo que es y lo que debe ser', *Química e Industria*, I(2) (1924), 39–41, p. 41.

[57] For a general overview, closely linked to medicine, see: Enrique Perdiguero, José Pardo-Tomás, Àlvar Martínez-Vidal, 'Physicians as a Public for the Popularisation of Medicine in Interwar Catalonia: The Monografies Mèdiques Series', in Faidra Papanelopoulou et al. (eds.), *Popularizing Science and Technology in the European Periphery, 1800–2000* (Aldershot: Ashgate, 2009), pp. 195–215. See also: Joseph Harrison, 'An Espanya Catalana: Catalanist Plans to Transform Spain into a Modern Capitalist Economy, 1898–1933', *Journal of Iberian and Latin American Studies*, 7(2) (2001), 143–56.

continued with Ramon Peypoch's chemistry papers in the journal *Ciència*, reinforcing the European character of modernisation of scientific Catalanism.[58] That liberal spirit also continued in the Republican 1930s and even subtly survived after the Civil War. Despite the public intervention of the INI, totalitarian projects for the organisation of chemistry and the strong control of the autarkic *Sindicato Vertical de Industrias Químicas*, some private companies and a few 'liberal' professionals operated in the chemical and pharmaceutical industries during a good part of Franco's dictatorship. Benefitting from pre-war experiences, liberal values permeated during the times of totalitarianism and autarchy, and in practice acted as shelters for chemists.

Drug shortages during the First World War and the damage caused by the 1918–19 Spanish flu led to a quest for self-sufficiency of the country in terms of pharmaceutical drugs. Financial support for a new company came from engineer Nicolás María Urgoiti (1869–1951), who was the director of the paper factory *La Papelera Española* and who had founded several newspapers[59] and the Calpe publishing house. Close to Republican values, Urgoiti stood for liberal, anti-monarchical ideas. In 1919, he convinced a group of prestigious scientists, trained at Ramón y Cajal's *Instituto Nacional de Higiene*, to found the *Instituto de Biología y Sueroterapia* (Ibys) (Figure 7.2),[60] which originally counted on the

---

[58] In 1926, chemist Ramon Peypoch (1898–1984) had founded a new scientific journal in the Catalan language: *Ciència. Revista Catalana de Ciència i Tecnologia*. The journal was, in fact, in opposition to Primo de Rivera's dictatorship. In 1932, Peypoch became the director of the physics and chemistry section of a new *Societat Catalana de ciències físiques, químiques i matemàtiques* (SCCFQM; Catalan Society for Physical, Chemical and Mathematical Sciences), in tune with the postulates of Catalanism aiming to build up independent institutions beyond the Spanish framework of the RSEFQ. The new society was set up under the banner of the *Institut d'Estudis Catalans* (IEC), a cultural project that began in 1907. Ramon Peypoch, 'Notes històriques sobre la Societat Catalana de Ciències Físiques, Químiques i Matemàtiques', *Butlletí de la Societat Catalana de Ciències Físiques, Químiques i Matemàtiques*, 1(1) (1977), 7–10.

[59] Such as *El Sol, La Voz, Crisol* and *Luz*.

[60] Ibys was founded in 1919 by a group of biologists: Gustavo Pittaluga, Lorenzo Ruiz de Arcaute, Adolfo Cervera, José Sanchís, Sadi de Buen, Gutiérrez Arrese, Tomás Garmendia, among others, and the entrepreneur Nicolás María Urgoiti. In 1936, Ibys had a technical board that included Jorge Francisco Tello, who also directed the Instituto Cajal during the period (1934–9), the prestigious physician Gregorio Marañón, Gustavo Pittaluga, Manuel Tapia, Luis Rodríguez and Antonio Madinaveitia. The director was Antonio Ruiz and the technical secretary was Adolfo Cervera. Ibys had six main sections: bacteriology (Jorge Ramón, Heliodoro del Castillo, L. Gonzalo Urgoiti, José Manuel Gómez), pharmacology (Tomás Alday), veterinary (Julio Hidalgo, Pascual Lucas), diagnostic (Lorenzo Ruiz), pharmaceutical chemistry (Julio Dávila, Nicolás María Urgoiti) and opotherapy (Oriol Utande, Antonio de la Vega). *Ibys, 1919–1944* (Madrid: Gráficas Reunidas, 1944); Javier Puerto, 'Los Laboratorios Ibys. Una excepción científica e industrial durante la dictadura franquista', *História da Saúde. Estudos do Século XX*, 12 (2012), 253–69.

Figure 7.2 Advertisement for the antacid Alugel. *Ibys*, 3(5), 1943 (leaflet).
Reproduced with permission of the *Biblioteca de Catalunya*, Barcelona.

collaboration of prestigious doctors such as Gregorio Marañón and August Pi Sunyer. After playing a relevant role at the UPUEE and in *Ciencia*, while in exile, epidemiologist Gustavo Pittaluga (1876–1956) led the scientific research at Ibys into serums, vaccines and endocrine gland extracts. Medical doctor and socialist militant José Madinaveitia (1897–?), the brother of the leading organic chemist, was another founding member of Ibys.

Ibys resisted the political purging of some of its members and became a sheltering ground for liberal scientists, with a certain acquiescence from Franco's authorities. In fact, Ibys hired a good number of scientists who, in the 1940s and 1950s, had no other professional opportunities and could not return to academia. This was the case for Moles, who, in spite of his painful prosecution, deep humiliation and cruel academic marginalisation, was welcomed unreservedly onto the technical staff.[61] Moles began his collaboration with Ibys in 1943, which lasted until his death in 1953. For a short period, Ibys also sheltered Miguel Catalán during the post-war period in which he suffered marginalisation for his political ideas, before obtaining a position at the CSIC in 1950. Adolfo González, Madinaveitia's right-hand man at the INFQ in the 1930s and a pupil of Staudinger in Zurich, also benefitted from the shelter of Ibys.[62] In 1944, the year of the 25th anniversary of Ibys, Moles belonged to Ibys's technical board and pharmaceutical chemistry section, where he probably had some awkward encounters with Gumersindo García, the former Falangist president of the *Sindicato Nacional de Industrias Químicas* (see Chapter 4).[63]

In the 1950s, Ibys produced a huge range of drugs for the Spanish market, including vitamins, hormones, serums, vaccines, products for diagnosis and other specialist pharmaceutical products. In the same period, and following recent research from the Warfare Service of the US Army, Ibys's journal welcomed papers such as a study of nitrogenated 'yperite' – the mustard gas that was used in the First World War – as a chemotherapy treatment for some kinds of tumour.[64]

---

[61] Antonio Gallego, one of the distinguished members of Cajal's research school, also suffered severe repression after the war in concentration camps and a long, painful military service in Africa, but he managed to make his living at the *Compañía Española de Penicilinas y Antibióticos* (CEPA). Raúl Rodríguez Nozal, *Uriach, Cambronero, Gallego. Farmacia e Industria. La producción de los primeros medicamentos en España* (Tres Cantos: Nívola, 2004). For the case of Moles, see: Enrique Moles, 'El peroxhidróxido de magnesio'. *Revista Ibys, Publicación trimestral del Instituto de Biología y Sueroterapia*, 6 (1946), 1–20.

[62] Canales, Gómez-Rodríguez, 'La depuración franquista', p. 474.

[63] *Ibys, 1919–1944* (Madrid: Gráficas Reunidas, 1944), pp. viii–ix.

[64] J. Andreu, J. Rivero, 'Nuestra experiencia com la mostaza nitrogenada', *Revista Ibys*, 4 (1950), 121–31.

In 1939, the company Zeltia emerged as a specialised laboratory, again as a response to the shortage of drugs for the Spanish market. Very much embedded within the rhetoric of autarchy, Zeltia advertised its products on the back cover of the journal *Ión* as a laboratory for producing insulin and devoted to the physical–chemical exploitation of Spanish raw materials.[65] Zeltia focused on the chemistry of natural products and on the extraction of active substances from local flora, as well as animal extracts of insulin and vitamins (Figure 7.3). In the turbulent post-war years, Fernando Calvet, who had acquired international training in biochemistry (see Chapter 3), but had lost his chair as a result of his liberal profile, soon became the scientific leader of the company. His research into the extraction of alkaloids from rye was the first successful enterprise of the company and provided a way by which he could protect himself from Francoist repression. Extractions from animal glands from slaughterhouses also became key raw materials for Calvet's chemistry of natural products on an industrial scale. Equally, organic chemist Ignacio Ribas was one of Zeltia's closest collaborators. Despite his marginalisation at the end of the Civil War, Ribas's expertise in the chemistry of natural products was of great value to the company. Zeltia also welcomed Andrés León, who, after receiving international training with Robert Robinson, ended up working in slaughterhouse labs, but still managed to publish some academic papers.[66] León worked at Zeltia on sulphonamides.[67]

In 1949, Zeltia contributed to the creation of *Antibióticos S. A.*, a joint entrepreneurial venture.[68] After the visit of Alexander Fleming to Francoist Spain in 1948 – he gave three lectures in Barcelona in June that year before a large local audience of medical doctors and local authorities[69] – the penicillin business gave way to the private initiatives of chemical and pharmaceutical firms, contributing to a certain liberal

---

[65] 'Laboratorio especializado, Insulina. Explotación químico-física de materias primas españolas' *Ión*, 8(2) (1942) (back cover). See also: Enrique Raviña, *The Evolution of Drug Discovery: From Traditional Medicines to Modern Drugs* (Weinheim: Wiley-VCH, 2010).

[66] See, for instance: Andrés León, Robert Robinson, 'Experiments on the Synthesis of Anthocyanins. Part XIV. Cyanenin Chloride, Malvenin Chloride, and an Indication of the Synthesis of Cyanin Chloride', *Journal of the Chemical Society* (1932), 2221–4.

[67] F. Giral, *Ciencia española en el exilio*, p. 55.

[68] In the 1960s, Zeltia diversified its production to agriculture and veterinary and human drugs, and later, by the 1980s, it exploited marine antitumour chemicals through the new firm Pharmamar.

[69] 'Sir Alexander Fleming's visit to the capital of Catalonia as the guest of its municipality has proved a great success [...] He delivered three lectures [...] "Some Problems of a Sceptic Wound", "Lysozome", and "The Use of Penicillin"'. 'Sir Alexander Fleming in Spain', *British Medical Journal*, 26 (1948), p. 1248.

Figure 7.3 Zeltia advertisement in the journal *Ión*, 8 (1942) (back cover).
Reproduced with permission of the *Biblioteca de Catalunya*, Barcelona.

industrial culture in the face of the official autarchy.[70] Zeltia's drugs – ephedrine, digitaline, hepatic extracts, foliculine, insulin, purpuripan and all kinds of vitamins, together with the production of the pesticide DDT – became key in the Spanish pharmaceuticals market, providing Calvet with a remarkable scientific reputation that allowed him to recover his university chair, first in Salamanca in 1944, and later in Barcelona (see Chapter 3).[71]

Other liberal responses appeared in the universities. In 1957, the chemist Francisco Buscarons (1906–89) resigned as rector of the University of Barcelona over disagreements about student repression.[72] Trained as a chemist in Zaragoza, Buscarons had been director of the chemical analysis laboratory of the Barcelona customs and excise building for 16 years (1927–43) and later obtained a chair in analytical chemistry, from which he achieved a considerable reputation in terms of research, teaching and collaboration with the chemical industry.[73] Buscarons' opposition to repression probably cloaked a hidden desire for academic autonomy in a very hostile political context. The tram strike of 14 January 1957 had the support of the university students, and it resulted in fierce police repression. The following day, students gathered at the campus of the University of Barcelona and the police entered the premises. Buscarons in person tried to stop the turmoil, but the police dismissed his authority and proceeded to make more than 40 arrests. As a result, the rector resigned. Despite strong official reluctance, the ministerial authorities finally accepted Buscarons' resignation, and the student protests gained further strength over the following months.[74] The event took place while Christian Democrat Joaquin Ruiz Giménez (1913–2009) was Minister for Education, and he had tried to appoint young rectors in a timid opening up of the Francoist universities.

---

[70] María Jesús Santesmases, *Antibióticos en la autarquía: banca privada, industria farmacéutica, investigación científica y cultura liberal en España, 1940–1960* (Madrid: Fundación Empresa Pública, 1999); María Jesús Santesmases, 'Distributing Penicillin: The Clinic, the Hero and Industrial Production in Spain, 1943–1952', in Viviane Quirke, Judy Slinn (eds.), *Perspectives on Twentieth-Century Pharmaceuticals* (Oxford: Oxford University Press, 2010), pp. 91–118. See also: María Jesús Santesmases, *The Circulation of Penicillin in Spain: Health, Wealth and Authority* (London: Palgrave Macmillan, 2018).

[71] 'Fernando Calvet Prats', *Anthropos*, 35 (1984), 1–27 (special issue); Joaquim Sales, *La química a la Universitat de Barcelona*, pp. 159–62.

[72] Buscarons accumulated considerable scientific prestige in Spain and abroad as an analytical chemist. Among his most important publications, the following are worth mentioning: Francisco Buscarons, *Análisis y arancel de productos químicos* (Barcelona: Herder, 1936); Francisco Buscarons, *Análisis inorgánico cualitativo sistemático* (Barcelona: Herder, 1942).

[73] *Universidad de Navarra. Archivo General. Fondo José María Albareda*. Correspondence Albareda-Buscarons, G-62-68 (1–5).

[74] Sales, *La química a la Universitat de Barcelona*, pp. 37–8, 166–8.

Buscarons could possibly be perceived as being far removed from any explicit liberal tradition, being too busy with his academic research and teaching at the university and his customs laboratory. However, his resignation placed him in no man's land between the right-wing Catholicism of Vitoria, Lora-Tamayo and Pascual and the totalitarian dreams of Jimeno and Albareda.

The 1960s brought major waves of academic dissent, and some liberal voices began to gain influence in the censored public sphere. Sociologist Salvador Giner described the demands of these group as 'classic liberal': free students unions, modernisation of the educational system, more support for research, freedom of speech and political, religious, cultural and national plurality.[75] The Spanish university crisis of 1965, for example, which led to police repression and the expulsion of a large number of students and distinguished professors, again brought to the fore a lively debate about lost liberal values in academia and in society and the need to recover political pluralism.[76] In 1966, the year of Albareda's death, students and professors gathered 'illegally' in a monastery in Barcelona to agree on the nature of a new democratic students association (*la Caputxinada*), but repression by the *Ministerio de Gobernación* immediately followed. Police entered several campuses and Lora-Tamayo, the Minister of Education and Science, expelled a good number of university students and professors.[77]

From outside the university system, other civic, liberal shelters appeared. Miguel Masriera sought out industrial shelters, but also used science popularisation as a strategy for survival in Franco's regime (Figure 7.4). Back from exile, and with no hope of recovering his pre-war academic university position, he established useful contacts with local and foreign chemical entrepreneurs. His scientific reputation attracted several firms for employment as a technical assessor, as well as foreign companies for consultancy work on the potential exploitation of Spanish

---

[75] Salvador Giner, 'Universitat, moviment estudiantil i poder polític a l'estat espanyol, 1939–1975', in Guillermo Lusa, Antoni Roca-Rosell (eds.), *Fem memòria per fer futur. La Universitat sota el franquisme. III Jornades Memorial Democràtic a la UPC. 16 i 17 de novembre de 2005* (Barcelona: UPC, 2005), pp. 27–67, 45. See also: Salvador Giner, 'Power, Freedom and Social Change in the Spanish University, 1939–1975', in Paul Preston (ed.), *Spain in Crisis: The Evolution and Decline of the Franco Regime* (New York: Barnes & Noble, 1976), pp. 183–211.

[76] Agustí Nieto-Galan, 'Reform and Repression: Manuel Lora-Tamayo and the Spanish University in the 1960s', in Ana Simões, Kostas Gavroglu, Maria Paula Diogo (eds.), *Sciences in the Universities of Europe, Nineteenth and Twentieth Centuries: Academic Landscapes* (Dordrecht: Springer, 2015), pp. 159–74.

[77] Solé Tura, Bricall, Raventós, Roca, Fontana, Lluch, Molas, Boigas, etc. Josep Maria Colomer, *Els estudiants de Barcelona sota el franquisme* (Barcelona: Curial, 1978), pp. 257–8; Elena Hernández et al., *Estudiantes contra Franco (1939–1975)* (Madrid: La Esfera de los Libros, 2007), p. 184.

Figure 7.4 Miguel Masriera in 1956 at the BBC.
A. Nieto-Galan, 'From Papers to Newspapers: Miguel Masriera (1901–1981) and the Role of Science Popularization under the Franco Regime', *Science in Context*, 26(3) (2013), 527–49 © Cambridge University Press 2013. Private collection of the author.

raw materials. Always involved on the margins of the academic system, Masriera never worked in industry full time. However, his joint projects with private entrepreneurs ranged from focusing on carbides and ketone derivatives, to the food supply, water purifiers, tanning chemicals and spectroscopic measures.[78] In addition, as discussed earlier (see Chapter 3), Masriera was a prolific scientific journalist at *La Vanguardia Española*,[79] but his articles also appeared in the newspaper *ABC* and the periodical *Destino*, the latter being considered one of the most 'liberal' publications of the late 1950s.[80]

---

[78] Agustí Nieto-Galan, 'From Papers to Newspapers: Miguel Masriera (1901–1981) and the Role of Science Popularization under the Franco Regime', *Science in Context*, 26(3) (2013), 527–49.
[79] Ibid.
[80] Isabel de Cabo, *La resistencia cultural bajo el franquismo. En torno a la revista 'Destino' (1957–1961)* (Barcelona: Altera, 2001). See also: Ramon Civit, *Destino i la cultura*

In 1948, in a letter he sent to Fréderic and Iréne Joliot-Curie, Masriera wrote, '*Je travaille ici isolé et tout à fait en dehors de la vie officielle espagnole*'.[81] His active role as a scientific journalist and his multiple publishing businesses allowed him a certain intellectual autonomy.[82] His excellent article on Dodge's chemical engineering course at the University of Barcelona early in 1954 (see Chapter 5) was a tacit endorsement of Calvet's cosmopolitan, liberal and open mind. In Masriera's words:

> [...] I know the short-sighted who believe that calling professors from outside is to lower the category of our own. For that reason, I admire more the determination with which the Chair of Technical Chemistry of the Faculty of Sciences [Calvet], with the enthusiastic support of the Rectorate and the University [Buscarons], has carried out this feat, overcoming the economic and academic difficulties to launch that international event. Good scholars do not lose out when they brush shoulders with the best foreigners, those who seek to bring fresh ideas to their chair and put their students in contact with the environment of other countries which are technically more advanced than us.[83]

Masriera represented a soft liberal resistance and a moderate conversion to the values of Francoism, in a context in which the press, his popular science articles and books and his cosmopolitanism from the pre-war times acted as an ideal shelter for his professional strategy. He was marginalised in the academic system, but he was able to earn his living through alliances with the Francoist elite. Old links with university colleagues, together with new projects with emergent industrialists and publishers, completed a subtle network of sociability in which his scientific background and popularising skills played crucial roles. Masriera's science popularisation helped him to survive in a regime that looked with suspicion on his liberal past, but that he served nonetheless. His popular science spread a cosmopolitanism that was tolerated by the regime's censorship and, at the same time, reinforced the anti-communist values in the Cold War.

Other chemists sought to bring fresh ideas to their chairs in other institutions outside the Francoist university system. During the Civil War, Moles's pupil, Augusto Pérez Vitoria, became Secretary of the 'Z'

---

catalana a les acaballes del franquisme (1966–1975). PhD thesis (Barcelona: University of Barcelona, 2012).

[81] 'I work in isolation here, outside the Spanish official life'. Fons Masriera. 3.2.C. Correspondence. Letter: 22 December 1948. IEC. Barcelona.

[82] For ambiguities regarding collaboration with the Nazi regime, see, for instance, the case of the chemist Peter Debye. Jurrie Reiding, 'Peter Debye: Nazi Collaborator or Secret Opponent', *Ambix*, 57(3) (2010), 275–300.

[83] Miguel Masriera, 'El curso del profesor Dodge', *La Vanguardia Española*, 12 March 1954, p. 5.

Services division of gunpowder and explosives, which controlled the chemical industry of the Republican side for a good part of the Civil War (see Chapter 3). He went into exile to Paris and survived professionally for years working in several chemical industries. Pérez Vitoria later joined UNESCO, and, in the 1960s, became Head of the Division of International Co-operation in Scientific Research, Department of Natural Science in Paris, as well as an editor of the journal *Impacto*.[84] Just like Bargalló in Mexico and Masriera in Barcelona, Pérez Vitoria worked on the popularisation of chemistry, a topic that since the 1955 International Conference in Madrid had constituted a useful tool for the legitimation of Franco's regime (see Chapter 6), but also was an ideal shelter for marginalised, liberal chemists. As with the case of F. Giral, in 1976, Pérez Vitoria recovered his chair in inorganic chemistry at the University of Murcia that he had gained in 1935 before the Civil War. In the late 1970s and 1980s, he campaigned for the recuperation of the memories of exiled chemists through the *Aula de Cultura Científica* in Santander.[85] F. Giral and Pérez Vitoria were two very exceptional cases, however, and the only two Spanish chemists who, having been repressed in the Civil War, finally recovered their old university chairs. In spite of that, their liberal ethos dramatically clashed with the remaining Francoist culture of the Spanish university system at the end of the dictatorship.

***

The exiled Latin American chemical community protected the Republican values of internationalism and pacifism and combined them with a liberal, flexible relationship (in economic terms) with private chemical firms, but frequent commitments in favour of public companies (oil, hormones and the extraction of natural products). Although prestigious names such as F. Giral and Madinaveitia carried out their previous research projects in Republican Spain, they adapted them to the new raw materials and built up new chemical institutions in a new country – Mexico – that welcomed them for their political affinities, but demanded new ways of defining pure and applied chemistry research in new institutional settings. Their ethos was constructed to a large extent in opposition to the values represented by

---

[84] Augusto Pérez Vitoria, 'Las políticas de ciencia y de la tecnología de la UNESCO', *Arbor*, 495 (1987), 67–92.
[85] In recent decades, under the leadership of Francisco González Redondo and Francisco González de Posada, *Amigos de la Cultura Científica* has published popular booklets, exhibition catalogues and books on Moles, Cabrera, Palacios, Catalán, del Campo, and the inventor Leonardo Torres Quevedo. www.todostuslibros.com/editorial/amigos-de-la-cultura-cientifica/page:1 (last accessed 10 September 2018).

Francoist chemistry. They also became chemist–diplomats for the survival of the Spanish Republicans in exile.

For years after the Civil War, industrial shelters continued to be the only professional option for the marginalised and repressed, such as Moles, Calvet, Catalán and Ribas. But other shelters appeared in the universities and in the public sphere for cases such as Buscarons and Masriera, respectively. Although they experienced and built on the same dictatorial regime that Lora-Tamayo enthusiastically endorsed, their political positions differed substantially. Whereas Lora-Tamayo epitomised a deep alliance between chemistry and power under a public rhetoric of technocracy and apoliticism (see Chapter 6), Buscarons set his own limits of academic power as university rector at the time of the student upheavals of the 1950s, and Masriera constructed his own survival shelters in the press and through his collaboration with private companies.

Literary historian Jordi Gracia has discussed in depth the 'silent resistance' to the regime of some intellectuals and the awakening of a critical consciousness during the dictatorship.[86] Gracia links the Spanish liberal tradition to the intellectual milieu of the ILE, Krausist philosophy and, as a logical result, to the Silver Age of the JAE. Attempts to recover liberal values after the catastrophic consequences of the Civil War inevitably sought out references to the early decades of the century. In Gracia's view, the long-standing liberal tradition was rooted in the Enlightenment and in intellectuals such as Miguel de Unamuno (1864–1936), Pío Baroja (1872–1956) and Juan Ramón Jiménez (1881–1958), but also Isaiah Berlin and John Rawls (1921–2002). It evolved in three different branches during Francoism: those with some early liberal hopes;[87] those trained as fascists, who attempted to integrate liberalism into the regime;[88] and the young Falangists, who moved towards liberalism and Marxism in late Francoism.[89]

This framework may also apply in our case. Following the university crisis of the 1960s, and in spite of the official optimism regarding economic growth, some chemical shelters had begun to challenge the values of the dictatorship, which formally ended after Franco's death on 20 November 1975. The *Caudillo* passed away just two years after the

---

[86] Jordi Gracia, *La resistencia silenciosa. Fascismo y cultura en España* (Barcelona: Anagrama, 2004). See also: Jordi Gracia, *Estado y cultura. El despertar de una conciencia crítica bajo el franquismo (1940–1962)* (Toulouse: Presses universitaires du Mirail, 1996); Javier Tusell, *Franco y los católicos. La política interior española entre 1945 y 1957* (Madrid: Alianza, 1984).

[87] Ortega, Marañón, Pla and Azorín. Gracia, *La resistencia silenciosa*, p. 34.

[88] D'Ors, Torrente Ballester, Laín Entralgo, Aranguren and Maravall. Ibid.

[89] Valverde, Sacristán, Castellet and Ferraté. Ibid.

oil crisis of 1973, which for the first time had seriously jeopardised the liberal values of welfare states during the Cold War and their mission to fight communism, but also to provide the necessary material well-being for all citizens in order to preserve their individual freedom. That oil crisis opened a new period in the history of Western liberalism – a return to the classical economy, which opposed the redistribution of wealth and state intervention in private business. This also opened a new period for Spain in which an economic crisis jeopardizing the developmental policies of the 1960s ran in parallel with a very complex political process of transition from a dictatorship to a liberal democracy.[90] At the beginning of this new period – with hopes for a better democratic future, but fears of the old, long-standing dictatorship – F. Giral's disappointing experience in the post-Franco Spanish university sadly brought him back to his Mexican shelter.

[90] Albert Carreras, Xavier Tafunell, *Historia económica de la España contemporánea* (Barcelona: Crítica, 2004).

Conclusion: The Moral Ambiguity of Chemistry

Although readers might claim that the previous chapters of this book talk about a particular case – the role of chemistry in twentieth-century Spain – my feeling is that having reached this point, this particular case tells us a lot about the status and the nature of chemistry as a profession in the last century and the ways in which chemists as experts in academia and industry co-constructed different political regimes. In a way, this is a book about a paradox: with chemistry being frequently presented as an apolitical, neutral, objective, technocratic field, it has played a significant political role in our age of extremes, from the German science-based heritage that preceded the First World War to the petrochemical and instrumental revolution in chemistry during the Cold War. Synthetic molecules in the laboratory and on an industrial scale, the analytical quality control of raw materials and finished products, standard chemical training in specialisations such as inorganic, organic, physical, medical, industrial and technical chemistry (all of them emerging fields in universities during the nineteenth century), mass consumption of colourants, drugs and plastics and a new culture of artificiality and material well-being constituted a complex sociotechnical network[1] in close alliance with political and economic elites, but apparently detached from any moral responsibility – at least until the 1970s, the period in which this book ends.

In our case, it was perhaps Enrique Moles, one of the key actors in the previous chapters, who most clearly epitomised that *moral ambiguity of chemistry* throughout the century. As we have discussed, Moles managed to develop his research projects in physical chemistry and particularly in the experimental determination of atomic weights under several political regimes, from the Monarchic Restoration of the early twentieth century and the seminal research policies of the JAE, Primo

---

[1] The theoretical framework that historians of technology, and science and technology studies (STS) scholars have developed in recent decades could be extrapolated to chemistry here. Wiebe E. Bijker, Thomas P. Hughes, Trevor Pinch (eds.), *The Social Construction of Technological Systems* (Cambridge, MA: MIT Press, 1987).

de Rivera's dictatorship – in which the project of the INFQ finally took shape under the industrialist aims of the regime, as reflected in the 1929 Chemistry Pavilion (*Palacio de la Química*) – the cosmopolitan dream of the Republic, the Civil War and Moles's involvement in the production of war weapons, and even at the beginning of Franco's dictatorship, when he tried to return from exile assuming that his chemistry would be compatible with the new power. Similarly, Lora-Tamayo's diene synthesis and its industrial applications in Franco's dictatorship seemed perfectly compatible, in his view, with his Ministry for Education and Science under the rhetoric of reform, modernisation and progress. Equally, the new petrochemical complexes of the 1960s sealed the old alliance between Spanish chemical entrepreneurs, foreign corporations and the 'developmentism' of Franco's regime in a context of demographic growth and increasing demand for chemicals for the material improvement of everyday life for the newly emerging middle classes. Here again, all chemicals seemed to circulate uncritically through society as innocent and efficient black boxes that brought about progress and tacitly legitimised power.

But the moral ambiguity of chemistry also had other faces. First, the never-ending tension between pure and applied chemistry – that is to say, the continuous and often conflicting relationship between academic research and industrial skills – has deeply shaped the nature and status of chemistry throughout history, but it was particularly idiosyncratic in the twentieth century, and more specifically in twentieth-century Spain. Second, the apparently apolitical character of chemistry became an ideal tool for those who wanted to appropriate it for varied and often conflicting projects of modernisation, from the cosmopolitan anti-war campaigns to the enthusiastic collaboration with totalitarian regimes. Third, the history of chemistry as a profession in the twentieth century is a history of a dynamic, changing and, in the Spanish case, troubled identity. It is a history of chemical societies, university faculties, research institutes, research schools, frequently damage jeopardy from wars (the First World War, the Second World War, the Spanish Civil War and the Cold War), the geopolitics of raw materials, the internationalisation of the chemical industry, the new consumer society of synthetic products and, again, the chemists' more or less explicit collaboration with anti-liberal regimes. Finally, an overview of chemistry as a profession in the twentieth century probably brings us to a deeper reflection on the role of chemists as *intellectuals* – in the broad sense of the word – from the glamorous, pacifist, anti-nuclear campaigns of double Nobel Prize winner Linus Pauling to the more obscure, routine, everyday practices of chemistry teachers and industrial chemists. This concluding chapter will close

Pure–Applied Chemistry

The pure–applied status of chemistry has been a topic of huge historiographical debate for decades.[2] In terms of the modern history of chemistry, we could begin with Charles Gillispie's seminal and illuminating paper on the 'natural history' of the chemical industry in the Enlightenment[3] and continue with Archibald and Nan Louise Clow's coinage of the term 'chemical revolution' for a history that was, in practice, the evolution of the chemical industry in the eighteenth century,[4] to later travel to the science-based industrial pattern of the late nineteenth-century German chemical industry[5] and the twentieth-century framework of 'unit operations' for the new chemical engineering.[6] As discussed in several chapters of the book, actors' claims for the pure and applied nature of chemistry have carried much political weight, both in terms of professional interests, but also as a broader political strategy in the construction of a particular regime. Moreover, what makes the pure–applied dichotomy particularly useful in the Spanish case is its persistence throughout the century with different terms and priorities, often in tune with specific values and world views.

---

[2] Christoph Meinel, 'Theory or Practice? The Eighteenth-Century Debate on the Scientific Status of Chemistry', *Ambix*, 30 (1983), 121–32; Robert F. Bud, Gerrylyn K. Roberts, 'Chemistry and the Concepts of Pure and Applied Science in Nineteenth-Century Britain', in Franco Calascibetta, Eugenio Torracca (eds.), *Storia e fondamenti della chimica: Atti del II Convegno Nazionale di Storia e Fondamenti della Chimica* (Rome: Accademia Nazionale delle Scienze detta dei XL, 1988), pp. 19–33. For a more general reflection on pure–applied science, the collective volume *The Mindful Hand* (2007) aims to overcome the pure–applied, science–technology, hand and thought dichotomy through sites and practices and by tracing back these kinds of categories through the voices of the actors. Lissa Roberts, Simon Schaffer, Peter Dear (eds.), *The Mindful Hand: Inquiry and Invention from the Late Renaissance to Early Industrialisation* (Amsterdam: Koninklijke Nederlandse Akademie van Wetenschappen, 2007). See also: Robert F. Bud, 'Introduction', *Isis*, *Focus: Applied Science*, 103 (2012), 515–17.

[3] Charles Gillispie, 'The Natural History of Industry', *Isis*, 48 (1957), 398–407.

[4] Archibald Clow, Nan Louise Clow, *The Chemical Revolution: A Contribution to Social Technology* (London: Batchworth Press, 1952).

[5] Georg Meyer-Thurow, 'The Industrialization of Invention: A Case Study from the German Chemical Industry', *Isis*, 73(3) (1982), 363–81.

[6] Clive Cohen, 'The Early History of Chemical Engineering: A Reassessment', *The British Journal for the History of Science*, 29 (1996), 171–94. See also: Jeffrey Allan Johnson, 'The Academic–Industrial Symbiosis in German Chemical Research, 1905–1939', in John Lesch (ed.), *The German Chemical Industry in the Twentieth Century* (Dordrecht: Kluwer, 2000), pp. 15–56.

In spite of the campaign by the Francoist chemists against their JAE colleagues for their supposed overly theoretical, 'pure' research, this book has provided multiple examples of 'applied' chemistry projects before the Civil War. Of particular note is the early dissemination of industrial chemistry journals such as *Química e Industria*. This is the justification for using the term 'industrial Silver Age' to describe the Spanish chemical culture of the JAE years, which had been, until now, poorly considered by historians of science and economic historians. The significant role of chemical industry projects in the early decades of the twentieth century therefore calls into question an overly simplistic image of a supposedly poorly developed country and a latecomer to industrialisation.[7] It was again Moles, in his address on the relations between university and industry, who, in 1929, defended the German tradition of the compatibility of pure, basic chemistry research – in his sophisticated experimental atomic weight measures – and industrial innovation.[8] In political terms, Moles was suggesting the need for a certain freedom in research – in a kind of Mertonian ethos – of a 'pure' chemistry that could ideally exist independently from government constraints; a basic, pure chemistry training as a platform for further industrial applications while retaining the intellectual value and prestige of the pure knowledge in itself. But Moles's dream in the 1930s collapsed during the Civil War in a context in which, as was the case in the two World Wars, research priorities, laboratory synthesis and large-scale production soon became the handmaidens of military needs at the front and led to the mass production of chemical weapons. The traumatic experience of the war acted as a training arena for the new spirit of 'applied' chemistry that emerged strongly from the beginnings of Franco's dictatorship.

It is not worth discussing here the main consensus among economic historians on the 1960s being a turning point for the industrialisation of Spain. However, from the chemistry perspective, there is no doubt that the Silver Age saw multiple efforts to strengthen links between academic and industrial chemistry. The industrial Silver Age helps us to revisit the Francoist criticism of the JAE for its elitist, centralised, 'pure' (that is to say, useless) research in chemistry, but also to question the old prejudice of considering science and technology (chemistry in our case) as a backwards science – or even a pseudoscience – in dictatorships.[9] In

---

[7] Jordi Nadal, 'La debilidad de la industria química española en el siglo XIX. Un problema de demanda', *Moneda y Crédito*, 176 (1986), 33–70.
[8] Enrique Moles, 'La Universidad y la Industria', *Química e Industria*, 7 (1930), 148–53.
[9] See, for instance: Robert Proctor, 'Nazi Science and Nazi Medical Ethics: Some Myths and Misconceptions', *Perspectives in Biology and Medicine*, 43(3) (2000), 335–46. See also: Jon Agar, *Science in the Twentieth Century and Beyond* (Cambridge: Polity Press, 2012).

fact, as discussed in several chapters, applied science – and applied chemistry in particular – became one of the key aims of Franco's dictatorship, and it was a useful way of co-constructing the regime in which many chemists played an important role, from Albareda's soil chemistry stations to Pascual's applied chemistry projects at the PJC in collaboration with his university research group. Moreover, in the autarchic 1940s and early 1950s, raw materials and their spread across the country acted as driving forces for political projects, which were frequently in the hands of the chemists. In a way, the spirit of chemistry that we find in articles in *Ión* and *Revista de Ciencia Aplicada* are good reflections of the ambitions of an euphoric totalitarian regime following its victory in the Civil War, which aimed to shape the country's raw materials in terms of its own political interests. In that context, chemistry became an ideal science for autarchy due to its economic but also political weight. In a way, pure chemistry and basic fundamental research – such as Moles's work on the determination of atomic weights, Catalán's spectroscopy and Madinaveitia's synthesis of natural products – were too often associated with liberal values, which the dictatorial, anti-liberal contexts were keen to suppress. The chemical culture of the INFQ clashed after the Civil War with the atmosphere of the new research institutes of the PJC.

In 1960, Manuel Lora-Tamayo published a paper in *Revista de Ciencia Aplicada* based on a lecture he gave in honour of Kurt Alder at the German Institute of Culture in Madrid. Lora-Tamayo wanted to emphasise the applied science agenda of diene synthesis for the production of hormones (steroids synthesis), morphine and insecticides.[10] But claims for a stronger pure chemistry also reappeared. Some pupils from Pascual's research school defended it again – tacitly recovering the 1929 spirit of Moles, but obviously without mentioning his name. In the 1960s, with the establishment of international chemical corporations across Spain, and the emergence of the new *polos químicos* or chemical plants, foreign chemical expertise – a new 'pure' chemistry – passed into foreign hands in the central premises of companies such as BASF, Hoechst, Bayer, DuPont and the ICI, leaving research at the PJC, the CSIC and the universities in a subsidiary position. The liberalisation of the chemical industry in the 1960s can therefore also be analysed again under the pure–applied divide.

Vannevar Bush's dream of a better world, both materially and morally, through science (and chemistry) at the end of the Second World War legitimised the applied chemistry programme of the PJC during Franco's

---

[10] Manuel Lora-Tamayo, 'La obra científica de Kurt Alder', *Revista de Ciencia Aplicada*, 74 (3) (1960), 193–205.

dictatorship, but did not bring democracy to Spain. In a way, the applied chemistry programme hindered individual political freedom and scientific creativity and reinforced totalitarian political structures. Paradoxically, the old Mertonian utilitarianism, once appropriated through chemistry in Francoist Spain, did not contribute to the democratisation of the country, which only formally occurred in the late 1970s after Franco's death.

Modernisation Paradoxes

Spanish chemists contributed to varied and often opposed programmes of modernisation in different political regimes. In the early decades of the century, the emerging chemical community reified its modernity in projects of improved chemical training, well-equipped laboratories and the creation of research schools, industrial links and international networks. They adapted their professional ambitions pragmatically, first to the last decades of the Restoration, then to Primo de Rivera's dictatorship – which can be considered as a late authoritarian result of the monarchy itself – and finally to the liberal values of the Second Republic. In a way, however, the novelty of the INFQ was particularly uncomfortable for the group of right-wing conservative chemists, such as Rocasolano's school, which harboured opposed ideas of modernity. The rhetoric of modernisation, modernity and progress was, however, in the hands of actors from both sides of the Civil War, and also in the propaganda of the 1934 International Chemistry Conference in Madrid and later in the grandiose CSIC celebrations in the early years of Franco's dictatorship. It is doubtful, though, that a very big difference can be established between the chemistry of natural products at the INFQ led by socialist Madinaveitia and the Catholic conservative Rocasolano's early biochemical school. After the Civil War, reactionary forms of modernity provided expertise and an ambitious agenda of applied chemistry.

The multiple faces of modernity can be also traced back through chemical diplomacy. In our case, what is really striking is the way in which Spanish chemists played a significant diplomatic role in Republican times, such as in the organisation and evolution of the 1934 International Conference, but at the same time there is solid evidence of Lora-Tamayo, Albareda, Pascual and others' use of the international chemical network, mainly at the end of the Second World War, to re-establish old scientific relations for the sake of the legitimation of the new regime. This allows us to revisit standard accounts of isolation and pure autarchic policies in early Francoism. Cosmopolitanism, therefore, which is frequently associated with modernity, had multiple faces and opposing political aims in the hands of Spanish chemists.

Again, the case of Moles describes very well the clash of opposing modernities and the construction of a historical narrative of discontinuity, which reinforced the identity of the Francoist chemists. After years of imprisonment and persecution, in 1949, Moles managed to get permission to travel abroad. Specifically, he attended the 15th IUPAC Conference in Amsterdam and he rejoined the atomic weights commissions. In April 1950, he enjoyed the public recognition of the *Societé chimique de France* and gave a lecture at '*La maison de la chimie*'; in 1951, he travelled to the IUPAC Conference in Washington, DC, and later received public recognition by the *Academia de Ciencias de Cuba*. In Spain, the press reported on these events, presenting Moles as a 'Spanish scientist', but never mentioning his repression and marginalisation under the new regime.[11] After his death in 1953, *Anales* reported on the event in a very cold, bureaucratic way, with a very short biographic profile that ends abruptly in 1934.[12]

In April 1953, only 15 days after Moles's death, the RSEFQ lavishly commemorated its 50th anniversary. Its president, Lora-Tamayo, acting once again as a chemist–historian, addressed an international audience, describing the evolution of the RSEFQ from its foundation in 1903 to its present state at the heart of the new regime. He mentioned the International Conference of 1934 in Madrid, an *annus mirabilis* that was difficult to forget, but, astonishingly, said not a single word on Moles.[13] In fact, Lora-Tamayo, Pascual and Moles met at the 1949 IUPAC Conference in Amsterdam. Unfortunately, there is no written evidence of the kind of dialogue (or cold silence) that the three chemistry luminaries might have established during their coffee breaks. However, what seems obvious to me is the inevitable clash of opposing modernities, in political terms, which the three men had to face in those circumstances under the banner of the apparently 'technical' topics for discussion of the IUPAC.

Although Lora-Tamayo admitted that the period of 1940–50 was one of isolation and convalescence from the Civil War, he envisaged the need to re-establish old international contacts through travels abroad

---

[11] The conservative newspaper *ABC* (10 May 1950) devoted half a page to the trip to France, with an article entitled '*Conferencia de un sabio español en la Maison de la chimie*', and the periodical *Semana* devoted the section '*Gente de la que apenas se habla*' (People we seldom talk about) to coverage of his trip to different European capitals. In all these articles, Moles is presented as a Spanish scientist who is famous abroad, but never referring to the marginalisation and repression he suffered.

[12] 'Necrológica de D. Enrique Moles Ormella', *Anales de la Real Sociedad de Química*, 49(B) (1953), 327–8.

[13] Manuel Lora-Tamayo, *Cincuenta años de Física y Química en España, 1903–1953. Bodas de Oro de la Real Sociedad Española de Física y Química* (Madrid: C. Bermejo, 1954).

and invitations from numerous foreign chemists in an appropriation – now for other political purposes – of the old cosmopolitanism of the JAE.[14] In practice, Francoist chemists maintained a good part of the scientific culture of the pre–Civil War generation: a research school model, international training and publications and an emphasis on experimental culture and well-equipped laboratories. Their research appeared in public as serving the aims and values of the new totalitarian regime: applied chemistry as a handmaiden of the chemical industry. This was indeed part of the modernisation paradox that Jimeno had also created at the end of the Civil War with his totalitarian project for chemistry in the new Francoist Spain – a plan that combined better training and a stimulus for chemical entrepreneurs and skilled workers under the rigid control of the fascist state.

A Troubled Identity

This book also reflects on the ways in which Spanish chemists behaved and identified themselves as a community, at least according to the voices of the most significant figures. This is obviously a very troubled story containing dissidence, revenge and profound reorganisations, which will probably require further research in the near future. What needs to be brought to the fore, however, is precisely that long-term conflict and the mechanisms the different groups used to build and preserve their own identities. As described in the previous chapters, the 'silver' generation of chemists suffered a traumatic split. Exile, internal repression and marginalisation dramatically reduced the number of academic chemists,[15] and so promising research schools of solid international reputation lost their momentum and could only partially reappear in exile and later in the dictatorship. The Francoist obsession with erasing any trace of the scientific culture of the JAE and the defeat of 'Republican' Spain did not, however, hinder the partial survival of the Silver Age community mainly in exile in Latin America, but also in a series of cases of internal liberal dissent and subtle resistance to the dictatorship. In exile, part of the community adapted their skills to new local demands and took advantage of the new land's natural resources and political, academic and industrial support.

---

[14] Manuel Lora-Tamayo, *Discurso leído en el acto de su recepción el día 21 de enero de 1948 en la Real Academia de Ciencias Exactas, Físicas y Naturales* (Madrid: Real Academia de Ciencias Exactas, Físicas y Naturales, 1948).
[15] Francisco Giral, *Ciencia española en el exilio (1939–1989)* (Barcelona: CIERE/Anthropos, 1994), p. 21.

Many accommodated themselves to Franco's regime, but retained aspects of the old scientific culture and reused their former international connections to run the new groups. Although in public they ignored or even dismissed the achievements of their masters' generation, in practice they attempted to reproduce in part their old scientific identity. After the Civil War, there were numerous attempts to emphasise a genealogy of radical discontinuity and rupture, but both groups – victims and victors – shared feelings of weakness and a peripheral status, which catalysed frequent, desperate attempts for international recognition. Equally, experimental techniques, research schools, material practices and research projects quietly continued. Despite the clashes over identity and hegemony, personal memories played a crucial role in the construction of chemists' collective identity.

In terms of publications, what had unified the whole community in the early twentieth century – *Anales* of the RSEFQ – was replaced after the Civil War by *Ión* (the main journal of the Falangist trade union *Sindicato Vertical*), *Revista de Ciencia Aplicada* (a leading voice of the PJC policies) and *Afinidad* (the journal of the Jesuit chemists), among a huge range of applied chemistry journals and original and translated chemistry textbooks that reflected the powerful chemical agenda of early Francoism.[16] *Anales* continued as the main journal of the RSEFQ, but now under a completely reformed scientific leadership. In exile, *Ciencia, Boletin del Instituto de Química* and chemistry textbooks published by Editorial Atlante in Mexico offered a good representation of the ambitious attempt to create a new chemical identity overseas.

From the Republican summer schools and international conferences in the 1930s to the Francoist display of the Butenandt Festival in the 1960s, the events of the community had changed, but they retained many of the tacit rules. In liberal and totalitarian regimes – and here again we see the moral ambiguity of chemistry – Nobel Prizes winners for chemistry, from Willstätter to Butenandt, were systematically appropriated by the JAE, the INFQ, the PJC, the CSIC and the university system as great foreign luminaries who legitimised the scientific practice of a peripheral scientific community in terms of international standards. In fact, in the early twentieth century, Spanish chemists published their research in international journals – often in joint papers with their Nobel Prize-winning mentors – but, once back from abroad, although they were already established in the Spanish system, they found matters much more difficult. Similarly, during the Francoist period, Lora-Tamayo, Pascual and

---

[16] *Catálogo del Libro Español de Química (1920–1955), con motivo del homenaje al P. E. Vitoria S.J., Afinidad (nº extraordinario, fuera de serie)*. May 1955.

others combined an autarchic-like strategy in tune with the regime with the intellectual ambition of publishing aspects of their research abroad.

Obsession with placing research schools in international networks transcended political regimes, and the pre-war, Republican dream of cosmopolitanism was, in one way or another, pursued by Francoist chemists such as Lora-Tamayo and Pascual, often proud of their role as co-constructers of the regime while also being able to publish their research in prestigious international journals. Albareda's soil chemistry fitted in very well with the autarchic projects of early Francoism, but at the same time brought together prestigious foreign chemists who were invited to Spain as part of the foreign policies of the new regime. Madinaveitia's Republican chemistry of natural products continued during the Franco period with Lora-Tamayo's applied organic chemistry projects, whereas García Banús's free radicals had their counterparts in Pascual's school in Barcelona. Moles's physical chemistry was appropriated to be used for other goals by Rius Miró, whereas the technical chemistry and industrial projects that had certainly already begun in the 1910s and 1920s were later continued under the banner of the Francoist 'applied chemistry', but never detached from the old trends of the industrial Silver Age.

In the struggle for the identity of the chemical community, networks of solidarity based on religious and political values played very important roles. In the same way that networks of Catholic, Protestant, Monarchic, Republican scientists, craftsmen and entrepreneurs had shaped strategies for dealing with industrial chemistry secrets and espionage in the nineteenth century, in the twentieth century the Society of Jesus, Opus Dei, Masonic anticlerical circles, Republican parties and the polarised political ideology of many of local actors played important roles in the ways in which the whole community of Spanish chemists evolved.[17]

In terms of the identity of the chemical community, the heroes of the Enlightenment – the Elhúyar brothers, Andrés del Río and even the Catalan chemist Martí Franquès (who was potentially an ideal candidate to strengthen political Catalanism) – continued to play key roles. Similarly, the grand old heroes of mining and metallurgy in colonial Spain – Álvaro Alonso Barba and Pedro de Medina, among many others – were crucial figures for the historical construction of the exiled chemists' identity, as Bargalló showed very clearly in his own research and who

---

[17] As described by George Wagner-Kyora for the history of the German industrial chemists: 'Identity construction [...] is a virtual political aspect of everyday acting and cultural self-imaging [...] it defines a kind of communicative social discourse, which derives not always from images but also from individual and collective memories'. George Wagner-Kyora, 'Continuities in the Identity Construction of Industrial Chemists, 1949–1970', *German Studies Review*, 29(3) (2006), 611–19, p. 611.

F. Giral often referred to in public addresses as a convincing argument in favour of the memory and the dignity of the 'university in exile'.

Spanish chemists were active chemist–historians, and frequently diffused historical accounts in their own professional interests, but also in tune with their political ideologies and the main goals of a specific regime. They developed a rich body of literature that reflected their changing place in international networks and their different self-perceptions as modernising agents very well. Their identity was strongly expressed through multiple commemorations. Laboratories and industries became two key sites through which they could assess progress and modernisation, as well as being a measure of the health of institutions (science faculties and research centres). External recognition through *pensionados*, international training and joint research projects acted as driving forces for internal legitimation close to political power and for gaining scientific authority within the chemical community in itself.

Chemists as Intellectuals

From cold molecules to hot historical actors, from apolitical subjects to active ideological agents – citizen scientists[18] – the history of a scientific community and its role in the co-construction of a specific political regime surely deserves further attention. As discussed in the Introduction of this book, there are numerous case studies of the role of liberal nineteenth-century German chemists and biologists under Hitler, chemists in Mussolini's Italy, engineers and physicists in Franco's Spain and many other cases in which the relations between science and power have been called into question by historians in recent decades. Nevertheless, what this book adds to that long-standing discussion is a historical assessment of the ways in which Spanish chemists in the twentieth century benefitted from the moral ambiguity of chemistry to fulfil their own professional ambitions and ideological priorities. If we go beyond an overly simplistic, perhaps naïve definition of intellectuals as supposedly free, objective observers of their present situation[19] and move closer to Antonio Gramsci's definition of *organic intellectuals* as servants of a specific social group with its own cultural hegemony and therefore being actors in the co-construction of a political regime, we could consider for instance: the

---

[18] Mark Walker (ed.), *Science and Ideology: A Comparative History* (London: Routledge, 2003).

[19] Edward Said, *Representations of the Intellectual. The 1993 Reith Lectures* (New York: Vintage Books, 1996). For an analysis on science as cultural hegemony under a Gramscian framework, see: Agustí Nieto-Galan, 'Antonio Gramsci Revisited: Historians of Science, Intellectuals, and the Struggle for Hegemony', *History of Science*, 46 (2011), 453–78.

Girals as organic intellectuals of the political programme of the Second Republic; Albareda as a key organic intellectual of the totalitarian policies of early Francoism through the CSIC and his soil chemistry programme; Lora-Tamayo as the ideal organic intellectual for the 'developmentism' of the 1960s and technocratic chemistry, which endorsed economic freedom but not political liberty; and Moles as a more multifaceted intellectual serving different political regimes but clashing dramatically with the Francoist cultural hegemony after the Civil War. As described by Gramsci in his *Prison Notebooks*:

Every social group, coming into existence on the original terrain of an essential function in the world of economic production, creates together with itself, organically, one or more strata of intellectuals which give it homogeneity and an awareness of its function not only in the economic but also in the social and political fields.[20]

Under a supposed rhetoric of neutrality, all biographies of the aforementioned names (and many others appearing in the chapters of this book) can be analysed as organic intellectuals of different and often opposing social groups. More or less consciously, they contributed to spreading the specific political values of their own time in a close alliance with their professional ambitions and career prospects.

In one way or another, our prosopography of Spanish chemists provides useful and sound data for the analysis of the role of professionals chemists (in academia and in industry) serving a particular ideology. It can obviously include, for instance, García Banús's bourgeois liberalism, Madinaveintia's committed socialism, F. Giral's admiration for Cardenas' Mexican leftist policies, the hot, pro-fascist proclamations of names such as Rocasolano in the 1930s and Vitoria's uses of acetylene molecules and their industrial applications in the 1950s – just before the emergence of the new petrochemistry – as ideal tools for the particular agenda of the Society of Jesus. Or we can take, for instance, the little-known names of the anarchist chemists who were in charge of the industrial effort during the Civil War,[21] or the Falangist chemists who maintained totalitarian political values through the development of a centralised organisation of the chemical industry, and – why not – the so-called Catalanist chemists who used the journal *Ciència* to fight in

---

[20] Quoted by David Forgacs (ed.), *The Antonio Gramsci Reader: Selected Writings 1916–1935* (New York: New York University Press, 2000), p. 301.

[21] Francisco Javier de Madariaga, 'La Comissió d'Indústries de Guerra de la Generalitat', in Guillermo Lusa, Antoni Roca-Rosell (eds.), *Fem memòria per fer futur. La tècnica i la Guerra Civil. IV Jornades Memorial Democràtic a la UPC. 30 de novembre de 2006* (Barcelona: UPC, 2006), pp. 123–35; Pelai Pagès, 'La Comisión de la Industria de Guerra de Cataluña (1936–1938)', *Ebre*, 38 (2008), 43–64.

favour of a cultural, linguistic counter-hegemony opposed to Spanish nationalism.[22] That framework of analysis does not exclude either public 'apolitical' statements of distinguished chemists such as del Campo at the end of the Civil War, or technocratic statements such as those of Lora-Tamayo in the 1960s during his years in Franco's cabinet as Minister for Education and Science, or Pascual's supposedly 'neutral' research projects to link the PJC and the universities.

Following the prosopographic sketches of this book, further research might reveal some clues for a more ambitious deontological reflection on the role of chemistry as a profession today in the context of Western liberal democracies that are seriously challenged by global state–corporate power and a growing public demand for environmental sensitivity and public health policies. Approaching chemists as potential intellectuals, then, could be a particularly rewarding research enterprise for the near future.

History and Memory

In his opening speech for the 1980–81 academic year in Salamanca, F. Giral passionately stressed the importance of what he called the 'university in exile', and the way in which that scientific community wanted to find paths towards reunification in the emerging democratic period. F. Giral was very explicit in paying homage to exiled colleagues for defending the liberal legacy of the Second Republic.[23] In private, he expressed profound disappointment after his return to Spain in 1977, and he missed his research school and the identity of the chemical community in Mexico.[24] F. Giral's idealisation of Republican academic life clashed with Lora-Tamayo's technocratic heritage, which was still embedded in the values of the dictatorship. In spite of both chemists' prestige for their work on natural products and their building of research schools and fruitful industrial links, their political disagreements drove

---

[22] In the first volume of *Ciència*, which appeared in February 1926, Peypoch stressed the need to introduce the Catalan language and culture into local academic life. 'Pòrtic', *Ciència. Revista Catalana de Ciència i Tecnologia*, 1(1) (1926), 1. See: Àngela Garcia Lledó, *La revista Ciència (1926–1933). Una eina al servei del catalanisme*. Master's thesis (Bellaterra: Universitat Autònoma de Barcelona, 2012).
[23] Francisco Giral, *Comentarios químico-farmacéuticos a la historia española en América* (Salamanca: Universidad de Salamanca, 1980), p. 99.
[24] Elena Aub, 'Entrevista realizada por Elena Aub en Salamanca, Departamento de Estudios Contemporáneos, IHAH-SEP, México INAH, 1981, PHO/10/Esp. 27', in Mª de la Soledad Alonso, Elena Aub, Marta Barnada (eds.), *Palabras del Exilio. De los que volvieron (Coordinación Eugenia Meyer)* (Mexico City: Secreataría de Educación Pública, 1988), p. 151.

them irreversibly apart. F. Giral's attempt to find a liberal reconciliation between the two universities – one in exile, the other the product of a 40-year dictatorial regime – became, in practice, a naïve proposal that was unable to filter into the ethos of the Spanish academic community.

In fact, signs of continuity, tacit practices and the continuing social acceptance of the old Francoist values seem to have pervaded to the present day. Even in the post-Franco democratic period, some of the chemists who contributed to the political project of the dictatorship have streets and teaching and research institutions named after them. Others have had statues to their memory erected in prominent public places and university classrooms. They enjoy public campaigns in support of them, opposing attempts to reassess their moral and political responsibilities. By contrast, as F. Giral's bitter disappointment showed, there is not much in the public sphere to remember the exiled, repressed, marginalised chemists.[25] It is true that Moles, among others, has recently received public recognition on several occasions, but the memory of many of the other names appearing in the previous chapters has not yet entered the Spanish public sphere.[26]

Today, one of the most prestigious research institutes of the CSIC occupies the former building of the INFQ in Madrid. It is still known as the 'Rocasolano', in honour of Antonio de Gregorio Rocasolano, who, as discussed in the previous chapters, was a distinguished chemist, but at the same time was one of the leaders of the repression of many of his academic fellows after the Civil War and an enthusiastic supporter of Franco's coup d'état in line with his right-wing ideology. Albeit lacking the pre-eminence of the sculptures of Nobel Prize winners Santiago Ramón y Cajal and Severo Ochoa, José María Albareda's bust is still exhibited today in a corner of the CSIC headquarters in Madrid. Equally, 'Juan de la Cierva' is the name of one of the most prominent programmes of research fellowships in Spain today, in honour of the famous Spanish engineer and supporter of Franco's dictatorship. His name was also given to the *Patronato Juan de la Cierva* (PJC), a key institution for science policies during the regime.[27] José Pascual gives name today to a street next to the Faculty of Chemistry of the University of Barcelona. Pharmacist Rafael Folch, one of the repressors in the Faculty of Pharmacy

---

[25] Recent attempts to dignify Enrique Moles in his own neighbourhood in Barcelona have been officially supported. In 2013, Barcelona City Council hung a commemorative plaque in honour of Moles at 25 Aulèstia i Pijoan street, at the entrance of the house where Moles was born in 1883.

[26] See, for instance, the commemorative books and exhibitions organised by the *Aula de Cultura Científica*, in which Augusto Pérez Vitoria played an important role.

[27] In 1936, Juan de la Cierva died in a plane crash in England, just at the beginning of the Civil War.

after the war, has had an avenue named after him in his hometown of Montblanc, Catalonia. Until recently, the chemist and industrialist Juan Abelló held the title of honorary pharmacist in the city of Reus and had a street named after him. Similarly, the *Fundación José Casarés Gil*, which was named after another chemist who actively supported Franco's dictatorship, is hosted today at the *Real Academia Nacional de Farmacia* in Madrid, with the support of the pharmaceutical industry, and producing regular publications.

There have been other reactions in favour of retaining the memory and honours of the Francoist chemists. In 2008, the *Real Academia Sevillana de Ciencias* (a local scientific society) sent a letter to the director of the *Diario de Sevilla* newspaper in a highly critical tone. It wanted to express in public its unease with the intention of the regional government of Andalusia to remove Lora-Tamayo's name from a secondary school. The *Academia* expressed its reluctance in the following terms:

In 1962 he [Manuel Lora-Tamayo] was appointed Minister for National Education and carried out ambitious cultural, educational, research policies, strengthening the links between teaching and research. As far as science policy was concerned, he notably expanded the number of research positions at the *Consejo Superior de Investigaciones Científicas* [CSIC], and enlarged our university system with the creation of university colleges, the origins of many of our universities. Professor Lora-Tamayo could not conclude his task; he resigned as minister because of his disagreement with the Government on how to tackle university problems. Seville's science is indebted to him.[28]

In the letter, Lora-Tamayo was perceived by his local colleagues as a great scientist and a local luminary who had gained prestigious academic positions throughout his long life. Nevertheless, the Seville Academy paid little attention to Lora-Tamayo's Francoist past. In accordance with the Historical Memory Act – passed by the Spanish Parliament in 1995 – all images and texts with explicit references to the dictatorship had to disappear from public spaces in Spain.

Even more recently, on 24 October 2016, a local newspaper in the city of Segovia (near Madrid) published an article against the removal of Fernández-Ladreda's name for a street in the city. Biographical data served to endorse that position:

After the proclamation of the Second Republic [... Fernández-Ladreda] organised the CEDA political party in Asturias, and became MP in 1933. The CEDA

---

[28] *Diario de Sevilla*, 4 August 2008, p. 4. Letters to the director, Lora-Tamayo. The *Junta de Gobierno* (government body) of the *Real Academia Sevillana de Ciencias* was against removing the name Lora-Tamayo from a secondary school in Seville.

History and Memory

was a 1930s right-wing Catholic party. In 1936, when the Civil War broke out, Republican loyalist troops and militia besieged Oviedo. Commander Fernández-Ladreda organised a volunteer defence battalion, which was later named after him. After the war, Franco appointed him as Minister for Public Works, to develop the 'stone and cement' policies [for large infrastructure projects], which so pleased the regime [...] Why [...] do you want to remove the name of such a distinguished benefactor of Segovia?[29]

Here again, the moral ambiguity of chemistry plays an important role in the public sphere, in the public perception of material and scientific progress as often being detached from any political responsibility. All of these mixed views on the ways in which historical memory should be preserved are yet more signs of the long-term wounds to the chemical community that have been described throughout the chapters of this book. Obviously, there is still a lot to be done in terms of recovering the biographies of many other chemists throughout the twentieth century. From that perspective, this book is just a first step to further prosopographies and bigger pictures.

In her book, *Historia Patria* (1997) – a masterclass in twentieth-century Spanish history through the lens of the teaching and popularisation of history – Carolyn P. Boyd rightly chose a quotation by Manuel Azaña. Azaña was prime minister of the Republic and one of the most prominent intellectuals of the liberal tradition. In 1920, he expressed his fears of a right-wing government as follows:

> Let us imagine the horrendous nightmare of the permanent accession to power of the Spanish right. Would they blow up the dams, block the canals, prohibit the use of *chemical fertilisers*? In no way. They would close the paths of spiritual progress, brutalise (even more) the sensibility of the *pueblo*, in order to avoid its arriving at the point where an attack on the liberal conscience is less tolerable than death itself.[30]

If Azaña was right, *chemical fertilisers* could have represented that moral ambiguity of chemistry and chemists' 'chameleonic' roles in different political regimes. The main question lies, however, in the way in which chemists spoke, experimented and published on fertilisers and on any other molecule, and the politics that inevitably embedded their professional careers and research projects; in a nutshell, in the way in which they constructed a particular cultural hegemony and became, more or less explicitly, powerful ideological agents of their time.

---

[29] *El Adelantado de Segovia*, 24 October 2016.
[30] Quoted by Carolyn P. Boyd, *Historia Patria, Politics, History and National Identity in Spain, 1875–1975* (Princeton: Princeton University Press, 1997), pp. 194–5 (emphasis added).

Addendum: Juan Julio Bonet Sugrañes (1940–2006)

Juan Julio Bonet Sugrañes was born in Barcelona in 1940, just at the end of the Civil War. He received a conservative, religious education in the context of National Catholicism during Franco's dictatorship.[31] As a natural result of his brilliant secondary education and his interest in chemistry, Bonet entered the IQS in 1956 as an undergraduate, obtaining a degree in chemical engineering in 1961. These were precisely the years in which the IQS received US funds for the modernisation of its laboratory equipment and research projects and achieved the first stage of the internationalisation of its research lines in the context of a significant growth of the chemical industry. In the same year, Bonet was awarded a fellowship by the Juan March foundation to do a PhD at the organic chemistry department of the ETH – the prestigious polytechnic school of Zurich and one of the key nodes of the international chemical network of the JAE years. He worked under the supervision of Professor Oskar Jeger, a pupil of the 1939 Nobel Prize winner for chemistry, Leopold Ruzicka, for his work on polymethylenes and higher terpenes, a prize shared that same year with Adolf Butenandt for his work on sex hormones.

On Bonet's arrival in Zurich, Jeger was working on the photochemistry of steroids – a new research field at that time. In 1965, Bonet submitted his PhD, entitled 'Sovolitic Reactions in the Steroid Field', which provided solid training for his future in steroid chemistry. In his own autobiography,[32] Bonet said he missed his years at the organic chemistry lab of the ETH, a site he perceived as being full of modern glassware, with easy access to reagents, a workshop for the making of new instruments and receiving generous federal funds for research as well as huge economic support from industry – again, one of the old dreams of the Silver Age. But what Bonet considered the turning point of the ETH was the institutional ability to attract top intellectual talent and true leaders of research schools.[33] One of his colleagues in Zurich, Kurt Schaffner, who years later became one of the directors of the Max-Planck Institute for Photochemistry in West Germany, described Bonet's perception of the ETH as follows:

[Bonet] became increasingly fascinated with the Zurich School of Chemistry which he learned to consist of an academic network extending over two local institutions, the chemistry departments of ETH and the University of Zurich,

---

[31] Eva Sevigné, *Juan Julio Bonet Sugrañes y el Laboratorio de Esteroides del Instituto Químico de Sarriá (IQS). 1965–1986*. Master's thesis (Bellaterra: Universitat Autònoma de Barcelona, 2008).
[32] Juan Julio Bonet Sugrañes, *Viaje al Reino de Saturno* (Madrid: Nívola, 2004).
[33] Ibid., p. 20.

located within short walking distance from each other. Both laboratories have a notable history of impressive scientific success. They form [...] a landscape of learning and research not readily found elsewhere. This author, possibly being biased as an off-spring of the largest unit in this landscape (the *Laboratorium für Organische Chemie* of ETH – *Eidgenössische Technische Hochschule*), it may suffice to recall here that international rating of the Zurich School accumulated altogether nine Nobel awards, two to Werner and Karrer at the university and three (to Ruzicka, Prelog and Ernst at ETH) to chemistry professors still in office, and four (to Willstätter, Kuhn, Reichstein and Staudinger) after they had moved from ETH to positions elsewhere.[34]

In 1965, after his ETH doctorate, Bonet soon became a lecturer in natural products at the IQS, and he founded a laboratory of steroids that would become the crucial site of his research school. In the early 1970s, he introduced organic photochemistry in teaching and research as a new, innovative field. From 1977 to 1986, Bonet was director of the Department of Organic Chemistry and deputy director of the whole institute (Figure 8.1). In all those years, he maintained international contacts and research collaborations and progressively built up a dynamic and fairly liberal research school for the synthesis of steroid molecules, in which he supervised 17 PhDs and close to 150 Master's dissertations, mainly working on the photochemistry of heterocyclic steroids.[35] After 20 years of leading his steroids research school at the IQS, Bonet left his position to join the chemical industry, and later, in 1988, he became director of technology transfer at a public institute for food and agriculture research for the Catalan Government. After becoming a chemist–historian with his novel *Viaje al Reino de Saturno* (A Journey to the Kingdom of Saturn), in which he attempted to establish a genealogy of chemists from A. L. Lavoisier to Oskar Jeger and Bonet's own research school on steroid chemistry, he died in 2006.

In my view, Bonet's biography is a good reflection of the main thesis and focus of this book and of the complex role of chemists as political actors in history. Bonet's institutional setting was apparently conservative, as was the whole of the Jesuit chemistry school led by Vitoria, and it was very close to industrial power, upper-class elites and support for Franco's dictatorship. But despite that framework, Bonet acquired a cosmopolitan training at the ETH in the early 1960s. This was a key period in which the young post–Civil War generation of Spanish chemists

---

[34] Kurt Schaffner, 'Juan Julio Bonet Sugrañes and His Relation with the Zurich School', *Afinidad*, 528 (2007), 103–4, p. 104.
[35] Lactames, lactones, oximes, diazasteroids, glicirretic acid and piperazines. Carme Brossa, 'El professor Joan Juli Bonet: Una revisió històrica de la seva recerca', *Afinidad*, 528 (2007), 108–46.

Figure 8.1 Juan Julio Bonet Sugrañes in his office at the Steroids Laboratory of the IQS in the 1980s.
Reproduced with permission of the *Institut Químic de Sarrià*, Barcelona.

began to re-establish regular international contacts and tacitly linked the old dreams of modernity of the early twentieth century with the new economic trends in the 1960s. As was the case with the JAE generation, once international training abroad ended and the young chemists had to establish themselves in the Spanish university setting, top international publications slowed down. However, the culture of chemistry in international research schools such as those of the ETH filtered into Bonet's peripheral academic system and provided a fresh, cosmopolitan education for future generations. García Banús remained at the ETH in the 1910s, as did Bonet in the 1960s, half a century later, and lines of continuity and common chemical values from both schools can be easily traced back. Educated in the totalitarian atmosphere of the 1940s and 1950s, but in close contact with international chemical trends in the 1960s, Bonet managed to build his own research school that paradoxically combined foreign collaborations and tolerant attitudes towards students and colleagues, but at the same time close ties with the Jesuit chemists' official programme at the IQS. With its light and dark sides, his steroids lab epitomised – like many other examples appearing in this book – the aforementioned moral ambiguity of chemistry.

If, at this point, the reader will allow me a few words on reflexivity, in the early 1980s, I was a Master's degree student in Bonet's research school, struggling with the chemistry of carbo-cations in steroid structures. At that time, in the transition to democracy after Franco's death in 1975, the old institutional structure of the IQS seemed outmoded, being still too deeply rooted in the times of the dictatorship. The steroid lab, however, provided a fresh atmosphere, open discussion and even political debates among Master's and PhD students. It included frequent visits by foreign experts, research seminars and informal gatherings of the group, as well as a true spirit of the community of chemists alongside a very definite identity. In addition, Bonet had maintained his international contacts abroad, mainly in Switzerland, but also in Germany. That was the case for the Max-Planck Institute for Photochemistry, one of Bonet's most innovative fields, representing a new conjuncture in the international nodes of chemistry to which to send his own students abroad. I was one of his modern *pensionados* working hard again in the lab to discover unexpected behaviours of steroid molecules in the presence of light. Some years later, I decided to move on to concentrate on history and the history of science, but my years in Bonet's research school and in his 'republic of chemistry' irreversibly marked my life and helped me to understand the precious value of research in a truly cosmopolitan environment.

In spite of the hard times in which he lived, Bonet became a master of a sort of 'liberal' chemistry, which I still feel needs to be further discussed and debated in our present troubled times. This is a major reason for why I have written this book and for my humble homage to Juan-Julio's memory.

Bibliography

Archives and Libraries

Archive of the Society for the Promotion of Science and Learning. Bodleian Library. Oxford.
Archivo de la Junta para Ampliación de Estudios e Investigaciones Científicas. Residencia de Estudiantes. Madrid.
Archivo del Ministerio de Asuntos Exteriores. Madrid.
Archivo del Ministerio de Defensa. Madrid.
Archivo del Ministerio del Interior. Madrid.
Archivo General de la Administración. Alcalá de Henares. Madrid.
Archivo Gráfico Carta de España. Dirección General de Migraciones. Ministerio de Empleo y Seguridad Social. Madrid.
Archivo Histórico de la Universidad Complutense de Madrid.
Archivo Histórico Nacional. Madrid. (*Diversos José Giral*).
Archivo José F. Demaría "Campúa."
Arxiu Fotogràfic de Barcelona.
Arxiu General. Diputació de Barcelona.
Arxiu Històric de l'Escola Tècnica Superior d'Enginyeria Industrial. Barcelona.
Arxiu Històric. Universitat de Barcelona.
Arxiu Montserrat Tarradellas i Macià. Monestir de Poblet. Tarragona.
Arxiu. Institut d'Estudis Catalans. Barcelona. (*Fons Masriera*).
Arxiu. Reial Acadèmia de Ciències i Arts de Barcelona. (*Fons Josep Pascual Vila*).
Biblioteca de Catalunya. Barcelona.
Biblioteca de Ciència i Tecnologia. Universitat Autònoma de Barcelona.
Biblioteca de l'Ateneu Barcelonès. Barcelona.
Biblioteca de l'Institut Químic de Sarrià. Barcelona.
Biblioteca de la Facultad de Farmacia. Universidad Complutense de Madrid.
Biblioteca de la Facultat de Química. Universitat de Barcelona.
Biblioteca de la Reial Acadèmia de Ciències i Arts de Barcelona.
Biblioteca Nacional de España. Madrid.
Bodleian Library. Oxford.
IUPAC. Historical Archive. Science History Institute. Philadelphia.
Pavelló de la República. Universitat de Barcelona.
Queen's College. University Archives, Oxford.

Bibliography

Rockefeller Archive Center. International Education Board. New York.
Universidad de Navarra. Archivo General. Pamplona. *(Fondo José María Albareda, Fondo Manuel Lora-Tamayo)*.

Scientific Journals (Spanish Primary Sources)

Afinidad. Revista de química teórica y aplicada
Anales de Edafología, Ecología y Fisiología Vegetal
Anales de Física y Química
Anales de la Junta para Ampliación de Estudios e Investigaciones Científicas
Anales de la Asociación Española para el Progreso de las Ciencias
Anales de la Real Sociedad Española de Física y Química
Anales del Centro de Investigaciones Vinícolas
Arbor
Boletín Informativo de la Unión de Profesores Universitarios Españoles en el Extranjero
Boletín del Instituto de Química
Butlletí de la Societat Catalana de Ciències Físiques, Químiques i Matemàtiques
Ciència. Revista Catalana de Ciència i Tecnologia
Ciencia. Revista hispano-americana de ciencias puras y aplicadas
Diario Oficial de la Exposición Internacional de Barcelona
Grasas y Aceites
Ión. Revista española de química aplicada
La industria química
Memorias de la Junta para Ampliación de Estudios e Investigaciones Científicas
Metalurgia y Construcción Mecánica
Monografies Mèdiques
Química e Industria
Residencia. Revista de la Residencia de Estudiantes
Revista de Ciencia Aplicada
Revista de Información de Química Analítica
Revista de Plásticos
Revista Ibys

Books and Articles (Primary and Secondary Sources)

This section includes, in alphabetical order, the main printed sources consulted and cited in the footnotes throughout the chapters, as well as some works of reference to provide complete information for the reader.

50 años de investigación en física y química en el edificio Rockefeller de Madrid, 1932–1982. Noviembre de 1982. Recuerdos históricos (Madrid: CSIC, 1982).

Abir-Am, Pnina G., 'Women in Research Schools: Approaching an Analytical Lacuna in the History of Chemistry and Allied Sciences', in Seymour

Mauskopf (ed.), *Chemical Sciences in the Modern World* (Philadelphia: University of Pennsylvania Press, 1993), pp. 375–91.

Abir-Am, Pnina G. (ed.), *La mise en mémoire de la science: Pour une ethnographie historique des rites commemoratifs* (Paris: Editions des Archives Contemporaines, 1998).

Abir-Am, Pnina G. and Clark A. Elliot, 'Introduction', in *Commemorative Practices in Science, Osiris (2nd series)*, 14 (1999), pp. 1–33.

Abollado, Carlos, 'La industria química y la química industrial', *Ión*, 45 (1945), 248.

Agar, Jon, *Science in the Twentieth Century and Beyond* (Cambridge: Polity Press, 2012).

Agell, José, 'El curso del professor Ostwald en el Instituto de Química Aplicada de Barcelona (Mayo 1924)', *Química e Industria*, I(9) (1924), 221–5.

Albareda, José María, *El Suelo: Estudio físico-químico y biológico de su formación y constitución; prólogo del Dr. Antonio de G. Rocasolano* (Madrid: Sociedad Anónima Española de Traductores y Autores, 1940).

Albareda, José María, *Consideraciones sobre la investigación científica* (Madrid: CSIC, 1951).

Albareda, José María, *Aumento de la población y producción agrícola* (Madrid: Real Academia de Farmacia, 1958).

Albareda, José María, 'Organization et tendances de la recherche scientifique en Espagne. Le Conseil supérieur de la recherche scientifique', *Impact. Science et Société*, 15(1) (1965), 43–57.

Alexander, Jerome (ed.), *Colloid Chemistry: Theoretical and Applied* (New York: The Chemical Catalo Co., 1926).

Allen, John, *Opus Dei. An Objective Look behind the Myths and Reality of the Most Controversial Force in the Catholic Church* (New York: Doubleday, 2005).

Alonso, María de la Soledad, Elena Aub and Marta Barnada, *Palabras del exilio. De los que volvieron* (Mexico City: Secreataría de Educación Pública, 1988).

Antequera, Juan José (ed.), *Huelva: Los peligros del cielo. Prevenciones del bando faccioso frente a ataques republicanos en 1936* (Seville: Facediciones, 2008).

Arendt, Hannah, *The Human Condition* (Chicago: University of Chicago Press, 1958).

Arendt, Hannah, *The Origins of Totalitarism* (London: G. Allen & Unwin, 1958) (1st edition 1951).

Arendt, Hannah, *Eichmann in Jerusalem: Ein Bericht von der Banalität des Bösen* (Leipzig: Reclam-Verlag, 1990) (1st edition 1963).

Aronowitz, Stanley, *Science as Power: Discourse and Ideology in Modern Society* (Minneapolis: University of Minnesota Press, 1988).

Arribas, Siro, *Introducción a la historia de la química analítica en España* (Oviedo: Universidad de Oviedo, 1985).

Artigas, Miguel et al. (eds.), *Una poderosa fuerza secreta. La Institución Libre de Enseñanza* (San Sebastian: Editorial Española, 1940).

Artigas, José Antonio, *El presente español de los vehículos con gasógeno. Memorias del Primer Congreso de Ingeniería Industrial* (Madrid: Asociación Nacional de Ingenieros Industriales, 1942).

Artís, Gaziella and Mireia Artís, 'Francesc Novellas i Roig (1874–1940). La química industrial', in Antoni Roca-Rosell, Josep Maria Camarasa (eds.),

Ciència i Tècnica al Països Catalans. Una aproximació biogràfica. 2 vols. (Barcelona: Fundació Catalana per a la Recerca, 1995), pp. 937–66.

Ash, Mitchell, 'Essay Review: Science, Technology and Higher Education under Nazism', *Isis*, 85 (1995), 458–62.

Ash, Mitchell, 'Scientific Changes in Germany, 1933, 1945, 1990: Towards a Comparison', *Minerva*, 37 (1999), 329–54.

Aub, Max, *La gallina ciega. Diario español. Edición, estudio introductorio y notas de Manuel Aznar Soler* (Barcelona: Alba editorial, 1995; 1st edition, 1971).

Aubin, David and Patrice Bret, *Le sabre et l'éprouvette. L'invention d'une science de guerre, 1914–1939* (Paris: Editions Noésis/Agnès Viénot, 2003).

Ausejo, Elena, *Por la ciencia y por la patria: La institucionalización científica en España en el primer tercio del siglo XX: La Asociación Española para el Progreso de las Ciencias* (Madrid: Siglo XXI, 1993).

Ausejo, Elena, 'La Asociación Española para el Porgreso de las Ciencias en el centenario de su creación', *Revista Complutense de Educación*, 19(2) (2008), 295–310.

Ausejo, Elena (ed.), *La ciencia española en la II^a República*. (Madrid: Fundación de Investigaciones Marxistas, 2008).

Azaña, Manuel, *Mi rebelión en Barcelona* (Madrid: Espasa Calpe, 1935).

Baldwin, N. J., 'The Archive of the Society for the Protection of Science and Learning', *History of Science*, 27 (1989), 103–5.

Balfour, Sebastian, *Deadly Embrace: Morocco and the Road to the Spanish Civil War* (Oxford: Oxford University Press, 2002) (Spanish version: Barcelona, Península, 2002).

Balfour, Sebastian and Paul Preston (eds.), *Spain and the Great Powers in the Twentieth Century* (London: Routledge, 1999).

Ball, Phillip, *Serving the Reich: The Struggle for the Soul of Physics under Hitler* (London: Vintage Books, 2013).

Ballestero, Alfonso, *Juan Antonio Suanzes 1891–1977. La política industrial de la postguerra* (León: LID Editorial Empresarial, 1993).

Baratas, Alfredo, 'El exilio científico español de 1939. Diáspora y reconstrucción de una comunidad científica', in Manuel Llusia and Alicia Alted (eds.), *La cultura del exilio republicano español de 1939. Actas del Congreso internacional celebrado en el marco del Congreso plural: Sesenta años despues* (Madrid-Alcalá-Toledo, Diciembre de 1999). (Madrid: UNED, 2003), pp. 707–14.

Baratas, Alfredo and Manuel Lucena, 'La 'Society for the Protection of Science and Learning' y el exilio republicano español', *Arbor*, 588 (1994), 25–48.

Barceló, José, 'Un siglo de espectroscopía: El centenario del descubrimiento del cesio por Bunsen y Kirchhoff', *Ión*, 20 (1960), 195–201.

Bargalló, Modesto, *La minería y la metalurgia en la América española durante la época colonial* (Mexico City: FCE, 1955).

Bargalló, Modesto, *Tratado de química inorgánica, fundamental y sistemática; para universidades, escuelas técnicas profesionales e institutos tecnológicos superiores* (Mexico City: Editorial Porrúa, 1962).

Barona, Josep Lluís, 'Los laboratorios de la Junta para Ampliación de Estudios e Investigaciones Científicas (J.A.E.) y la Residencia de Estudiantes (1912–1939)', *Asclepio*, 59(2) (2007), 87–114.

Barona, Josep Lluís (ed.), *El exilio científico republicano* (Valencia: Universitat de València, 2010).
Bayod, Ángel (ed.), *Franco visto por sus ministros* (Barcelona: Planeta, 1983).
Bensaude-Vincent, Bernadette, 'In the Name of Science', in John Krige and Dominque Pestre (eds.), *Science in the Twentieth Century* (Amsterdam: Harwood Publishers, 1997), pp. 319–38.
Berlin, Isaiah, *Four Essays on Liberty* (Oxford: Oxford University Press, 1958).
Bermejo, Luis, 'El problema de los carburantes en España', in Emilio Jimeno (ed.), *Universidad de Barcelona. Aspectos y problemas de la nueva organización de España. Ciclo de Conferencias organizado por la Universidad de Barcelona* (Barcelona: Universidad de Barcelona, 1939), pp. 231–53.
Bermejo, Luis, 'El Instituto Rockefeller', in Miguel Artigas et al. (eds.), *Una poderosa fuerza secreta. La Institución Libre de Enseñanza* (San Sebastian: Editorial Española, 1940), pp. 197–202.
Berrojo, Raúl, *Enrique Moles y su obra*. PhD thesis (Barcelona: Universitat de Barcelona, 1980).
Bertomeu-Sánchez, José Ramón, Duncan Thorburn Burns and Brigitte Van Tiggelen (eds.), *Neighbours and Territories: The Evolving Identity of Chemistry. The 6th International Conference on the History of Chemistry* (Louvain-la-Neuve: Mémosciences, 2008).
Bijker, Wiebe E., Thomas P. Hughes and Trevor Pinch (eds.), *The Social Construction of Technological Systems* (Cambridge, MA: MIT Press, 1987).
Blas, Luis, *Química de guerra* (Barcelona: Salvat, 1940).
Blas, Luis, *Agenda del químico* (Madrid: Aguilar, 1942).
Blas, Luis, 'Presente y futuro de las industrias químicas derivadas del acetileno', *Ión*, 78 (1948), 3–6.
Blas, Luis, *Química de los plásticos* (Madrid: Aguilar, 1948).
Bonet, Juan Julio, *Viaje al Reino de Saturno. Un viaje de ida y vuelta a los orígenes de la química moderna* (Madrid: Nívola, 2004).
Bonnín, Pere, *Fèlix Serratosa. Químic, humanista i mestre de generacions d'universitaris* (Barcelona: Fundació Catalana per a la Recerca, 1995).
Bosch Giral, Pedro, Juan Francisco García de la Banda, Joaquín Pérez-Pariente and Manoel Toural Quiroga, *Protagonistas de la química en España: Los orígenes de la catálisis* (Madrid: CSIC, 2010).
Boyd, Carolyn P., *Historia Patria. Politics, History and National Identity in Spain, 1875–1975* (Princeton: Princeton University Press, 1997).
Boyd Rayward, W. and Mary Ellen Bowden (eds.), *The History and Heritage of Scientific and Technological Information Systems* (Philadelphia: Chemical Heritage Foundation, 2002)
Bretislav, Friedrich, Dieter Hoffmann, Jürgen Renn, Florian Schmaltz and Martin Wolf (eds.), *One Hundred Years of Chemical Warfare: Research, Deployment, Consequences* (Cham: Springer Open, 2017).
Bricall, Josep Maria, *Política econòmica de la Generalitat (1936–1939)* (Barcelona: Edicions 62, 1979).
Brock, William H., *The Fontana History of Chemistry* (London: Fontana Press, 1992).
Brooke, John H., 'Methods and Methodology in the Development of Organic Chemistry', *Ambix*, 34 (1987), 147–55.

Brossa, Carme, 'El professor Joan Juli Bonet: Una revisió històrica de la seva recerca', *Afinidad*, 528 (2007), 108–46.
Bud, Robert, 'Introduction', *Isis, Focus: Applied Science*, 103 (2012), 515–17.
Bud, Robert F. and Gerrylyn K. Roberts, 'Chemistry and the Concepts of Pure and Applied Science in Nineteenth-Century Britain', in Franco Calascibetta and Eugenio Torracca (eds.), *Storia e fondamenti della chimica: Atti del II Convegno Nazionale di Storia e Fondamenti della Chimica* (Rome: Accademia Nazionale delle Scienze detta dei XL, 1988), pp. 19–33.
Budd, Adam (ed.), *The Modern Historiography Reader: Western Sources* (London: Routledge, 2009).
Burriel, Fernando, Felipe Lucena and Siro Arribas, *Química analítica cualitativa: Teoría y semimicrométodos* (Madrid: Paraninfo, 1959).
Bustinza, Florencio, 'Obra docente y científica y perfil humano del Profesor D. Obdulio Fernández Rodríguez', in *Libro de Homenaje al Profesor D. Obdulio Fernández Rodríguez, con motivo del cincuentenario de su ingreso en la Real Academia de Ciencias Exactas, Físicas y Naturales* (Madrid: Real Academia de Ciencias Exactas, Físicas y Naturales, 1969), pp. 7–30.
Caballero, Ernesto and María Carmen Azcuénaga, *La Junta para Ampliación de Estudios e Investigaciones Científicas. Historia de sus centros y protagonistas (1907–1936)* (Gijón: Ediciones Trea, 2010).
Cabana, Francesc, *Fàbriques i empresaris. Els protagonistes de la revolució industrial a Catalunya. Vol. I. Metal·lúrgics. Químics* (Barcelona: Enciclopèdia Catalana, 1992).
Callon, Michel (ed.), *La science et ses réseaux. Genèse et circulation des faits scientifiques* (Paris: La Découverte, 1988).
Calvet, Fernando, *Bioquímica para médicos, químicos y farmacéuticos* (Madrid: Alhambra, 1956).
Calvo, Felipe, *Emilio Jimeno Gil. Semblanza de un maestro* (Santander: Amigos de la Cultura Científica, 1984).
Calvo, Felipe and José María Gulemany, 'Contribución del profesor Emilio Jimeno Gil al prestigio de la Universidad de Barcelona: Rector de 1939 a 1941', in *Història de la Universitat de Barcelona. I Simposium del 150 Aniversari de la Restauració* (Barcelona: Publicacions de la Universitat de Barcelona, 1990), pp. 471–83.
Calvo, Lluís (ed.), *El CSIC en Cataluña (1942–2012): Siete décadas de investigación científica* (Barcelona: CSIC, 2012).
Camprubí, Lino, 'One Grain, One Nation Rice Genetics and the Corporate State in Early Francoist Spain (1939–1952)', *Historical Studies in the Natural Sciences*, 40(4) (2010), 499–531.
Camprubí, Lino, *Engineers and the Making of the Francoist Regime* (Cambridge, MA: MIT Press, 2014).
Camprubí, Lino, *Los ingenieros de Franco. Ciencia, catolicismo y Guerra fría en el Estado franquista* (Barcelona: Crítica, 2017).
Camprubí, Lino, Xavier Roqué and Francisco Sáez de Adana (eds.), *De la Guerra Fría al calentamiento global. Estados Unidos, España y el nuevo orden científico mundial* (Madrid: La Catarata, 2018).
Camps, Francesc, 'La química en el CID-CSIC: 40 años de labores científicas', in Lluís Calvo (ed.), *El CSIC en Cataluña (1942–2012): Siete décadas de investigación científica* (Barcelona: CSIC, 2012), pp. 143–56.

Canales, Antonio, 'La política científica de posguerra', in Amparo Gómez and Antonio Canales (eds.), *Ciencia y fascismo. La ciencia española de postguerra* (Barcelona: Laertes, 2009), pp. 105–36.

Canales, Antonio and Amparo Gómez, 'La depuración franquista de la Junta para la Ampliación de Estudios e Investigaciones Científicas (JAE): Una aproximación cuantitativa', *Dynamis*, 37(2) (2017), 459–88.

Cano Pavón, José Manuel, 'La investigación química en Granada en el siglo actual (1900–1975)', *Dynamis*, 16 (1996), 317–67.

Capella, Santiago, José Antonio Chamizo, Julián Garritz and Andoni Garritz, 'La huella en México de los químicos del exilio español de 1939', in Antonio Bolívar (ed.), *Científicos y humanistas del exilio español en México* (Mexico City: Academia Mexicana de Ciencias, 2006), pp. 155–72.

Capitán, Luis Fermín (ed.), *Un siglo de estudios de química en Granada (1913–2013)* (Granada: Universidad de Granada, 2014).

Carreras, Albert and Xavier Tafunell, *Historia económica de la España contemporánea* (Barcelona: Crítica, 2004).

Carreras, Juan José and Miguel Ángel Ruiz Carnicer (eds.), *La Universidad española bajo el régimen de Franco (1939–1975)* (Zaragoza: Institución Fernando el Católico, 1991).

Casado, Santos, 'Ciencia y conciencia bajo los tilos. Los laboratorios de la Residencia de Estudiantes y el exilio de 1939' in *Los científicos del exilio español en México* (Mexico City: Universidad Michoacana de San Nicolás de Hidalgo, Sociedad Mexicana de Historia de la Ciencia y de la Tecnología, Sociedad Española de Historia de las Ciencias y de las Técnicas, 2001), pp. 125–46.

Casanova, José V., 'The Opus Dei Ethic, the Technocrats and the Modernization of Spain', *Social Science Information*, 22(1) (1983), 27–50.

Casanova, Julián, *República y Guerra Civil*. Vol. 8, in Josep Fontana and Ramón Villares (eds.), *Historia de España* (Barcelona/Madrid: Crítica/Marcial Pons, 2007).

Casares, José, *El problema de la investigación científica en España. Discurso inaugural: Sección 3ª, Ciencias Físicoquímicas* (Madrid: Asociación Española para el Progreso de las Ciencias, 1911).

Castells, Josep and Félix Serratosa (eds.), *Obra científica del professor Dr. José Pascual Vila (1895–1979)* (Barcelona: EUNIBAR, 1982).

Castillo, Adolfo, *Albareda fue así: Semilla y surco* (Madrid: CSIC, 1971).

Catálogo del Libro Español de Química (1920–1955), con motivo del homenaje al P. E. Vitoria S.J, Afinidad (nº extraordinario, fuera de serie) (Barcelona: Asociación de Químicos del Instituto Químico de Sarriá, 1955).

Cazorla-Sánchez, Antonio, *Políticas de la victoria. La consolidación del Nuevo Estado franquista (1938–1953)* (Madrid: Marcial Pons, 2000).

Cebollada, José Luis, 'Antonio de Gregorio Rocasolano y la Escuela Química de Zaragoza', *Llull*, 11 (1908), 189–216.

Chadwick, Owen, *The Secularization of the European Mind in the Nineteenth Century* (Cambridge: Cambridge University Press, 1975).

Chaves Palacios, Julián, 'Oposición política a la monarquía de Alfonso XIII. José Giral y los republicanos en la Dictadura de Primo de Rivera', *Hispania*, 252 (2016), 159–87.

Civit, Ramon, *Destino i la cultura catalana a les acaballes del franquisme (1966–1975)*. PhD thesis (Barcelona: Universitat de Barcelona, 2012).
Clara, Fernando, Cláudia Ninhos and Sasha Grishin (eds.), *Nazi Germany and Southern Europe, 1933–45: Science, Culture and Politics* (Basingstoke: Palgrave Macmillan, 2016).
Claret, Jaume, *La repressió franquista a la Universitat catalana* (Vic: Eumo/Institut Universitari d'Història Jaume Vicens i Vives, 2003).
Claret, Jaume, *El atroz desmoche. La destrucción de la universidad española por el franquismo* (Barcelona: Crítica, 2006).
Clow, Archibald and Nan Louise Clow, *The Chemical Revolution: A Contribution to Social Technology* (London: Batechworth Press, 1952).
Cobos, José M., 'Luces y sombras del apoyo de la II^a República española a la ciencia', in Elena Ausejo (ed.), *La ciencia española en la II^a República*. (Madrid: Fundación de Investigaciones Marxistas, 2008), pp. 9–39.
Cobos, José M. et al., 'El Boletín Informativo de la Unión de Profesores Univesitarios Españoles en el Extranjero', *Llull*, 27 (2004), 27–60.
Cohen, Clive, 'The Early History of Chemical Engineering: A Reassessment', *The British Journal for the History of Science*, 29 (1996), 171–94.
Coleman, Kim, *A History of Chemical Warfare* (London: Palgrave, 2005).
Colomer, Josep María, *Els estudiants de Barcelona sota el franquisme* (Barcelona: Curial, 1978).
Connelly, John and Michael Grüttner (eds.), *Universities under Dictatorship* (University Park: The Penn State University Press, 2005).
Cornwell, John, *Hitler's Scientists: Science, War, and the Devil's Pact* (London: Penguin Books, 2003).
Coverdale, John F., 'José María González Barredo. An American Pioneer', *Studia et Documenta Rivista dell'Istituto Storico San Josemaría Escrivá*, 10 (2016), 23–44.
Crawford, Elizabeth, *Nationalism and Internationalism in Science, 1880–1930* (Cambridge: Cambridge University Press, 1992).
Crawford, Elisabeth T., John L. Heilbron and Rebecca Ullrich, *The Nobel Population 1901–1937: A Census of the Nominators and Nominees for the Prizes in Physics and Chemistry* (Berkeley: Office for History of Science and Technology, University of California, 1987).
Crowther, James Gerald, *La ciencia en el país de los Soviets. Traducción directa del inglés de Francisco Giral* (Madrid: Cenit, 1931).
Cueli, José, 'Matemáticas, física y química', in *El exilio español en México 1932–1982* (Mexico City: Salvat/FCE, 1982), pp. 531–43.
de Cabo, Isabel, *La resistencia cultural bajo el franquismo. En torno a la revista 'Destino' (1957–1961)* (Barcelona: Altera, 2001).
de Diego García, Emilio, *Historia de la industria en España. La química* (Madrid: Escuela de Organización industrial. Actas Editorial, 1996).
De l'Espagne chimique (bref aperçu). Notes sur l'enseignement et la pratique de la chimie en Espagne. Avril 1934 (Madrid: IX Congreso Internacional de Química, 1934).
de Madariaga, Francisco Javier, *Las industrias de guerra de Cataluña durante la Guerra Civil*. PhD thesis (Tarragona: Universitat Rovira i Virgili, 2003).

de Madariaga, Francisco Javier, 'La Comissió d'Indústries de Guerra de la Generalitat', in Guillermo Lusa and Antoni Roca-Rosell (eds.), *Fem memòria per fer futur. La Tècnica i la Guerra Civil. IV Jornades Memorial Democràtic a la UPC. 30 de novembre de 2006* (Barcelona: UPC, 2006), pp. 123–35.

de Madariaga, Francisco Javier, 'Las industrias de guerra de Cataluña durante la Guerra Civil', *Ebre*, 38 (2008), 43–64.

de Riquer, Borja, *Historia de España (IX): La dictadura de Franco* (Barcelona/Madrid: Crítica/Marcial Pons, 2010).

de Riquer, Borja, 'Viure i historiar el franquisme', *L'Avenç*, 417 (2015), 24–33.

Deichmann, Ute, *Biologists under Hitler* (Cambridge, MA: Harvard University Press, 1996) (1st German edition 1992).

Deichmann, Ute, 'The Expulsion of Jewish Chemists and Biochemists from Academy in Nazi Germany', *Perspectives on Science*, 7(1) (1999), 1–86.

Delgado, Lorenzo, 'Cooperación cultural y científica en clave política. Crear un clima de opinión favorable para las bases USA en España', in Lorenzo Delgado and María Dolores Elizalde (eds.), *España y Estados Unidos en el siglo XX* (Madrid: CSIC, 2005), pp. 207–44.

Diario del XXXII Congreso Internacional de Química Industrial (Barcelona: 1960).

Divall, Colin, 'Education for Design and Production: Professional Organization, Employers, and the Study of Chemical Engineering in British Universities, 1922–1976', *Technology and Culture*, 35 (1994), 258–88.

Dodge, Barnett F., 'A Chemical Engineer in Spain', *Chemical Engineering Progress*, 50(9) (1954), 36, 62, 64, 66, 68–9.

Donnelly, James F., 'Representations of Applied Science: Academics and Chemical Industry in Late-Nineteenth-Century England', *Social Studies of Science*, 16 (1986), 195–234.

Donnelly, James F., 'Chemical Engineering in England, 1880–1922', *Annals of Science*, 45(6) (1988), 555–90.

Donnelly, James F., 'Defining the Industrial Chemist in the United Kingdom, 1850–1921', *Journal of Social History*, 29 (1996), 779–96.

Duarte, Ángel and Pere Gabriel, 'Una sola cultura política republicana ochocentista en España?' *Ayer. Revista de Historia Contemporánea*, 39 (2000), 11–35.

Duran, Ricard, 'Colorants Artificials, 1856–1936: Josep Prats i Aymerich', in Josep Batlló et al. (eds.), *Actes de la VIII Trobada d'Història de la Ciència i de la Tècnica* (Barcelona: SCHCT, 2006), pp. 487–94.

Edgerton, David, *Warfare State. Britain, 1920–1970* (Cambridge: Cambridge University Press, 2006).

Edgerton, David, 'Not Counting Chemistry: How We Misread the History of Twentieth-Century Science and Technology', *Distillations*, Spring 2008, Science History Institute. www.sciencehistory.org/distillations/magazine/not-counting-chemistry-how-we-misread-the-history-of-20th-century-science-and (last accessed 29 September 2018).

Enríquez, Alberto (ed.), *Exilio español y ciencia mexicana: Génesis del Instituto de Química y del Laboratorio de Estudios Médicos y Biológicos de la Universidad Nacional Autónoma de México 1939–1945* (Mexico City: Colegio de México, 2000).

Erickson, Paul, Judy L. Klein, Lorraine Daston, Rebecca Lemov, Thomas Sturm and Michael D. Gordin, *How Reason Almost Lost Its Mind: The Strange Career of Cold War Rationality* (Chicago: University of Chicago Press, 2013).

Estrada, Humberto 'Francisco Giral González', in *Profesores eméritos. Semblanzas* (Mexico City: UNAM, 1989), pp. 15–19.

Estruch, Joan, *Saints and Schemers: Opus Dei and Its Paradoxes* (Oxford: Oxford University Press, 1995).

Fell, Ulrike (ed.), *Chimie et industrie en Europe. L'apport des sociétés savantes industrielles du XIXè sièvle à nos jours* (Paris: Editions des archives contemporaines, 2003).

Fennell, Roger, *History of IUPAC, 1919–1987* (Oxford: Blackwell Science, 1994).

Fernández, Obdulio, *Discurso leído ante la Real Academia de Ciencias Exactas, Físicas y Naturales por el Sr. _____ en su recepción pública y contestación del Excmo. Sr. D. José Rodríguez Carracido el día 29 de junio de 1918* (Madrid: Imprenta clásica española, 1918).

Fernández, Obdulio, 'La evolución de la química desde el VIII Congreso Internacional', in *IX Congreso Internacional de química pura y aplicada. Madrid 5–11 April 1934.* (Madrid: C. Bermejo, 1934), pp. 3–48.

Fernández, Obdulio, *Recuerdos de una vida* (Madrid: C. Bermejo, 1973) (reprinted in 1992).

Fernández Alonso, José Ignacio, *Hacia una teoria cuántica de las sustancias que muestran actividad cancerígena. Lección inaugural del curso 1955–56* (Valencia: Tipografía Moderna, 1955).

Fernádez-Ladreda, José María, *Discurso leído por el Excmo. Sr. D. José María Fernández-Ladreda y Menéndez-Valdés, en la solmenne sesión celebrada el día 27 de novembre de 1947 para tomar posesión de una plaza de académico de número y Contestación en nombre de la Academia por el Sr. D. Luis Blas y Álvarez, académico de número* (Madrid: Real Academia Nacional de Farmacia, 1947).

Fernández-Ladreda, José María, *El Doctorado en Química Industrial y la formación de los químicos para la industria. Discurso correspondiente a la apertura del curso académico 1949–1950* (Madrid: Universidad, 1949).

Fernández Terán, Rosario E., *El profesorado del 'Instituto Nacional de Física y Química' ante la Guerra Civil, el proceso de depuración y el drama del exilio.* PhD thesis (Madrid: Universidad Complutense de Madrid, 2005).

Fernández Terán, Rosario. E. and Francisco González Redondo, 'Antonio Madinaveitia y el Laboratorio de Química Biológica de la Junta para Ampliación de Estudios' in José Manuel Sánchez Ron and José García-Velasco (eds.), *100 años de la JAE. La Junta para Ampliación de Estudios e Investigaciones Científicas en su Centenario* (Madrid: Publicaciones de la Residencia de Estudiantes, 2010), pp. 743–61.

'Fernando Calvet Prats', *Anthropos*, 35 (1984), 1–27.

Florensa, Clara, 'James Bond, Pepsi-Cola y el accidente nuclear de Palomares (1966)', in Lino Camprubí, Xavier Roqué and Francisco Sáez de Adana (eds.), *De la Guerra Fría al calentamiento global. Estados Unidos, España y el nuevo orden científico mundial.* (Madrid: La Catarata, 2018), pp. 17–38.

Font, Josep, 'Química 1950–2000', *Anales de Química*, 111(1) (2015), 21–4.

Forgacs, David (ed.), *The Antonio Gransci Reader: Selected Writings 1916–1935* (New York: New York University Press, 2000).

Forgan, Sophie, 'Atoms in Wonderland', *History and Technology*, 19(3) (2003), 177–96.

Formentín, Justo and Esther Rodríguez, *La Fundación Nacional para Investigaciones Científicas (1931–1939)* (Madrid: CSIC, 2009).

Foster, Mary Louise 'The Education of Spanish Women in Chemistry', *Journal of Chemical Education*, 8(1) (1931), 30–4.

Foster, Mary Louise, 'Chemistry in Spain', *Journal of Chemical Education*, XI(7) (1934), 426–7.

Fourneau, Ernest and Antonio Madinaveita, *Síntesis de medicamentos orgánicos* (Madrid: Calpe, 1921).

Fourneau, Jean-Pierre, 'Ernest Fourneau, fondateur de la chimie thérapeutique francaise: feuillets d'albums', *Revue d'histoire de la pharmacie*, 275 (1987), 335–55.

Fox, Robert, *Science Without Frontiers: Cosmopolitanism and National Interests in the World of Learning, 1870–1940* (Corvallis: Oregon State University Press, 2016).

Freemantle, Michael, *The Chemists' War, 1914–1918* (London: Royal Society of Chemistry, 2015).

Furukawa, Yasu, *Inventing Polymer Science: Staudinger, Carothers and the Emergence of Macromolecular Chemistry* (Philadelphia: University of Pennsylvania Press, 1998).

Furter, William F. (ed.), *History of Chemical Engineering* (Washington, DC: American Chemical Society, 1980).

Galán, Fernando, 'Sobre la callada desaparición de un viejo y excelente maestro de la Universidad de Salamanca: El professor Carlos Nogareda Doménech. In Memoriam', *Salamanca. Revista de Estudios*, 33–4 (1994), 349–58.

Galí, Alexandre, *Història de les Institucions i del moviment cultural a Catalunya, 1900–1936* (Barcelona: Fundació Alexandre Galí, 1981).

García Banús, Antonio, 'Los radicales libres en química orgánica', *Anales de la Real Sociedad Española de Física y Química*, 23 (1925), 147–63.

García Lledó, Àngela, *La revista Ciència (1926–1933): Una eina al servei del catalanisme*. Master's thesis (Bellaterra: Universitat Autònoma de Barcelona, 2012).

García Naharro, Fernando, *El papel de la ciencia. Publicaciones científico-técnicas durante el franquismo (1939–1966)*. PhD thesis (Madrid: Universidad Complutense de Madrid, 2017).

García Pérez, Rafael, *Franquismo y Tercer Reich: las relaciones económicas hispano-alemanas durante la Segunda Guerra Mundial* (Madrid: Centro de Estudios Constitucionales, 1994).

Garritz, Andoni, 'Obituario. Francisco Giral González: Un verdadero maestro', *Revista de la Sociedad Química de México*, 46(2) (2002), 193–5.

Garritz, Andoni, 'Breve historia de la educación química en México', *Boletín de la Sociedad Química Mexicana*, 1(2) (2007), 3–24.

Garritz, Andoni and Ricardo Valdez, 'Modesto Bargalló Ardévol. Un químico español que se transformó en mexicano', *Educación Química*, 19(1) (2008), 3–8.

Gavroglu, Kostas et al., 'Science and Technology in the European Periphery: Some Historiographical Reflections', *History of Science*, 46 (2008), 1–23.

Geison, Gerald L., 'Scientific Change, Emerging Specialties and Research Schools', *History of Science*, 10 (1981), 20–40.

Geison, Gerald L. and Frederic L. Holmes (eds.), *Research Schools. Historical Reappraisals. Osiris. Second Series*, 8 (1993).
Gil, Salvador, *Discurso pronunciado por el R. P. Dr. Salvador Gil Quinzá, S.J. Director del Instituto Químico de Sarriá, en el homenaje al R. P. Eduardo Vitoria en su 90° aniversario y 50° de la fundación del Instituto Químico. Barcelona 7 de mayo de 1955* (Barcelona: Instituto Gráfico Oliva de Vilanova, 1955).
Gillispie, Charles, 'The Natural History of Industry', *Isis*, 48 (1957), 398–407.
Giner, Salvador, 'Power, Freedom and Social Change in the Spanish University, 1939–1975', in Paul Preston (ed.), *Spain in Crisis: The Evolution and Decline of the Franco Regime* (New York: Barnes & Noble, 1976), pp. 183–211.
Giner, Salvador, 'Universitat, moviment estudiantil i poder polític a l'estat espanyol, 1939-1975' in Guillermo Lusa and Antoni Roca-Rosell (eds.), *Fem memòria per fer futur. La Universitat sota el franquisme. III Jornades Memorial Democràtic a la UPC. 16 i 17 de novembre de 2005* (Barcelona: UPC, 2005), pp. 27–67.
Giral, Francisco, *Comentarios químico-farmacéuticos a la historia española en América* (Salamanca: Universidad de Salamanca, 1980).
Giral, Francisco, 'Química orgánica (1932–1939)', in *50 años de investigación en física y química en el edificio Rockefeller de Madrid, 1932–1982. Noviembre de 1982. Recuerdos históricos* (Madrid: CSIC, 1982), pp. 39–42.
Giral, Francisco, *Ciencia española en el exilio (1939–1989). El exilio de los científicos españoles* (Barcelona: CIERE/Anthropos, 1994).
Giral, Francisco, *Vida y obra de José Giral Pereira* (Mexico City: UNAM, 2004).
Gleason, Abbott, *Totalitarianism: The Inner History of the Cold War* (Oxford: Oxford University Press, 1995).
Glick, Thomas, *The Comparative Reception of Relativity* (Dordrecht: Kluwer, 1987).
Glick, Thomas, *Einstein in Spain: Relativity and the Recovery of Science* (Princeton: Princeton University Press, 1988).
Glick, Thomas, 'La Fundación Rockefeller en España: Augustus Trowbridge y las negociaciones para el Instituto Nacional de Física y Química, 1923–1927' in José Manuel Sánchez Ron (ed.) *La Junta para Ampliación de Estudios e Investigaciones Científicas 80 años después* (Madrid: CSIC, 1988), pp. 281–300.
Glick, Thomas, 'Dictating to the Dictator: Augustus Trowbridge, the Rockefeller Foundation, and the Support of Physics in Spain, 1923–1927', *Minerva*, 43(2) (2005), 121–45.
Golinski, Jan, *Science as Public Culture: Chemistry and Enlightenment in Britain, 1760–1820* (Cambridge: Cambridge University Press, 1992).
Gómez, Amparo and Antonio Canales (eds.), *Ciencia y fascismo. La ciencia española de postguerra* (Barcelona: Laertes, 2009).
Gómez, Amparo, Antonio Canales and Brian Palmer (eds.), *Science Policies and Twentieth-Century Dictatorships: Spain, Italy, and Argentina* (Farnham: Ashgate, 2015).
González, Carlos and Antonio Santamaría (eds.), *Física y química en la Colina de los Chopos. 75 años de investigación en el edificio Rockefeller del CSIC (1932–2007)* (Madrid: CSIC, 2009).
González Redondo, Francisco and Rosario Fernández Terán, 'Cajal y la nueva senda de la química orgánica en España. En torno a Antonio Madinaveitia Tabuyo', in *Actas del III Simposio 'Ciencia y Técnica en España de 1898 a 1945:*

Cabrera, Cajal, Torres Quevedo' (Madrid: Amigos de la Cultura Científica, 2004), pp. 127–42.

González de Posada, Francisco, *Julio Palacios: Físico español, aragonés illustre* (Madrid: Amigos de la Cultura Científica/Fundación General de la Universidad Politécnica de Madrid, 1993).

González de Posada, Francisco, 'Enrique Moles Ormella (1883–1953): Farmacéutico, químico y artista', *Anales de la Real Academia Nacional de Farmacia*, 71 (2005), 673–702.

González de Posada, Francisco et al., *Ángel del Campo y Cerdán. Eminente químico español. Cuenca 1881; Madrid 1944* (Madrid: Amigos de la Cultura Científica. Academia de Ciencias e Ingeniería de Lanzarote, 2006).

Gooday, Graeme, '"Vague and Artificial": The Historically Elusive Distinction between Pure and Applied Science', *Isis*, 103(3) (2012), 546–54.

Gracia, Jordi, *Estado y cultura. El despertar de una conciencia crítica bajo el franquismo (1940–1962)* (Toulouse: Presses Universitaires du Mirail, 1996).

Gracia, Jordi, *La resistencia silenciosa. Fascismo y cultura en España* (Barcelona: Anagrama, 2004).

Graham, Helen, *The Spanish Republic at War* (Cambridge: Cambridge University Press, 2002).

Graham, Loren, *Science in Russia and the Soviet Union: A Short History* (Cambridge: Cambridge University Press, 1993).

Green, Mott, 'Writing Scientific Biography', *Journal of the History of Biology*, 40(4) (2007), 727–59.

Guia de les institucions científiques i d'ensenyança. Diputació de Barcelona (Barcelona: Publicacions del Consell de Pedagogia, 1916).

Guirao, Fernando, *Spain and the Reconstruction of Western Europe 1945–1957: Challenge and Response* (London: Macmillan, 1998).

Guldi, Jo and David Armitage, *The History Manifesto* (Cambridge: Cambridge University Press, 2014).

Gutiérrez Ríos, Enrique, 'Investigación y libertad', *Nuestro Tiempo*, 109 (1963), 8–20.

Gutiérrez Ríos, Enrique, *José María Albareda: Una época de la cultura española* (Madrid: CSIC, 1970).

Haber, Ludwig Fritz, *The Poisonous Cloud: Chemical Warfare in the First World War* (Oxford: Clarendon Press, 1986).

Hammerstein, Notke, 'National Socialism and the German Universities', *History of Universities*, 18(1) (2003), 170–88.

Harris, Steven J., 'Transposing Merton's Thesis: Apostolic Spirituality and the Establishment of the Jesuit Scientific Tradition', *Science in Context*, 3 (1989), 29–65.

Harrison, Joseph, 'An Espanya Catalana: Catalanist Plans to Transform Spain into a Modern Capitalist Economy, 1898–1933', *Journal of Iberian and Latin American Studies*, 7(2) (2001), 143–56.

Hayes, Peter, 'Carl Bosch and Carl Krauch: Chemistry and the Political Economy of Germany, 1925–1945', *Journal of Economic History*, 47(2) (1987), 353–63.

Hayes, Peter, *Industry and Ideology: IG Farben and the Nazi Era* (Cambridge: Cambridge University Press, 1989).

Herf, Jeffrey, *Reactionary Modernism. Technology, Culture and Politics in Weimar and the Third Reich* (Cambridge: Cambridge University Press, 1984).

Hernández, Elena, Miguel Ángel Ruiz Carnicer and Marc Baldó, *Estudiantes contra Franco (1939-1975). Oposición política y movilización juvenil* (Madrid: La Esfera de los Libros, 2007).

Herrán, Néstor, '"Science to the Glory of God": The Popular Science Magazine *Ibérica* and its Coverage of Radioactivity, 1914-1936', *Science & Education*, 21(3) (2012), 335-53.

Herrán, Nestor and Xavier Roqué, 'An Autarkic Science: Physics, Culture and Power in Franco's Spain', *Historical Studies in the Natural Sciences*, 43(2) (2013), 202-35.

Herrán, Néstor and Xavier Roqué (eds.), *La física en la dictadura. Físicos, cultura y poder en España (1939-1975)* (Bellaterra: Servei de Publicacions de la Universitat Autònoma de Barcelona, 2013).

Homburg, Ernst, 'Shifting Centres and Emerging Peripheries: Global Patterns in Twentieth-Century Chemistry', *Ambix*, 52(1) (2005), 3-6.

Homenaje al Prof. D. Manuel Lora-Tamayo. Presidente de esta Real Academia con motivo de su 70º aniversario (Madrid: Real Academia de Ciencias Exactas, Físicas y Naturales, 1975).

Homenaje a D. Manuel Lora-Tamayo. Sesión solemne celebrada el 20 de octubre de 2003 (Madrid: Instituto de España, 2003).

Howard, Frank A. 'Autarky in 1935. A Chemical Appraisal of the Self-Sufficient State', *Industrial and Engineering Chemistry*, 27(7) (1935), 770-3.

Ibáñez Martín, José, *Hacia un nuevo orden universitario. Discurso de inauguración del año académico 1940-41* (Valladolid: Universidad de Valladolid, 1940).

Ibáñez Martín, José, *La investigación española* (Madrid: Publicaciones Españolas, 1947).

Ibáñez Martín, José, *X años de servicios a la cultura española. Madrid, 1939-1949* (Vitoria/Madrid: Hijos de Heraclio Fournier/Editorial Magisterio, 1950).

Ignatieff, Michael, *The Warrior's Honor: Ethnic War and the Modern Conscience* (London: Chatto and Windus, 1998).

IV Reunión anual de la Real Sociedad Española de Física y Química y I de los Institutos 'Alonso Barba' y 'Alonso de Santa Cruz' del CSIC, San Sebastián, 11-15 junio 1945 (Madrid: Nuevas Gráficas, 1945).

IX Congreso Internacional de Química Pura y Aplicada. Madrid, 5-11 Abril 1934 (Madrid: Bermejo, 1934).

Jarausch, Konrad, *The Unfree Professions: German Lawyers, Teachers and Engineers, 1900-1950* (Oxford: Oxford University Press, 1990).

Jasanoff, Sheila, *States of Knowledge: The Co-Production of Science and Social Order* (London: Routledge, 2004).

Jimeno, Emilio (ed.), *Universidad de Barcelona. Aspectos y problemas de la nueva organización de España. Ciclo de conferencias organizado por la Universidad de Barcelona* (Barcelona: Universidad de Barcelona, 1939).

Jimeno, Emilio, *Ciencia y Técnica* (Madrid: Saeta, 1940).

Jimeno, Emilio and Facundo Rolf Morral, *La ciencia y la industria* (Barcelona: Publicaciones del Instituto de la Metalurgia y la Mecánica. Universidad de Barcelona, 1936).

Johnson, Jeffrey Allan, *The Kaiser's Chemists: Science and Modernization in Imperial Germany* (Chapel Hill: University of North Carolina Press, 1990).

Johnson, Jeffrey Allan, 'The Academic–Industrial Symbiosis in German Chemical Research, 1905–1939', in John Lesch (ed.), *The German Chemical Industry in the Twentieth Century* (Dordrecht: Kluwer, 2000), pp. 15–56.

Johnson, Jeffrey Allan, 'Chemical Warfare in the Great War', *Minerva*, 40(1) (2002), 93–106.

Johnson, Jeffrey Allan, 'Reflections on War and the Changing Perception of Chemistry', *Ambix*, 58 (2011), 99–100.

Johnson, Jeffrey Allan, 'Dilemmas of Nineteenth-Century Liberalism among German Academic Chemists: Shaping National Science Policy from Hofmann to Fischer, 1865–1919', *Annals of Science*, 72(2) (2015), 224–41.

Johnson, Jeffrey Allan and Roy MacLeod (eds.), *Frontline and Factory: Comparative Perspectives on the Chemical Industry at War, 1914–1924* (Dordrecht: Springer, 2006).

Josephson, Paul R., *Totalitarian Science and Technology* (New York: Humanity Books, 2005).

Judt, Tony, *Postwar: A History of Europe since 1945* (London: Penguin Books, 2005).

Julià Saurí, Ángel, 'El IX Congreso de Química Industrial', *Química e Industria*, 67 (1929), 202–3.

Karachalios, Andreas, *I chimici di fronte al fascismo. Il caso di Giovanni Battista Bonino (1899–1985)* (Palermo: Istituto Gramsci Siciliano, 2001).

Knight, David and Helge Kragh (eds.), *The Making of the Chemist: The Social History of Chemistry in Europe 1789–1914* (Cambridge: Cambridge University Press, 1998).

Krige, John, *American Hegemony and the Postwar Reconstruction of Science in Europe* (Cambridge, MA: MIT Press, 2006).

Krige, John, 'Atoms for Peace, Scientific Internationalism and Scientific Intelligence', *Osiris*, 21 (2006), 161–81.

Krige, John and Dominque Pestre (eds.), *Science in the Twentieth Century* (Amsterdam: Harwood Publishers, 1997).

Latour, Bruno, *Science in Action: How to Follow Scientists and Engineers through Society* (Cambridge, MA: Harvard University Press, 1987).

Laviana, Juan Carlos (ed.), *1962: Del Contubernio de Munich a la huelga minera* (Madrid: Unidad Editorial, 2006).

Laylin, James K. (ed.), *Nobel Laureates in Chemistry, 1901–1992* (Washington, DC: American Chemical Society and Chemical Heritage Foundation, 1993).

Leitz, Christian, *Economic Relations between Nazi Germany and Franco's Spain, 1936–1945* (Oxford: Clarendon Press, 1996).

Leitz, Christian, 'Nazi Germany and Francoist Spain, 1936–1945', in Sebastian Balfour and Paul Preston (eds.), *Spain and the Great Powers in the Twentieth Century* (London: Routledge, 1999), pp. 127–150.

Lesch, John E. (ed.), *The German Chemical Industry in the Twentieth Century* (Dordrecht: Kluwer, 2000).

León-Sanz, Pilar, 'Science, State and Society. José María Albareda's *Consideraciones sobre la investigación científica*', *Prose Studies*, 31(3) (2009), 227–40.

Lewis, Paul H., *Latin Fascist Elites: The Mussolini, Franco and Salazar Regimes* (London: Preager, 2002).
Liedtke, Boris, 'Spain and the US, 1945–1975', in Sebastian Balfour and Paul Preston (eds.), *Spain and the Great Powers in the Twentieth Century* (London: Routledge, 1999), pp. 229–44.
Lilla, Mark, Ronald Dworkin and Robert Silvers (eds.), *The Legacy of Isaiah Berlin* (New York: Review Books, 2001).
Llopis, Antonio, 'Primeras materias y productos químicos españoles', in *IX Congreso Interacional de Química Pura y Aplicada. Tomo IX, Grupo VIII, conferencias, memorias y comunicaciones de enseñanza y economia* (Madrid: C. Bermejo, 1934), pp. 5–12.
López, María Silvia, 'Aproximación al estudio de las publicaciones sindicales españolas desarrolladas durante el franquismo (1936–1975)', *Historia y Comunicación Social*, 8 (2003), 159–85
López, Santiago, 'El Patronato Juan de la Cierva (1939–1960), I. Parte: Las instituciones precedentes', *Arbor*, 619 (1997), 201–38.
López, Santiago, 'El Patronato Juan de la Cierva (1939–1960), II. Parte: La organización y la financiación', *Arbor*, 625 (1998), 1–44.
López, Santiago, 'El Patronato Juan de la Cierva (1939–1960), III. Parte: La investigación científica y tecnológica', *Arbor*, 637 (1999), 1–32.
López, Santiago, 'Las ciencias aplicadas y las técnicas: La Fundación Nacional de Investigaciones Científicas y Ensayos de Reformas y el Patronato Juan de la Cierva del CSIC (1931–1961)', in Ana Romero and María Jesús Santesmases (eds.), *Cien años de política científica en España* (Bilbao: Fundación BBVA, 2008), pp. 79–106.
López Rodó, Laureano, *Memorias* (Barcelona: Plaza y Janés, 1990).
López Sánchez, José María, 'El Ateneo Español de México y el exilio intelectual republicano', *Arbor*, 735 (2009), 41–55.
Lora-Tamayo, Manuel, *Química para médicos* (Madrid: Librería Victoriano Suárez, 1940) (1st edition 1934).
Lora-Tamayo, Manuel, 'Orientaciones para una posible reforma de la Facultad de Ciencias', *Revista Nacional de Educación*, 1(2) (1941), 40–50.
Lora-Tamayo, Manuel, *Organización actual de la investigación científica (Texto difundido de las conferencias dadas en el Instituto de Investigación Técnica de Barcelona y en la Feria de Muestras of Valencia. Mayo 1946)* (Madrid: CSIC, Patronato 'Juan de la Cierva Cordorniu', 1946).
Lora-Tamayo, Manuel, *Discurso leído en el acto de su recepción el día 21 de enero de 1948 en la Real Academia de Ciencias Exactas, Físicas y Naturales... Contestación por Antonio Rius Miró* (Madrid: Real Academia de Ciencias Exactas, Físicas y Naturales, 1948).
Lora-Tamayo, Manuel, 'El momento actual de la ciencia española', *Arbor*, 43 (1949), 381–93.
Lora-Tamayo, Manuel, *La investigación en química orgánica. Discurso pronunciado por Manuel Lora-Tamayo, en la sesión de clausura del XI pleno del Consejo Superior de Investigaciones Científicas* (Madrid: CSIC, 1951).
Lora-Tamayo, Manuel, *Cincuenta años de física y química en España (1903–1953)* (Madrid: Real Sociedad Española de Física y Química, 1953).

Lora-Tamayo, Manuel, *Moral profesional del investigador* (Madrid: CSIC, 1953).

Lora-Tamayo, Manuel, 'La obra científica de Kurt Alder', *Revista de Ciencia Aplicada*, 74(3) (1960), 193–205.

Lora-Tamayo, Manuel, *Investigación e Industria. Conferencia en la Sesión de Clausura de la X Reunión Bienal de la Real Sociedad Española de Física y Química y VIII de los Institutos de Física y Química del CSIC y de la Junta de Energía Nuclear (12–19 julio 1961)* (Madrid: Real Sociedad Española de Física y Química, 1961).

Lora-Tamayo, Manuel, *La investigación científica* (Madrid: Editora Nacional, 1963).

Lora-Tamayo, Manuel, 'Estructura de las facultades universitarias y su profesorado', *Revista de Educación*, 174 (1965), 17–26.

Lora-Tamayo, Manuel, 'La investigación al servicio de la defensa', *Arbor*, 281 (1969), 5–19.

Lora-Tamayo, Manuel, *Un clima para la ciencia* (Madrid: Gredos, 1969).

Lora-Tamayo, Manuel, *Sucinta historia de un instituto de investigación. Discurso leido en el acto de recepción como académico de honor por el Excmo. Sr. D. Manuel Lora-Tamayo y contestación del Excmo. Sr. D. Antonio Cortés Lladó, presidente de la Real Academia el dia 24 de mayo de 1973* (Seville: Real Academia de Medicina de Sevilla, 1973).

Lora-Tamayo, Manuel, *'Progresos en 50 años de química orgánica'. Conferencia inaugural por el prof. D. _____Real Sociedad Española de Física y Química. 75 Aniversario. Madrid 2–7 de octubre de 1978* (Madrid: Real Sociedad Española de Física y Química, 1978).

Lora-Tamayo, Manuel, *La investigación química española* (Madrid: Alhambra, 1981).

Lora-Tamayo, Manuel, 'Recuerdos del CSIC en su 50 aniversario', *Arbor*, 529 (1990), 99–115.

Lora-Tamayo, Manuel, *Lo que yo he conocido (Recuerdos de un viejo catedrático que fue ministro)* (Puerto Real: Federico Joly y Cia- Ingrasa, 2002).

'Los laboratorios de la Residencia', *Residencia. Revista de la Residencia de Estudianes*, 1 (1934), 26–30.

Lucier, Paul, 'The Origins of Pure and Applied Science in Gilded Age America', *Isis, Focus: Applied Science*, 103 (2012), 527–36.

Lusa, Guillermo, *La Escuela de ingenieros en guerra (1936–1939)* (Barcelona: Escola Tècnica Superior d'Enginyeria Industrial de Barcelona, UPC, 2007).

Lusa, Guillermo and Antoni Roca-Rosell (eds.), *Fem memòria per fer futur. La Tècnica i la Guerra Civil. IV Jornades Memorial Democràtic a la UPC. 30 de novembre de 2006* (Barcelona: UPC, 2006).

Lusa, Guillermo and Antoni Roca-Rosell (eds.), *Fem memòria per fer futur. La Universitat sota el franquisme. III Jornades Memorial Democràtic a la UPC. 16 i 17 de novembre de 2005* (Barcelona: UPC, 2005)

Lusa, Guillermo and Antoni Roca-Rosell (eds.), *Fem memòria per fer futur. Tècnica, Medicina i Guerra Civil. V Jornada Memorial Democràtic a la UPC. 29 de novembre de 2007* (Barcelona: UPC, 2007).

Macrakis, Kristie, *Surviving the Swastika: Scientific Research in Nazi Germany* (Oxford: Oxford University Press, 1993).

Madariaga, Benito and Celia Valbuena, *La Universidad Internacional de Verano en Santander (1933-1936)* (Guadalajara: Universidad Internacional Menéndez Pelayo. Ministerio de Universidades e Investigación, 1981).

Madinaveitia, Antonio, *La nueva química* (Mexico City: El Nacional, 1942).

Magallón, Carmen, 'Mujeres de las ciencias físico-químicas en España. El Instituto Nacional de Ciencias y el Insituto Nacional de Física y Química (1910-1936)', *Llull*, 20 (1997), 529-74.

Magallón, Carmen, *Pioneras españolas en las ciencias: Las mujeres del Instituto Nacional de Física y Química* (Madrid: CSIC, 1998).

Magallón, Carmen, 'El laboratorio Foster de la Residencia de Señoritas. Las relaciones de la JAE con el "International Institute for Girls in Spain", y la formación de las jóvenes científicas españolas', *Asclepio*, 59(2) (2007), 37-62.

Maier, Charles S., 'Between Taylorism and Technocracy: European Ideologies and the Vision of Industrial Productivity in the 1920s', *Journal of Contemporary History*, 5(2) (1970), 27-61.

Mainer, José Carlos, *La Edad de Plata (1902-1939). Ensayo de interpretación de un proceso cultural* (Madrid: Cátedra, 1983).

Maiocchi, Roberto, 'Scieziati italiani e scienza nazionale (1919-1939)', in Simonetta Saldani and Gabriele Turi (eds.), *Fare gli italiani. Scuola e cultura nell'Italia contemporanea. II. Una società di massa* (Bolonia: Il Mulino, 1993), pp. 41-86.

Maiocchi, Roberto, *Gli scieziati del Duce. Il ruolo dei ricercatori e del CNR nella política autarchica del fascismo* (Rome: Carocci, 2003).

Maiocchi, Roberto, *Scienza e fascismo* (Rome: Carocci, 2004).

Malet, Antoni, 'Las primeras décadas del CSIC: Investigación y ciencia para el franquismo', in Ana Romero and María Jesús Santesmases (eds.), *Cien años de política científica en España* (Bilbao: Fundación BBVA, 2008), pp. 211-56.

Malet, Antoni, *El paper polític de la Delegació del CSIC a Catalunya (1941-1956)* (Barcelona: Fundació Carles Pi i Sunyer, 2009).

Malet, Antoni, 'José María Albareda (1902-1966) and the Formation of the Spanish Consejo Superior de Investigaciones Científicas.' *Annals of Science*, 66 (2009), 307-32.

Mancebo, María Fernanda, *La Universidad de Valencia. De la Monarquía a la República (1919-1939)* (Valencia: Universitat de València/Instituto de Cultura Juan Gil-Albert, 1994).

Mancebo, María Fernanda, 'Tres vivencias del exilio en México: Max Aub, Adolfo Sánchez Vázquez y Francisco Giral', *Migraciones y Exilios*, 5 (2004), 85-102.

Manias, Chris, 'The Race Prussienne Controversy: Scientific Internationalism and the Nation, *Isis*, 100(4) (2009), 733-57.

Manrique, José María and Lucas Molina, *Las armas de la Guerra Civil española. El primer estudio global y sistemático del armamento empleado por ambos contendientes* (Madrid: La Esfera de los Libros, 2006).

Manrique, José María and Lucas Molina, *Guerra química en España, 1912-1945* (Madrid: Galland Books, 2012).

Martín-Aceña, Pablo and Francisco Comín, *INI, 50 años de industrialización en España* (Madrid: Espasa-Calpe, 1991).

Martínez Garroño, María Eugenia, 'Antonio García Banús en Colombia: una aportación del exilio español de capital importancia para la química colombiana', in *VII Congreso Internacional de Historia de América*, 3 vols. (Zaragoza: 1998), I, pp. 491–505.

Mascareñas, Eugenio, *Consideraciones generales acerca de la enseñanza y estudio particular del estado en que se halla la de las ciencias experimentales en España. Discurso inaugural leído en la solemne apertura del cusro académico de 1899 a 1900, ante el claustro de la Universidad de Barcelona por el Dr. D. Eugenio Mascareñas Hernández, Catedrático de la Facultad de Ciencias* (Barcelona: Hijos de Jaime Jepús, 1899).

Mateo, Pedro Luis et al., 'Jesús Yoldi Bereau (1894–1936), el profesor de química que no se presentó a su destino', *Anales de Química*, 110(4) (2014), 286–93.

Mauskopf, Seymour (ed.), *Chemical Sciences in the Modern World* (Philadelphia: University of Pennsylvania Press, 1993).

May, Larry and Jerome Kohn (eds.), *Hannah Arendt: Twenty Years Later* (Cambridge, MA: MIT Press, 1997).

Maya, Alfonso, 'Actividades productivas e innovaciones técnicas', in *El exilio español en México, 1932–1982* (Mexico City: Salvat/FCE, 1982), pp. 148–52.

McCartan, Tom (ed.), *Kurt Vonnegut: The Last Interview and Other Conversations* (New York: Melville House, 2012).

Medina-Doménech, Rosa María, Guillermo Olagüe and Juan Carlos Ortiz de Zárate, 'Ciencia y técnica en la Granada de principios de siglo: el impacto del descubrimiento de los rayos X (1897–1907)', *Llull*, 17 (1994), 103–16.

Meiksins, Peter F., 'The Myth of Technocracy: The Social Philosophy of American Engineers in the 1930s', *History of Political Thought*, 21(3) (2000), 501–23.

Meinel, Christoph, 'Theory or practice? The Eighteenth-Century Debate on the Scientific Status of Chemistry', *Ambix*, 30 (1983), 121–32.

Meissner, William, 'The Jesuit as a Priest–Scientist', *Bulletin of the American Association of Jesuit Scientist*, 40 (1963), 25–33.

Meyer-Thurow, Georg, 'The Industrialization of Invention: A Case Study from the German Chemical Industry', *Isis*, 73(3) (1982), 363–81.

Moles, Enrique, 'Un curso teórico-práctico de química física', *Anales de la Junta para Ampliación de Estudios e Investigaciones Científicas*, 4 (1911), 67–70.

Moles, Enrique, 'Dieci anni di ricerche sui gas', *Gazzetta Chimica Italiana*, 56 (1926), 915–47.

Moles, Enrique, 'Los nuevos laboratorios de química de la Facultad de Ciencias de Madrid', *Anales de la Real Sociedad de Física y Química*, 2(2) (1929), 33–49.

Moles, Enrique, 'La Universidad y la Industria', *Química e Industria*, 7 (1930), 148–53.

Moles, Enrique, *Del momento científico español, 1775–1825. Discurso leído en el acto de ingreso en la Academia de Ciencias Exactas, Físicas y Naturales* (Madrid: C. Bermejo, 1934).

Moles, Enrique, 'Veinte años de investigacions acerca de las densidades gaseosas', *Madrid. Cuadernos de la Casa de Cultura*, 1 (1937), 33–51.

Moles, Enrique, 'El peroxhidróxido de magnesio', *Revista Ibys*, 6 (1946), 1–20.

Moles, Enrique, *'Determinació de pesos moleculars de gasos pel mètode de les densitats límits', Enric Moles i Ormella. Traducció, introducció i comentaris de Joaquim Sales i Agustí Nieto-Galan* (Barcelona: Societat Catalana de Química, 2013).

Molinero, Carme, Manel Risques and Francesc Vilanova (eds.), *Sobre el franquisme i Catalunya. Homenatge a Borja de Riquer i Permanyer* (El Papiol: Edafós, 2015).

Molinero, Carme and Pere Ysàs, *Els industrials catalans durant el franquisme* (Vic: Eumo, 1991).

Molinero, Carme and Pere Ysàs, *El règim franquista: Feixisme, modernització i consens* (Vic/Girona: Eumo/Universitat de Girona, 1992).

Monge, Gerardo, *Memoria del XXII Congreso Internacional de Química Industrial Barcelona. Congrès international de chimie industrielle* (Barcelona: 1949).

Monge, Gerardo, *Figuras de la industria química española* (Barcelona: 1955).

Mora, Jordi, *Unidad de la materia y diversidad ideológica Discursos ontológicos en la España de la segunda mitad del siglo XIX*. PhD thesis (Bellaterra: Universitat Autònoma de Barcelona, 2014).

Moradiellos, Enrique, *La España de Franco (1939–1975). Política y sociedad* (Madrid: Síntesis, 2000).

Moradiellos, Enrique, *Franco: Anatomy of a Dictator* (London: I. B. Tauris, 2018).

Moreno González, Antonio, 'A grandes males, grandes remedios. Una Sociedad española para el adelanto de la física y la química', *Anales de la Real Sociedad Española de Física y Química*, 2 (2003), 244–65.

Morrell, Jack, 'The Chemist Breeders: The Research Schools of Liebig and Thomson', *Ambix*, 19 (1972), 1–46.

Morris, Peter (ed.), *From Classical to Modern Chemistry: The Instrumental Revolution* (London: Royal Society of Chemistry/Science Museum, 2002).

Morris, Peter, *The Matter Factory: A History of the Chemistry Laboratory* (London: Reaktion Books, 2015).

Morris, Peter and Anthony S. Travis, 'The Role of Physical Instrumentation in Structural Organic Chemistry', in John Krige and Dominique Pestre (eds.), *Science in the Twentieth Century* (Amsterdam: Harwood Publishers, 1997), pp. 715–39.

Moya Gago, María Ignacia, *La enseñanza de la química inorgánica en la Universidad de Barcelona (1900–1936)*. PhD thesis (Barcelona: Universitat de Barcelona, 1980).

Muñoz, Josep M., 'Paul Preston: El compromís emocional de l'historiador', *L'Avenç*, 338 (2008), 17–24.

Muñoz Soro, Javier, *Cuadernos para el diálogo (1963–1976). Una historia cultural del segundo franquismo* (Madrid: Marcial Pons, 2006).

Musil, Robert, 'There Must Be More to Love than Death: A Conversation with Kurt Vonnegut', *The Nation*, 231(4) (1980), 128–32.

Nadal, Jordi, 'La debilidad de la industria química española en el siglo XIX. Un problema de demanda', *Moneda y Crédito*, 176 (1986), 33–70.

Nielsen, Anita Kildebaek and Soňa Štrbáňová (eds.), *Creating Networks in Chemistry: The Founding and Early History of Chemical Societies in Europe* (Cambridge: RSC Publishing, 2008).

Nieto-Galan, Agustí, 'Seeking an Identity for Chemistry in Spain: Medicine, Industry, University, the Liberal State and the New "Professionals"', in David Knight and Helge Kragh (eds.), *The Making of the Chemist: The Social*

History of Chemistry in Europe 1789–1914 (Cambridge: Cambridge University Press, 1998), pp. 177–90.

Nieto-Galan, Agustí, 'The Images of Science in Modern Spain. Rethinking the "polémica"', in Kostas Gavroglu (ed.), *The Sciences in the European Periphery during the Enlightenment* (Dordrecht: Kluwer, 1998), 65–86.

Nieto-Galan, Agustí, 'Free Radicals in the European Periphery: "Translating" Organic Chemistry from Zurich to Barcelona in the Early Twentieth Century', *The British Journal for the History of Science*, 37 (2004), 167–91.

Nieto-Galan, Agustí, 'Antonio Gramsci Revisited: Historians of Science, Intellectuals, and the Struggle for Hegemony', *History of Science*, 49(4) (2011), 453–78.

Nieto-Galan, Agustí, 'A Republican Natural History in Spain around 1900: Odón de Buen (1863–1945) and His Audiences', *Historical Studies in the Natural Sciences*, 42(3) (2012), 159–89.

Nieto-Galan, Agustí, 'From Papers to Newspapers: Miguel Masriera (1901–1981) and the Role of Science Popularisation under the Franco Regime', *Science in Context*, 26(3) (2013), 527–49.

Nieto-Galan, Agustí, 'Reform and Repression: Manuel Lora-Tamayo and the Spanish University in the 1960s', in Ana Simões, Kostas Gavroglu and Maria Paula Diogo (eds.), *Sciences in the Universities of Europe, Nineteenth and Twentieth Centuries. Academic Landscapes* (Dordrecht: Springer, 2015), pp. 159–74.

Nieto-Galan, Agustí, *Science in the Public Sphere: A History of Lay Knowledge and Expertise* (London: Routeldge, 2016).

Nieto-Galan, Agustí and Joaquim Sales, 'Josep Pascual Vila (1895–1979): Una aproximación biogràfica', *Anales de Química*, 113(1) (2017), 10–47.

Nieto-Galan, Agustí and Joaquim Sales, 'Exilio y represión científica en el primer franquismo. El caso de Enrique Moles', *Ayer. Revista de Historia Contemporánea*, 114(2) (2019), 279–311.

Nye, Mary Jo, *Science in the Provinces: Scientific Communities and Provincial Leadership in France, 1860–1930* (Berkeley: University of California Press, 1986).

Nye, Mary Jo, *From Chemical Philosophy to Theoretical Chemistry: Dynamics of Matter and Dynamics of Disciplines, 1800–1950* (Berkeley: University of California Press, 1993).

Nye, Mary Jo, *Before Big Science: The Pursuit of Modern Chemistry and Physics, 1800–1940* (Cambridge, MA: Harvard University Press, 1996).

Nye, Mary Jo, 'Scientific Biography: History of Science by Another Means', *Isis*, 97(2) (2006), 322–32.

Olson, Richard, *Scientism and Technocracy in the Twentieth Century: The Legacy of Scientific Management* (London: Lexington Books, 2016).

Ord-Hume, Arthur W. J. G., *Juan de la Cierva and His Autogiros* (Catrine: Stenlake Publishing, 2011).

Ordóñez Alonso, María Magdalena, 'Los científicos del exilio español en México. Un perfil', http://clio.rediris.es/clionet/articulos/cientificos.htm#* (last accessed 21 November 2016).

Oreskes, Naomi, 'Why I Am a Presentist', *Science in Context*, 26 (2013), 595–609.

'Organización de las Naciones Unidad para la Educación, la Ciencia y la Cultura (UNESCO). Festival de Ciencias. Madrid. 19 a 22 octubre de 1955. Informe de la Secretaría', *Anales de la Asociación Española para el Progreso de las Ciencias*, 21 (1956), 212–23.

Osorio, Ramiro, *Historia de la química en Colombia* (Bogotá: Instituto Colombiano de Cultura Hispánica, 1985).

Otero Carvajal, Luis Enrique (dir.), *La Universidad Nacional Católica. La reacción antimoderna* (Madrid: Universidad Carlos III, 2014).

Otero Carvajal, Luis Enrique et al., *La destrucción de la ciencia en España: Depuración universitaria en el franquismo* (Madrid: Editorial Complutense, 2006).

Otero Carvajal, Luis Enrique and José María López Sánchez, *La lucha por la modernidad. Las ciencias naturales y la Junta para Ampliación de Estudios* (Madrid: Residencia de Estudiantes, 2012).

Pagés, Pelai, 'La Comisión de la Industria de Guerra de Cataluña (1936–1938)', *Ebre*, 38 (2008), 43–64.

Palao, Gerardo, 'Influencias extranjeras en la investigación química española (1904–1965), *Llull*, 13 (1990), 131–52.

París, Carlos, 'La pretensión de una Universidad Tecnocrática', in Juan José Carreras and Miguel Ángel Ruiz Carnicer (eds.), *La Universidad española bajo el régimen de Franco (1939–1975)* (Zaragoza: Institución Fernando el Católico, 1991), pp. 437–55.

París, Higinio, 'La industria química y la economía española', *Ión*, 1(1) (1941), 27–9.

Pascual, José, *La química en la Facultad de Ciencias de Barcelona. Discurso inaugural del año académico 1951–52* (Barcelona: Universidad de Barcelona, 1951).

Pascual, José, *La investigación y sus diferentes clases. Discurso leído en el aula magna de la Universidad de Barcelona el día 5 de mayo* (Barcelona: CSIC. Delgación de Barcelona, 1953).

Pascual, José, *'Configuración, mecanismos y conformación en química orgánica', Discurso leído en le acto de su recepción por el Exmo. Sr. Dr. D. Pascual Vila y contestación del Exmo. Sr. Dr. D. Manuel Lora-Tamayo* (Madrid: Real Academia de Ciencias Exactas, Físicas y Naturales, 1963).

Perdiguero, Enrique, José Pardo-Tomás and Àlvar Martínez-Vidal, 'Physicians as a Public for the Popularisation of Medicine in Interwar Catalonia: The *Monografies Mèdiques* Series', in Faidra Papanelopoulou et al. (eds.), *Popularizing Science and Technology in the European Periphery* (Aldershot: Ashgate, 2009), pp. 95–215.

Pérez, Pablo, 'José María Albareda: La ciencia al servicio de Dios', *Nuestro Tiempo*, 665 (2010), 52–7.

Pérez, Pablo, 'International Contacts in the First Years of the Spanish CSIC', in Fernando Clara, Cláudia Ninhos and Sasha Grishin (eds.), *Nazi Germany and Southern Europe, 1933–45: Science, Culture and Politics* (Basingstoke: Palgrave Macmillan, 2016), pp. 68–83.

Pérez-Pariente, Joaquín, 'El Jesuita Eduardo Vitoria: la química como apostolado', in Pedro Bosch Giral et al., *Protagonistas de la química en España: Los orígenes de la catálisis* (Madrid: CSIC, 2010), pp. 62–116.

Pérez-Pariente, Joaquín, Javier Soria Ruiz and Juan Francisco García de la Banda, 'The Origin of the Institute of Catalysis and Petroleum Chemistry of the Spanish National Research Council and Its Relationship with the Development of Catalysis in Spain', *Catalysis Today*, 259 (2015), 3–8.

Pérez Vitoria, Augusto, 'El IX Congreso Internacional de Química pura y aplicada y la XI Conferencia de la Unión Internacional de Química', *Anales de Física y Química*, 32(2) (1934), 195–207.

Pérez Vitoria, Augusto, 'Las políticas de ciencia y de la tecnología de la UNESCO', *Arbor*, 495 (1987), 67–92.

Peypoch, Ramon, 'Notes històriques sobre la Societat Catalana de Ciències Físiques, Químiques i Matemàtiques', *Butlletí de la Societat Catalana de Ciències Físiques, Químiques i Matemàtiques*, 1(1) (1977), 7–10.

Pollock, Ethan, *Stalin and the Soviet Science Wars* (Princeton: Princeton University Press, 2006).

Pohl-Valero, Stefan 'Thermodynamics, Social Thinking and Biopolitics in Spain under the Restoration', *Universitas Humanistica*, 69 (2010), 35–60.

Prats Aymerich, José, *Síntesis Química. Memoria que en cumplimiento de las disposicions vigentes eleva á la superioridad José Prats Aymerich, Profesor numerario de la Escuela Superior de Industrias y de Ingenieros Textiles de Tarrasa. Pensionaso en el Extranjero para el Estudio de dicha ciencia desde diciembre de 1908 hasta agosto de 1909* (Madrid: Fortanet, 1910).

Presas, Albert, 'Nota histórica: Una conferencia de José María Albareda ante las autoridades académicas alemanas', *Arbor*, 631–2 (1998), 343–57.

Presas, Albert, 'Science on the Periphery. The Spanish Reception of Nuclear Energy: An Attempt at Modernity?', *Minerva*, 43 (2005), 197–218.

Presas, Albert, 'La inmediata posguerra y la relación científica y técnica con Alemania', in Ana Romero and María Jesús Santesmases (eds.), *Cien años de política científica en España* (Bilbao: Fundación BBVA, 2008), pp. 173–209.

Preston, Paul (ed.), *Spain in Crisis: The Evolution and Decline of the Franco Regime* (New York: Barnes & Noble, 1976).

Preston, Paul, *Franco* (London: Fontana Press, 1995).

Preston, Paul, *The Spanish Holocaust: Inquisition and Extermination in Twentieth-Century Spain* (London: Harper Press, 2012).

Proctor, Robert, *Racial Hygiene: Medicine under the Nazis* (Cambridge, MA: Harvard University Press, 1988).

Proctor, Robert, *Value-Free Science? Purity and Power in Modern Knowledge* (Cambridge, MA: Harvard University Press, 1991).

Proctor, Robert, *Adolf Butenandt (1903–1995) Nobelpreisträger, Nationalsocialist und MPG-Präsident. Ein esrter Blick in der Nachlass* (Berlin: Max-Planck-Gesellschaft zur Förderung der Wissenschaften, 2000).

Proctor, Robert, 'Nazi Science and Nazi Medical Ethics: Some Myths and Misconceptions', *Perspectives in Biology and Medicine*, 43(3) (2000), 335–46.

Puerto, Javier, *Giral: El domador de tormentas. La sombra de Manuel Azaña* (Madrid: Corona Borealis, 2003).

Puerto, Javier, 'Los Laboratorios Ibys. Una excepción científica e industrial durante la dictadura franquista', *História da Saúde. Estudos do Século XX*, 12 (2012), 253–69.

Puerto, Javier, 'José Giral Pereira (1879–1962) profesor de bioquímica', *Panacea*, 6 (2015), 21–5.
Puerto, Javier, *Ciencia y política. José Giral Pereira* (Madrid: Real Academia de la Historia/BOE, 2016).
Puig, Núria, 'The Frustated Rise of Spanish Chemical Industry between the Wars', in Anthony S. Travis et al. (eds.), *Determinants in the Evolution of the European Chemical Industry, 1900–1939: New Technologies, Political Frameworks, Markets and Companies* (Dordrecht: Kluwer, 1998), pp. 301–20.
Puig, Núria, 'El crecimiento asistido de la industria química en España: Fabricación Nacional de Colorantes y Explosivos, 1922-1965', *Revista de Historia Industrial*, 15 (1999), 105–36.
Puig, Núria, *La nacionalización de la industria farmacéutica en España. El caso de las empresas alemanas, 1914–75* (Madrid: Fundación Empresa Pública, 2001).
Puig, Núria, *Bayer, Cepsa, Repsol, Puig, Schering y La Seda. Constructores de la Química española* (Madrid: LID, 2003).
Puig, Núria, 'La ayuda económica de Estados Unidos y la americanización de los empresarios españoles', in Lorenzo Delgado and María Dolores Elizalde (eds.), *España y Estados Unidos en el siglo XX* (Madrid: CSIC, 2005), pp. 181–206.
Puig, Núria and Santiago López, *Ciencia e industria en España: El Instituto Químico de Sarriá, 1916–1992* (Barcelona: Institut Químic de Sarrià, 1992).
Puig, Núria and Santiago López, 'Chemists, Engineers and Entrepreneurs: The Chemical Institute of Sarrià's Impact on Spanish Industry (1956–1992), *History and Technology*, 11 (1994), 345–59.
Puig-Samper, Miguel Ángel (dir.), *Tiempos de investigación JAE-CSIC. Cien años de ciencia en España* (Madrid: CSIC, 2007).
Putman, Robert, 'Elite Transformation in Advanced International Societies', *Comparative Political Studies*, 10 (1977), 383–412.
Quiroga, Alejandro, *Making Spaniards National Catholicism and the Nationalisation of the Masses during the Dictatorship of Primo de Rivera (1923–1930)*. PhD dissertation (London: London School of Economics and Political Science, 2004).
Ramberg, Peter, 'Chemical Research and Instruction in Zurich, 1831–1871', *Annals of Science*, 72(2) (2015), 170–86.
Ramos, Isabel, *Profesores, alumnos y saberes en la Universidad de Salamanca en el Rectorado de D. Antonio Tovar (1951–1956)* (Salamanca: Ediciones Universidad de Salamanca, 2009).
Ramsay, William, *Química moderna teórica y sistemática; traducida del inglés por Ángel del Campo Cerdán* (Madrid: Romo, 1912).
Real Academia de Ciencias Exactas, Físicas y Naturales, *Libro Homenaje al Profesor D. Obdulio Fernández y Rodríguez, con motivo del cincuentenario de su ingreso en la Real Academia de Ciencias Exactas, Físicas y Naturales ... Discurso leído por D. Florencio Bustinza Lachiondo en la solemne sesión celebrada el día 21 de diciembre de 1968 para hacer entrega al Excmo. Sr. D. O.F.R. de la medalla Echegaray* (Madrid: La Academia, 1969).
Real Sociedad Española de Física y Química, *V Reunión anual de la Real Sociedad Española de Física y Química y II de los Institutos de Física y Química del CSIC. Resúmenes de las comunicaciones científicas. Granada, 26 abril al 1 de mayo de 1948* (Madrid: 1948).

Real Sociedad Española de Física y Química, *Bodas de Oro de la Real Sociedad Española de Física y Química. Coloquio sobre problemas de la Sintesis de Diels y Alder. Madrid 16–17 de abril 1953. Sala de Conferencias del Instituto Alonso Barba de Química* (Madrid: C. Bermejo, 1953).

Redondo, Gonzalo, *Política, cultura y sociedad en la España de Franco (1939–1975)*. 3 vols. (Pamplona: Ediciones Universidad de Navarra, 1999).

Reiding, Jurrie, 'Peter Debye: Nazi Collaborator or Secret Opponent', *Ambix*, 57(3) (2010), 275–300.

Reinhardt, Carsten (ed.), *Chemical Science in the Twentieth Century: Bridging Boundaries* (Weinheim: Wiley-VCH, 2001).

Reinhardt, Carsten, *Shifting and Rearranging: Physical Methods and the Transformation of Modern Chemistry* (Sagamore Beach: Science History Publications, 2006).

Reinhardt, Carsten, 'Sites of Chemistry in the Twentieth Century', *Ambix*, 62(2) (2015), 109–13.

Reinhardt, Carsten, 'Focus: What's in a Name? Chemistry as a Nonclassical Approach to the World', *Isis*, 109(3) (2018), 559–607.

Reinhardt, Carsten and Harm Schröter, 'Academia and Industry in Chemistry: The Impact of State Intervention and the Effects of Cultural Values', *Ambix*, 51(2) (2004), 99–106.

Renneberg, Monika and Mark Walker (eds.), *Science, Technology and National Socialism* (Cambridge: Cambridge University Press, 1994).

Reol Tejada, Juan Manuel, *Homenaje a las grandes figuras de las ciencias farmacéuticas: Obdulio Fernández y Enrique Moles* (Madrid: Real Academia Nacional de Farmacia, 2005).

Reynolds, Terry S., 'Defining Professional Boundaries: Chemical Engineering in the Early Twentieth Century', *Technology and Culture*, 27 (1986), 694–716.

Rhees, David J., 'Corporate Advertising, Public Relations and Popular Exhibits: The Case of DuPont', *History and Technology*, 10 (1993), 65–75.

Ribas, Albert, *La Universitat Autònoma de Barcelona (1933–1939)* (Barcelona: Edicions 62, 1976).

Richards, Michael, *A Time of Silence (1936–45): Civil War and the Culture of Repression in Franco's Spain* (Cambridge: Cambridge University Press, 1998).

Riera, Ignasi, *Els catalans de Franco* (Barcelona: Plaza y Janés, 1998) (Spanish version, 1999).

Rius Miró, Antonio, *Introducción a la Ingeniería Química* (Madrid: Alfa, 1944).

Rius Miró, Antonio, *Discurso leído en el acto de su recepción por le Excmo. Sr. _____ y Contestación del Excmo. Sr. D. Francisco Navarro Borrás, el día 21 de novembre de 1945* (Madrid: Real Academia de Ciencias Exactas, Físicas y Naturales, 1945).

Roberts, Gerrylynn K. and Anna E. Simmons, 'British Chemist Abroad 1887–1971: The Dynamics of Chemists's Careers', *Ambix*, 60(1) (2009), 103–28.

Roberts, Lissa, Simon Schaffer and Peter Dear (eds.), *The Mindful Hand: Inquiry and Invention from the Late Renaissance to Early Industrialisation* (Amsterdam: Koninklijke Nederlandse Akademie van Wetenschappen, 2007).

Roca-Rosell, Antoni, 'Ciencia y sociedad en al época de la Mancomunitat de Catalunya (1914–1923)', in José Manuel Sánchez Ron (ed.), *Ciencia y sociedad en España: De la Ilustración a la Guerra Civil* (Madrid: El Arquero/CSIC, 1988) pp. 223–52.

Rocasolano, Antonio de Gregorio, *De la vida a la muerte. Trabajos del Laboratorios de Investigaciones Bioquímicas de la Facultat de Ciencias de Zaragoza* (Zaragoza: Editorial Gambón, 1937).

Rocasolano, Antonio de Gregorio, 'La investigación científica, acaparada y estropeada', in Miguel Artigas et al. (eds.), *Una poderosa fuerza secreta. La Institución Libre de Enseñanza* (San Sebastian: Editorial Española, 1940), pp. 149–60.

Rocasolano, Antonio de Gregorio, 'La táctica de la Institutción', in Miguel Artigas et al. (eds.), *Una poderosa fuerza secreta. La Institución Libre de Enseñanza* (San Sebastian: Editorial Española, 1940), pp. 125–30.

Rocasolano, Antonio de Gregorio and Luis Bermejo, *Química para médicos y naturalistas* (Madrid: Ramona Velasco, 1929).

Rodríguez Nozal, Raúl, 'Orígenes, desarrollo y consolidación de la industria farmacéutica española (ca. 1850–1936)', *Asclepio*, 52(1) (2000), 127–59.

Rodríguez Nozal, Raúl, *Uriach, Cambronero, Gallego. Farmacia e Industria. La producción de los primeros medicamentos en España* (Tres Cantos: Nívola, 2004).

Rodríguez Nozal, Raúl, 'Sanidad, farmacia y medicamento industrial durante la segunda República', *Papeles de la Fundación de Investigacions Marxistas*, 28(2) (2008), 163–85.

Rodríguez Nozal, Raúl and Antonio González Bueno (eds.), *El medicamento de fabricación industrial en la España contemporània* (Madrid: CERSA, 2008).

Rodrigo, Antonina, *Una mujer silenciada. María Teresa Toral, ciencia, compromiso y exilio* (Barcelona: Ariel, 2012).

Rojahn, Carl August, *Preparación de productos químicos y químico-farmacéuticos, traducido del original alemán y considerablemente ampliado por el profesor Francisco Giral, Catedrático de Farmacia de la Facultad de Santiago (España), Jefe de los Departamentos de síntesis orgánica en los Laboratorios Hormona, México*. 2 vols. (Mexico City: Atlante, 1942).

Romero, Ana, 'Un viaje de José María Otero Navascués. Los inicios de la investigación nuclear en España', *Arbor*, 659–60 (2000), 509–25.

Romero, Ana and José Manuel Sánchez Ron, *Energía nuclear en España. De la JEN al Ciemat* (Madrid: Doce Calles, 2001).

Romero, Ana and María Jesús Santesmases (eds.), *Cien años de política científica en España* (Bilbao: Fundación BBVA, 2008).

Romero, Francisco, 'Spain and the First World War', in Sebastian Balfour and Paul Preston (eds.), *Spain and the Great Powers in the Twentieth Century* (London: Routledge, 1999), pp. 32–52.

Romero, Francisco and Angel Smith (eds.), *The Agony of Spanish Liberalism: From Revolution to Dictatorship, 1913–23* (Basingstoke: Palgrave/Macmillan, 2010).

Rossiter, Margaret W., *Women Scientists in America before Afirmative Action, 1940–1972* (Baltimore/London: Johns Hopkins University Press, 1995).

Rovighi, Alberto and Filippo Stefani, *La partecipazione italiana alla Guerra Civile Spagnola* (Rome: Ufficio storico dello Stato maggiore dell'Esercito, 1993).

Ruiz Carnicer, Miguel Ángel, 'Spanish Universities under Franco', in John Connelly and Michael Grüttner (eds.), *Universities under Dictatorship* (University Park: The Penn State University Press, 2005), pp. 113–38.

Russell, Colin A., '"Rude and Disgraceful Beginnings": A View of History of Chemistry from the Nineteenth Century', *The British Journal for the History of Science*, 21(3) (1988), 273–94.

Russell, Colin A., Noel G. Coley and Gerrylynn K. Roberts, *Chemists by Profession: The Origins and Rise of the Royal Institute of Chemistry* (Milton Keynes: Oxford University Press and the Royal Institute of Chemistry, 1977).

Ryan, Alan (ed.), *The Idea of Freedom: Essay in Honour of Isaiah Berlin* (Oxford: Oxford University Press, 1979).

Said, Edward, *Representations of the Intellectual: The 1993 Reith Lectures* (New York: Vintage Books, 1996).

Sales, Joaquim, *La química a la Universitat de Barcelona* (Barcelona: Publicacions i Edicions de la Universitat de Barcelona, 2011).

Sánchez, Agustín and Silvia Figueroa (eds.), *De Madrid a México. El exilio español y su impacto sobre el pensamiento, la ciencia y el sistema educativo mexicano* (Madrid: Universidad Michoacana de San Nicolás de Hidalgo & Comunidad de Madrid, 2001).

Sánchez Ron, José Manuel (ed.), *Ciencia y sociedad en España: De la Ilustración a la Guerra Civil* (Madrid: El Arquero/CSIC, 1988).

Sánchez Ron, José Manuel (ed.), *La Junta para Ampliación de Estudios e Investigaciones Científicas 80 años después: 1907–1987*. 2 vols. (Madrid: CSIC, 1988).

Sánchez Ron, José Manuel, *Miguel Catalán. Su obra y su mundo* (Madrid: CSIC/ Fundación Ramón Menéndez Pidal, 1994).

Sánchez Ron, José Manuel, *Un siglo de ciencia en España* (Madrid: Residencia de Estudiantes, 1998).

Sánchez Ron, José Manuel, *Cincel, martillo y piedra. Historia de la ciencia en España (siglos XIX y XX)* (Madrid: Taurus, 1999).

Sánchez Ron, José Manuel, *50 años de cultura e investigación en España: La fundación Juan March (1955–2005)* (Barcelona: Crítica, 2005).

Sánchez Ron, José Manuel and Antoni Roca-Rosell, 'Spain's First School of Physics: Blas Cabrera's Laboratorio de Investigaciones Físicas', in Gerald L. Geison and Frederic L. Holmes (eds.), *Research Schools. Historical Reappraisals. Osiris. Second Series*, 8 (1993), pp. 127–55.

Sánchez Santiró, Ernest, *Científics i professionals: La Facultat de Ciències de València (1857–1939)* (Valencia: Universitat de València, 1998).

San Román, Elena, *Ejército e industria: El nacimiento del INI* (Barcelona: Crítica, 1999).

Santesmases, María Jesús, 'Scientific Organizations in Spain (1950–1970): Social Isolation and International Legitimation of Biochemists and Molecular Biologists in the Periphery', *Social Studies of Science*, 27 (1997), 187–219.

Santesmases, María Jesús, 'El legado de Cajal frente a Albareda. Las ciencias biológicas en los primeros años del CSIC y los orígenes del CIB', *Arbor*, 631-2 (1998), 305–32.

Santesmases, María Jesús, *Antibióticos en la autarquía. Banca privada, industria farmacéutica, investigación científica y cultura liberal en España, 1940–1960* (Madrid: Fundación Empresa Pública, 1999).

Santesmases, María Jesús, 'Severo Ochoa and the Biomedical Sciences in Spain under Franco, 1959–1975', *Isis*, 91 (2000), 706–34.

Santesmases, María Jesús, *Entre Cajal y Ochoa: Ciencias biomédicas en la España de Franco, 1939–1975* (Madrid: CSIC, 2001).

Santesmases, María Jesús, 'Orígenes internacionales de la política científica', in Ana Romero and María Jesús Santesmases (eds.), *Cien años de política científica en España* (Bilbao: Fundación BBVA, 2008), pp. 293–326.

Santesmases, María Jesús, 'Distributing Penicillin: The Clinic, the Hero and Industrial Production in Spain, 1943–1952', in Viviane Quirke and Judy Slinn (eds.), *Perspectives on Twentieth-Century Pharmaceuticals* (Oxford: Peter Lang, 2010), pp. 91–118.

Santesmases, María Jesús, *The Circulation of Penicillin in Spain: Health, Wealth and Authority* (London: Palgrave/Macmillan, 2018).

Santesmases, María Jesús and Emilio Muñoz, 'The Scientific Periphery in Spain: The Establishment of a Biomedical Discipline at the Centro de Investigaciones Biológicas', *Minerva*, 35 (1997), 27–45.

Saraiva, Tiago, *Fascist Pigs: Technoscientific Organisms and the History of Fascism* (Cambridge, MA: MIT Press, 2016).

Saraiva, Tiago and Norton Wise, 'Autarky/Autarchy: Genetics, Food Production, and the Building of Fascism', *Historical Studies in the Natural Sciences*, 40(4) (2010), 419–28.

Sarasua, Juan Manuel, *La protección colectiva de la población civil contra las armas químicas en el Guerra Civil española. Los Cursillos de Defensa Pasiva de la Junta de Defensa Pasiva de Cataluña (1937–1938)*. Master's thesis (Bellaterra: Univeristat Autònoma de Barcelona, 2007).

Saz, Ismael, *Las caras del franquismo* (Granada: Comares, 2013)

Schaffner, Kurt, 'Juan Juilo Bonet and His Relation with the Zurich School', *Afinidad*, 64(528) (2007), 103–4.

Schieder, Wolfgang and Achim Trunk (eds.), *Adolf Butenandt und die Kaiser-Wilhelm-Gesellschaft. Wissenschaft, Industrie und Politik im Dritten Reich* (Göttingen: Wallstein Verlag, 2004).

Schummer, Joachim, Bernadette Bensaude-Vincent and Brigitte Van Tiggelen (eds.), *The Public Image of Chemistry* (Singapore: World Scientific Publishing, 2007).

Segura, Manuel, Alberto Gomis and José María Sánchez, 'Modesto Bargalló Ardévol (1894–1981), Maestro de maestros e historiador de la ciencia', *Lull*, 34 (2011), 419–42.

Serratosa, Félix, '¿Qué investiga Vd? Carta abierta a D. Pedro Rocamora, Director de Arbor', *Arbor*, 295–6 (1970), 5–19.

Servos, John, *Physical Chemistry from Ostwald to Pauling: The Making of Science in America* (Princeton: Princeton University Press, 1990).

Sevigné, Eva, *Juan Julio Bonet Sugrañes y el Laboratorio de Esteroides del Instituto Químico de Sarriá (IQS). 1965–1986*. Master's thesis (Bellaterra: Universitat Autònoma de Barcelona, 2008).

Seward, Albert Charles (ed.), *Science and the Nation: Essays by Cambridge Graduates with an Introduction by Lord Moulton* (Cambridge: Cambridge University Press, 1917).

Shapin, Steven, *The Scientific Life: A Moral History of a Late Modern Vocation* (Chicago: University of Chicago Press, 2008).
Shapin, Steven, *Never Pure: Historical Studies of Science as If It Was Produced by People with Bodies, Situated in Time, Space, Culture and Society, and Struggling for Credibility and Authority* (Baltimore: Johns Hopkins University Press, 2010).
Shapin, Steven and Arnold Thackray, 'Prosopography as a Research Tool in History of Science: The British Scientific Community, 1700–1900', *History of Science*, 12 (1974), 1–28.
Shorten, Richard, 'Rethinking Totalitarian Ideology: Insights from the Anti-Totalitarian Canon', *History of Political Thought*, 36(4) (2015), 716–61.
Shortland, Michael and Richard Yeo, *Telling Lives in Science: Essays on Scientific Biography* (Cambridge: Cambridge University Press, 1996).
Siegmund-Schultze, Rienhard, 'The Problem of Antifascist Resistance of "Apolitical" German Scholars', in Monika Renneberg and Mark Walker (eds.), *Science, Technology and National Socialism* (Cambridge: Cambridge University Press, 1994), pp. 312–23.
Sirera, Carles, 'Neglecting the Nineteenth Century: Democracy, the Consensus Trap and the Modernization Theory in Spain', *History of the Human Sciences*, 28(3) (2015), 51–67.
Smyth, Denis, 'Franco and the Allies in the Second World War', in Sebastian Balfour and Paul Preston (eds.), *Spain and the Great Powers in the Twentieth Century* (London: Routledge, 1999), pp. 185–209.
Spitler, Ernst. G., 'The Priest–Scientist in the Church', *Bulletin of the American Association of Jesuit Scientists*, 39 (1962), 30–7.
Stoetzer, Otto Carlos, *Karl Christian Friedrich Krause and his Influence in the Hispanic World* (Vienna: Bohlau, 1998).
Stokes, Raymond G., *Opting for Oil: The Political Economy of Technological Change in the West Germany Chemical Industry, 1945–1961* (Cambridge: Cambridge University Press, 1994).
Suay-Matallana, Ignacio, *Análisis químico y expertos en la España contemporánea. Antonio Casares Rodríguez (1812–1888) y José Casares Gil (1866–1961)*. PhD thesis (Valencia: Universitat de València, 2014).
Suay-Matallana, Ignacio, 'Between Chemistry, Medicine and Leisure: Antonio Casares and the Study of Mineral Waters and Spanish Spas in the Nineteenth Century', *Annals of Science*, 73(3) (2016), 289–302.
Sureda, Josep, *Antologia Científica* (Palma de Mallorca: Govern de les Illes Balears. Conselleria d'Innovació i Energia, 2003).
Toca, Ángel, *La introducción de la gran industria química en España. Solvay y su planta de Torrelavega (1887–1935)* (Santander: Universidad de Cantabria, 2005).
Toca, Ángel, '"Dos profesiones para un solo cometido." La introducción de la ingeniería química en España durante el primer franquismo', *Dynamis*, 26 (2006), 253–85.
Toca, Ángel, 'Ingeniería Química en España: Los orígenes (1850–1936) (I)', *Anales de Química*, 103 (2) (2007), 47–53.
Toca, Ángel, 'Ingeniería Química en España: Los orígenes (1850–1936) (II)', *Anales de Química*, 103(3) (2007), 759–66.

Tomeo, Mariano, *Rocasolano. Notas bio-bibliográficas* (Zaragoza: La Académica, 1941).
Torrellas, Susana, *Química i indústria a l'Exposició Internacional de Barcelona de 1929*. Master's thesis (Bellaterra: Universitat Autònoma de Barcelona, 2013).
Toural Quiroga, Manoel, 'Antonio Madinaveitia, un científico republicano', in Pedro Bosch Giral et al., *Protagonistas de la química en España: Los orígenes de la catálisis* (Madrid: CSIC, 2010), pp. 223–85.
Trunk, Achim, 'Biochemistry in Wartime: The Life and Lessons of Adolf Butenandt, 1936–1946', *Minerva*, 44 (2006), 285–306.
Tusell, Javier, *Franco y los católicos. La política interior española entre 1945 y 1957* (Madrid: Alianza, 1984).
Udías, Agustín, *Jesuit Contribution to Science: A History* (Dordrecht: Springer, 2015).
Uhthoff, José, 'El químico en nuestro país. Lo que es y lo que debe ser', *Química e Industria*, 1(2) (1924), 39–41.
Uhthoff, Josep, 'La indústria química a l'Exposició', *Ciència: Revista catalana de ciència i tecnologia*, 36 (1930), 627.
Urbain, Georges, *Conferencias acerca de la químca física de los complejos minerales ... traducidas por Ángel del Campo. Catedrático de la Universidad de Madrid* (Madrid: Eduardo Arias, 1917).
Urbieta, María Teresa and José Llombart, 'Las actividades científicas del químico Eugenio Muñoz Mena en el País Vasco durante los años anteriores a la guerra civil', in *Los científicos del exilio español en México* (Mexico City: Universidad Michoacana de San Nicolás de Hidalgo, Sociedad Mexicana de Historia de la Ciencia y de la Tecnología, Sociedad Española de Historia de las Ciencias y de las Técnicas, 2001), pp. 235–58.
Valentines-Álvarez, Jaume, 'La introducció de la gran indústria de la química a Espanya i el nou règim de sabers a la perifèria', *Quaderns d'Història de l'Enginyeria*, 8 (2007), 361–75.
Valera, Miguel, 'Eduardo Vitoria S.J. A Contemporary Leader in the Spanish Chemical World', *Journal of Chemical Education*, 33(4) (1956), 161–6.
Valera, Manuel et al., 'La Guerra Civil española y la investigación científica en química. Estudio preliminar', in Javier Echeverría and Marisol de Mora (eds.), *Actas del III Congreso de la Sociedad Española de Historia de la Ciencia y de la Técnica. San Sebastián, 1–6 Octubre 1984* (San Sebastian: Editorial Guipuzcoana, 1986), Vol. III, pp. 395–407.
Valbuena Vázquez, Pablo, *Historia de la Escuela Técnica Superior de Ingenieros Industriales de Madrid desde 1901 hasta 1972* (Madrid: E.T.S.I. Industriales (UPM), 1996). http://oa.upm.es/5331/1/PFC_PABLO_VALBUENA_VAZQUEZ.pdf (last accessed 5 December 2016).
Vergara, Juan, *La química orgánica en España en el primer tercio del siglo XX*. PhD thesis (Valencia: Universitat de València, 2004).
Vessuri, Hebe, 'La ciencia académica en América Latina en el siglo XX', *Redes. Revista de Estudios Sociales de la Ciencia*, 1(2) (1994), 41–76.
Vián, Ángel, *Publicaciones científicas (1936–1984). Homenaje académico ... por sus 51 años de Trabajo en el campo de la química industrial* (Madrid: Técnicas Reunidas, 1985).

Vián, Ángel and Joaquín Ocón, *Elementos de Ingeniería Química (Operaciones Básicas)* (Madrid: Aguilar, 1952).

Victori, Lluís, *L'Institut Químic de Sarrià, 1905–2005* (Barcelona: Institut Químic de Sarrià, 2005).

Victori, Lluís, 'Eduard Vitoria i Miralles: pioner de la química al nostre país', *Revista de la Societat Catalana de Química*, 7 (2006), 60–64.

Victori, Lluís, P. *Salvador Gil. Creador del IQS moderno (1934–1957)* (Barcelona: Institut Químic de Sarrià, 2009).

Villa, Dana (ed.), *The Cambridge Companion to Hannah Arendt* (Cambridge: Cambridge University Press, 2000).

Villena, Leonardo, *Julio Palacios: Labor didáctica, confinamiento y proyección internacional* (Santander: Amigos de la Cultura Científica, 1985).

Viñas, Ángel, *Los pactos secretos de Franco con Estados Unidos: Bases, ayuda económica, recortes de soberanía* (Barcelona: Grijalbo, 1981).

Viñas, Ángel, *Guerra, dinero, dictadura. Ayuda fascista y autarquia en el España de Franco* (Barcelona: Crítica, 1984).

Vitoria, Eduardo, *Conferencias de química moderna dadas en el Laboratorio Químico del Ebro de la Compañía de Jesús por el P. Eduardo Vitoria S.J. Director de dicho establecimiento* (Tortosa: Biarnés y Foguet, 1907).

Vitoria, Eduardo, *Catálisis química. Sus teorías y aplicaciones en el laboratorio y en la industrias* (Barcelona: Tipografía Católica, 1911).

Vitoria, Eduardo, *Prácticas químicas para cátedras y laboratorio* (Barcelona: Tipografía R. Casals, 1914).

Vitoria, Eduardo, *La ciencia química y la vida social. Conferencias de vulgarización científica dadas en el paraninfo de la Universidade Literaria y en el Centro Escolar y Mercantil de Valencia, del 23 al 28 de noviembre de 1915, por el _____ Dr. en Ciencias. Director del Laboratorio Químico del Ebro* (Barcelona: Tipografía Católica Pontificia, 1916).

Vitoria, Eduardo, *Manual de quimica moderna teórica y experimertal con sus principales aplicaciones al comercio y a la industria* (Barcelona: Tipografía Católica Casals, 1929).

Vitoria, Eduardo, *Química del carbono. Teoría y práctica con vistas muy especiales a la síntesis en el laboratorio y en la industria* (Barcelona: Tipografía Católica Casals, 1940).

Vitoria, Eduardo, *El pan y el vino eucarísticos: Estudio químico-litúrgico* (Bilbao: La Editorial Vizcaína, 1944).

Vitoria, Eduardo, *Autobiografía, 1864–1958. Edición facsímil del original manuscrito con transcripción, introducción y notas al cuidado de Lluís Victori i Companys* (Barcelona: Institut Químic de Sarrià, 2005).

Vonnegut, Kurt, *Cat's Cradle* (Harmondsworth: Penguin, 1965) (1st edition 1963).

Wagner, Bernd C., *IG Auschwitz. Zwangsarbeit und Vernichtung von Häftlingen des Lagers Monowitz 1941–1945* (Munich: K. G. Saur Verlag, 2000).

Wagner-Kyora, Georg, 'Continuities in the Identity Construction of Industrial Chemists, 1949–1970', *German Studies Review*, 29(3) (2006), 611–19.

Walker, Mark, *German National Socialism and the Quest for Nuclear Power, 1939–1949* (Cambridge: Cambridge University Press, 1989).

Walker, Mark, *Nazi Science: Myth, Truth and the German Atomic Bomb* (Cambridge, MA: Perseus, 1995).
Walker, Mark (ed.), *Science and Ideology: A Comparative History* (London: Routledge, 2003).
Warleta, José, *Autogiro. Juan de la Cierva y su obra* (Madrid: Instituto de España, 1977).
Weinstein, David, 'Nineteenth- and Twentieth-Century Liberalism', in George Klosko (ed.), *The Oxford Handbook of the History of Political Philosophy* (Oxford: Oxford University Press, 2011).
Willstätter, Richard, *From My Life: The Memoirs of Richard Willstätter* (New York: W. A. Benjamin Inc., 1965).
Witkop, Bernhard, 'Remembering Heinrich Wieland (1877–1957), portrait of an organic chemist and founder of modern biochemistry', *Medicinal Research Reviews*, 12(3) (1992), 195–274.
Wizinger, Robert, *Carbón, aire y agua. Revisado, puesto al día y adaptado al caso de España por el Dr. J.M. Fernández-Ladreda y Menéndez-Valdés, Coronel de Artillería. Catedrático de la Universidad de Sevilla* (Madrid: Aguado, 1942).
Worscester, Thomas (ed.), *The Cambridge Companion to the Jesuits* (Cambridge: Cambridge University Press, 2008).
Ynfante, Jesús, *La prodigiosa aventura del Opus Dei* (Paris: Ruedo Ibérico, 1970).
Zimmerman, David, 'The Society for the Protection of Science and Learning and the Politicization of British Science in the 1930s', *Minerva*, 44(1) (2006), 25–45.

Index

α, β, and γ rays, 65
Abelló, Juan, 146, 150, 161, 183, 230
Abir-Am, Pnina G.
  commemorative practices, 6
Abollado, Carlos, 146, 161
Academia de Ciencias de Cuba, 222
Academia de Ciencias de Zaragoza, 38
Academia Pontificia de Ciencias, 127–8, 130
Acción Católica, 117, 126
Acción Española, 96. *See* Rocasolano
Acción Nacional, 96
Agell, José, 31, 35, 42–3, 146, 185
  Escola Industrial, Barcelona, 34
  industrial dream, 203
  Institut de Química Aplicada, 43, 203
Agreement
  on tariffs and trade (GATT), 165
  US–Spain (1953), 149, 154, 162, 175
  West Germany–Spain, 181
Agriculture, 11, 43, 45, 48, 121, 143, 178, 200, 233
  US Department, 178
agrochemical stations, 127. *See* Albareda
Albareda, José María, xi, xii, xiii, xviii, 14, 103, 110, 115–16, 129, 161
  Academia Pontificia de Ciencias, 127
  against Moles, 108
  against the INFQ, 117
  against the JAE, 118
  and Butenandt, 179
  and Ibáñez Martín, 112–13
  and Lora-Tamayo, 173
  and Opus Dei, 126, 128, 168
  and Rius Miró, 150
  and the CSIC, 112–13, 129, 135, 158
  as organic intellectual, 227
  bust, 229
  death, 174, 210
  leftist repression, 101
  network, 130, 156, 221
  right-wing Catholicism, 210
  school, 150

science policies, 8, 134
selection committees, 95
soil chemistry, xiii, 127, 129, 134, 143, 156, 220, 225
totalitarian ambitions, 15
Universidad de Navarra, 173
Alcalá Zamora, Niceto, 68
alcohol, 76, 79, 178, 185
  fermentation, 105
Alder, Kurt, 158, 160, 171, 220
Algerian War, 178
alkaloids, 16, 20, 31, 50, 61, 68, 98, 104, 150, 207
Allardt, Helmut, 180
Allmand, Arthur J., 97
alloys, 81, 121–2
ambassador(s), 177, 180
  chemists as, 15, 138, 156
American Chemical Society, 32, 159, 197
ammonia, 77, 91, 190
analytical chemistry, 30, 37, 101, 103, 209
anarchist, 170
  brigades, 102
  CNT, 89, 92
  groups, 84, 101–2, 192
  revolutionaries, 84
  violence, 103
  workers, 87
aniline black, 27
antacid Alugel, 205
anti-clericalism, 194, 225
antiknock agent, 87
anti-liberalism, 123, 138
anti-Semitism, 57, 96, 194
apoliticism, 10, 15, 18, 85, 92, 95, 103, 156, 164, 166–7, 181, 187–8, 216–17, 226, 228
appeasement policies, 100
applied chemistry, 14, 16, 32, 48–9, 55, 67, 69, 82, 96, 109, 110–11, 113, 116, 123, 125, 127, 134, 138, 143–6, 149,

Index

155–7, 170, 179, 185, 213, 217, 220–1, 224–5
Arendt, Hannah, 1
Argentina, 99
Aries, Robert S., 152
arms embargo, 100
Armstrong, Henry E., 71
Arrhenius, Svante, 24
artificial fibres, 78, 137, 142, 182
Artigas, José Antonio, 140
Asociación Católica Nacional de Propagandistas (ACNP), 92, 96, 126, 130
Asociación Nacional de Químicos de España (ANQUE), 159, 184
Asturias, 55, 78, 136
Atlante (publisher), 196, 199, 202, 224
atomic bomb project, 154
atomic weights, 16, 32, 34, 38, 49, 58, 61, 104, 109, 111, 118, 220, 222. *See* Moles
  commission, 32, 222
  experimental determination, 58, 61, 118, 216, 220
'atoms for peace', 8, 176
Aub, Max, xii, xiii
Aula de Cultura Científica, 213
Austria, 5
autarchy, 14–15, 121, 137–40, 155, 158–9, 161–2, 186, 204, 207, 209, 220
autarky, 10, 15, 134, 137–8, 158
Azaña, Manuel, 21, 52, 98, 231

backwardness, 10, 14, 23, 25, 27–8, 32, 39, 41, 44, 50–1, 56, 120, 154
Badische Anilin und Soda Fabrik (BASF), 142, 147, 186, 220
Bakelite Corporation (chemical company), 80
Balmer, Brian
Banco de México, 198
Banco Urquijo, 171
Barba, Álvaro Alonso, 118
Barceló, José, 125
Bargalló, Modesto, 195–6, 200, 202, 213, 225
Barger, George, 67, 72, 105
Barnés, Adela, 100, 194, 196, 200
Barnés, Dorotea, 100
Barnés, Francisco, xvii
Barnés, Pilar, 194
Barton, Derek, 199
Basel, 41, 150
Basque Country, 64
Battle of Annual, 84

Batuecas, Tomás, 176
Bayer (chemical company), 183, 220
Bayern Academy of Science, 180
Belgium, 5, 44, 46, 107
benzene, 27, 31, 139, 182
Bergmann, Ernst, 194
Berkeley, 70
Berlin, 12, 29, 31, 142
  Dahlem, 179, 200
  Wall, 168
Berlin, Isaiah, 189, 214
Bermejo, Luis, 14, 36, 38, 95–6, 103, 110, 118, 138–9
  Nobel Prize nominator, 72
biochemistry, 8, 15, 38–9, 68, 103, 105, 127–8, 179, 207. *See* Calvet, Rosasolano
  Adolf Butenandt, 179
  Hans von Euler, 105
  Kaiser Wilhelm Institute, 179
biography, xiii, xiv, xvii, 3, 18, 126, 195, 227, 231
biography (auto), 6
biological chemistry, 22, 58, 128–9, 201
bituminous slates, 141, 147, 182
Bladé, Artur, xvii, 200
Blas, Luis, 92, 141. *See* chemical weapons
Bogotá, 98
Bolívar, Ignacio, 20
Bolívar, José Ignacio, 199
bombing, 199
  Barcelona, 85
  civil population, 199
  Guernica, 85
Bonet, Juan Julio, v, xviii, 232–5
Bonino, Giovanni Battista, 3, 70, 140
Bonn, 157
Bosch, Carl, 190
Bourbon Restoration, 2, 14, 25, 33, 50
Britain, 36, 46, 83, 149, 156
British Council, 156
Bruni, Giuseppe, 31, 140
Bureau of Standards, Washington, DC, 40, 46, 128
Burgos
  government, 112, 126
  headquarters, 142
Buscarons, Francisco, 14, 158, 209–10, 212, 214
Bush, Vannevar, 164–5, 175, 220
Butenandt, Adolf, 179–81, 224, 232
Buxó, Joaquín, 145

Cabrera, Blas, 20, 32, 36, 38, 56–7, 61, 97, 202
calcium carbide, 141
calorimetry, 160
CALTECH, 70, 160
Calvet, Fernando, xvii, 30, 67, 99, 103–5, 110, 153, 207, 209, 212, 214
Calvo Serer, Rafael, 191
Cámara de Comercio, Madrid, 146
Cámara Nacional de Industrias Químicas, 44, 76, 80
Camprubí, Lino, 9
CAMPSA, 141, 182
Canales, Antonio, 8
Cannon, Walter, 100
Caputxinada, La, 210
Caracas, 98, 201–2
Carceller, Demetrio, 142, 145, 161, 183
Cárdenas, Lázaro, 193, 201
carotenoids, 157
Casamada, Ramón, 101
Casares, José, 26, 28, 33, 115
Castells, José, 186
Catalan
　chemical entrepreneurs, 45
　chemical industry, 203
　Front popular, 87
　government, 64–5, 90, 200. See *Generalitat de Catalunya*
　industry, 171
　journals, 200
　language, xiv, 96
　names, xiv
　separatism, 103
　university system, 96
Catalan Society of Biology, 72
Catalán, Miguel, 20, 30, 56, 62, 93, 103, 105–6, 110–11, 161, 199, 206–7, 214, 220
　and del Campo, 34
　multiplets, 30
　repression, 105–6
　spectroscopy, 57, 59, 62, 103
Catalanism, 23, 65, 90, 105, 203–4, 225
　scientific, 204
catalysis, 159–60
Catholic chemists, 16, 44, 80, 82, 112, 115, 221
Catholicism, 96–7, 101, 116, 126–7, 130–1, 134–5, 165, 172, 210, 225
cellulose, 66, 78, 89, 121, 136, 139, 170, 185
censorship, 153, 165, 174, 192, 194, 212
Centre de chimie théorique de France, 160

Centro de Estudios y Experiencias (La Marañosa), 86
Centro de Investigación y Desarrollo (CID), 171. See Pascual
Centro de Investigaciones Vinícolas, 79
Centro Español de Plásticos, 144
Chalonge, Daniel, 109
Chattaway, Frederick Daniel, 29, 99, 104–5. See Calvet
chemical community, 4–5, 9, 14, 24, 40, 68, 71, 84, 110, 189, 191, 221, 226
　and religion, 126
　division, 14, 191
　European, 5
　exile, 16, 197
　identity, 5, 8, 225
　international, 39
　Latin American, 213
　Mexico, 228
　public rituals, 20
　repression, 93
　Republican values, 103
　scientific culture, 111
　wounds, 231
chemical diplomacy, 155–62
chemical engineering, xvi, 46, 138, 149–54, 173, 201, 212, 218, 232
　and Ángel Vián, 151–2
　and Antonio Rius Miró, 150–1
　textbooks, 151
　Yale, 153
chemical gases, xv, 54, 86, 89–90
chemical industry, 40–50, 73–81
chemical training, 14, 24, 43, 50, 147, 149, 157, 216, 221
chemical war, xv, 54, 85, 92
chemical warfare, 85
chemical weapons, xv, 14, 16, 64, 68, 83–6, 89, 90, 92, 141, 219
chemist-historians, xvii, xviii, 6, 55, 202, 222, 226, 233
chemistry-based industry, 25, 167
chlorine, 85, 92, 182
citizen scientists, 13, 226. See Mark Walker
city mayors, 14
civil defence, 86–7
civil discourse, 10. See Thomas Glick
Civil War, xi, xii, xiv, xviii, 2, 4, 7, 9, 13–14, 16, 18, 20, 28, 38, 72, 76, 78–80, 82, 84, 93, 99, 103, 105, 109, 111, 116–7, 129, 130, 138, 217
　academic repression, 17, 93, 99, 110
　and the Jesuits, 102, 145
　chemical, 84–93
　exile, 16, 191–202

Index

German firms, 142
heroes, 142
industrial effort, 227
INFQ, 152
Lora-Tamayo, 136, 157
Madrid resistance, 100
Opus Dei, 113, 126, 128
victims, 18
war weapons, 217
cleansing
  commissions, 95
  committees, 96, 110–11
  process, 103
coal, 9, 47, 49, 78, 113, 121, 136, 139, 140–3, 147, 149, 182–3
  mines, 139
coal tar, 47, 49, 139
co-evolution, 137, 192
Cold War, 1, 9, 15, 165, 166–8, 216–7
  allies, 175–81
  anti-communism, 212, 215
  defence policies, 168
  logic, 155, 165
Colegio de Estudios Superiores de Deusto, 173
Colegio de México, 197
Coleman, Kim, 85
collectivisation, 84
Collège de France, 107
colloid chemistry, 32, 38–9, 127, 143, 199
Colombia, 99, 195, 200–1
colonial crisis, 7, 22
Coloquios Íntimos de Estudios Norteamericanos, 176
Columbia University, 34, 46
combustibles, 113, 136, 139–40, 183
Comisión Impulsora y Coordinadora de la Investigación Científica, 198
Comissió d'Indústries de Guerra, 87
Comité Conjunto Hispano Americano, 176
Comité Iberoamericano de Publicaciones Científicas, 198
communism, 9, 86, 91, 100–3, 155, 158, 168, 170, 175, 192, 212, 215
Confederación Española de Derechas Autónomas (CEDA), 65
Confederación Nacional del Trabajo (CNT), 87, 89, 92
Consejo de la Economía Nacional, 49
Consejo Nacional de la Industria Química, 121
Consejo Superior de Investigaciones Científicas. *See* CSIC
Consell d'Economia, 90
Consell de Sanitat de Guerra, 87–8
Consiglio Nazionale delle Ricerche, 3, 136
Constitution of Cádiz, 190

contraceptive pill, 199
Contubernio de Munich, 168
coordination complexes, 29, 94
co-production
  chemistry and power, xiii, 14
  physics and power, 9
  science and power, 17, 193
corporate chemistry, 181–7
corporations, 166, 176, 183, 186, 188, 217, 220
Cortes Españolas, 115, 123, 146, 172
cortisone, 195, 198
cosmochemistry, 104, 109
cosmopolitanism, 14, 16, 26–33, 54–5, 81, 103, 168, 202, 212, 221, 225. *See Junta para Ampliación de Estudios* (JAE), *See* League of Nations
Coulson, Charles Alfred, 160
craftsmen, 225
Crick, Francis, xv
Cros (chemical company), 48, 76, 145, 182–3
Crowther, James Gerald, 52
CSIC, 108, 112–3, 116–18, 126, 129, 134–5, 144–6, 149–50, 155–6, 158–9, 161, 170–82, 186, 206, 220–21, 224, 227, 229–30
Cuba, 22, 168, 222
customs laboratory. *See* Buscarons, Le Boucher
Czech Republic, 5

Daudel, Raymond, 160
DDT, 209
de Elhúyar, Fausto, 55, 225
de Elhúyar, Juan José, 55, 225
De Gaulle, Charles, 178
de la Cierva, Juan, 113, 229
Ynfiesta, Juan Luis de la, 144
de Riquer, Borja, xvi, 18
de Salas, Eduardo, 99
Debye, Peter, 171
Dehon Little, Arthur, 151
Deichmann, Ute, 3–4
del Amo, Gregorio, 98
del Campo, Ángel, 29–30, 32, 34, 62, 66–7, 103, 125, 228. *See* Catalán
  spectroscopy, 34
del Fresno, Carlos, 67, 78
del Río, Andrés, 55, 202, 225
Delarue, Dieter, 31
Delgado, Joaquín (anarchist), 170
demographic growth, 171–2, 217
Denmark, 5, 56

272   Index

depolitisation of the culture, 165. *See* Tony Judt
Deutsche Chemische Gesellschaft, 32
Deutsche Gesellschaft für Dokumentation, 181
Deutscher Akademischer Austauschdienst, 156
Development Plan, 166, 173–4, 183
developmentism, 166, 187, 217, 227
Dexter Award, 197
diene synthesis, xiii, 17, 171, 217, 220
diplomacy
   chemical, 138, 155–61, 221
Diputación de Barcelona, 129, 145
Dirección de Guerra Química de la Armada, 85
dissent, 16
   academic, 210
   liberal, 16, 53, 174, 192, 200, 202, 223
distillation, 151, 182
   bituminous coal, 49
   bituminous slates, 147, 182
   coal tar, 49, 121
   lignite, 139, 142
   vacuum, 35
   wool, 77
Djerassi, Carl, 198
Dodge, Barnett F., 152–4, 212. *See* chemical engineering
Dow (chemical company), 80, 183, 185
Drucker, Carl, 62
DuPont (chemical company), 80–1, 99, 183, 220
dyestuffs, 20, 24, 27, 31, 41, 47–8, 61, 77, 89, 186

economic growth, 116, 138, 166, 169, 214
economic historians, 76, 155, 168, 219
edaphology, 136, 143, 156
Edgerton, David, 11, 83
Edinburgh, 29
   Royal Society of, 159
   University of, 72, 105
Education and Science
   Minister, 13, 168–9, 174, 217, 228. *See* Lora-Tamayo
Eidgenössische Technische Hochschule. *See* ETH
Einstein, Albert, xv, 12, 100
Eisenhower, Dwight D., 162
electricity, 34–6, 48, 57, 121
electrochemistry, 62, 92, 122, 128, 150, 160
   chair, 62
   laboratory, 59
   section at the INFQ, 57
Electroquímica de Flix, 182

Empresa Nacional Calvo Sotelo, 147
Energía e Industrias Aragonesas, 91
Enlightenment (new), 55–64
Enlightenment, 6, 55, 214, 218, 225
ENPETROL, 182
Escola d'Enginyers Industrials, Barcelona, 89
Escola d'Enginyers Tèxtils de Terrassa, 41
Escola de Directors d'Indústries Químiques, Barcelona, 35
Escola Industrial, Barcelona, 34, 42
Escrivá de Balaguer, Josemaría, 113, 126, 128
Escuela de Oficiales de Guerra Química de Toledo y Salamanca, 92
Escuela Industrial, Madrid, 150
Escuela Nacional de Ciencias Biológicas, Mexico, 194
Escuela Nacional de Ciencias Químicas, Mexico, 197
Estatut d'Autonomia (Catalonia), 64
ETH, 12, 29, 32, 35–6, 41, 50, 57, 156–7, 160, 202, 232–3. *See* García Banús. *See* Bonet
Ethiopia, 140
ethos, xviii, 9–10, 13, 16, 53, 116, 126, 134, 191–2, 199, 213, 219, 229
   chemical community, 9, 13, 116, 126, 191, 229
   exiled chemical community, 16, 195, 197, 213
   liberal, xviii, 10, 16, 192, 213
   Mertonian, 219
   religious, 134
   republican, 53
ethyldichlorarsine, 85, 87
ethylene, 132, 159, 182–3
exile, xii, xiii, 14–15, 18, 21, 93, 95, 97–101, 106, 109–11, 115, 171, 213, 224, 229
   British aid, 98
   chemical community, 4, 16, 192–202, 225
   historiography, xvi
   Nazi Germany, 100
   Republican government, 13
   university, 228. *See* Francisco Giral
explosives, 44–5, 76, 83, 89–101, 137, 213

Fábrica Nacional de Colorantes y Explosivos, 77, 91, 145
Fábrica Nacional de Productos Químicos, 84
Fabbriche Riunite Agricoltori Italiani
Fagés, Juan, 24
Falange, 123, 125, 142, 172

Index

fascism, 3, 9, 65, 84, 100, 103, 116, 137, 139–40, 189, 202, 214, 223, 227
  antifascist manifestoes, 98, 103
  antifascist resistance, 100
  fascist chemistry, 116–26
  fascist Italy, 3, 54, 65, 136–40, 155, 162
  Nazi Germany, 65, 100, 138, 155, 162
fellow travellers (scientists as), 3, 156, 188
Fernández, José Ignacio, 160, 176
Fernández, Obdulio, 27–8, 40, 66–7, 81, 102
Fernández-Ladreda, José María, 13, 92, 96, 126, 142, 151–2, 159, 230
Ferran, Antonio, 34
Ferrer Pi, Pedro, 177
fertilisers, 76–7, 89, 137, 140, 143, 182, 231
  phosphate, 154
  soil chemistry, 127
First World War, 29, 66, 68, 76–7, 131, 216–17
  chemical weapons, xv, 14, 54, 65, 83, 92
  drug shortages, 204
  Fritz Haber, 84
  Hugo Stolzenberg, 84
  mustard gas, 85, 206
  nationalism, 3
  Spanish neutrality, 26
  technocracy, 163
Fischer, Emil, 29, 190
Fischer–Tropsch process, 139, 142–3
Fleming, Alexander, 207
Folch, Rafael, 229
Font, Josep, 176
Food and Agriculture Organization (FAO), 200
food chemistry, 193, 200
Fordism, 163
Foster, Mary Louise, 63, 73
Fourneau, Ernest, 29–30, 48, 58, 71, 98, 108
Fowler, Alfred, 30, 62
France
  diplomacy, 155
  Front populaire, 193
  Third Republic, 53, 192
Franco, Francisco, 97, 121, 171, 231
  and Adolf Butenandt, 179–80
  and Antonio Rius Miró, 150
  and Benito Mussolini, 91
  and Emilio Jimeno, 118
  and José Antonio Suanzes, 137
  and Juan Abelló, 146, 150
  and Juan de la Cierva, 113
  and the Count of Godó, 110
  and the IQS, 177
  and the Society of Jesus, 178
  award (*premio*), 123
  cabinet, 14, 17, 107, 168, 170, 228
  chemical entrepreneurs' support, 183
  coup d'état, 82–5, 103, 118, 166, 229
  death, 166, 173, 200, 214, 221, 235
  victory in 1939, 129, 165, 170, 191
Franco's dictatorship, xi, xii, xiv, 2, 7–8, 10, 13–18, 73, 80, 155, 161, 166, 168, 179, 187, 190–3, 203–4, 210, 213, 217, 219–21, 224, 229–30, 232, 234
Francoism, xvi, 9, 17, 19, 110, 123, 126–7, 138, 147, 212, 214, 221, 224, 225, 227
Francoist chemists, 38, 115–16, 126, 134, 138–9, 156, 187, 214, 219, 222–5, 230
free radicals, 35, 41, 225
Freiburg, 31, 36
fuel, xv, 78, 139–40, 142, 154, 171
  for planes, 90
  gasogen, 141
  planes, 87, 89
  synthetic, 77, 147
full professors (*catedráticos*), 172
Fundación del Amo, 98
Fundación Nacional para Investigaciones Científicas y Ensayos de Reformas (FNICER), 78, 81, 193

Galicia, 105
García Banús, Antonio, xi–xiii, xvii, 32, 50, 64, 105, 110
  and Fernando Calvet, 105
  and Juan Julio Bonet, 234
  and Julius Schmidlin, 41
  and Miguel Masriera, 36, 87
  Colombia, 200
  exile, 98, 115, 157
  free radicals, xiii, 35, 41, 225
  Institut de Química, 64
  research school, 39, 98
García Lorca, Federico, 93, 98
García, Gumersindo, 123. See *Falange*
gas masks, 83, 85
gasogen, 18, 140–1
gasoline, 137, 139, 141, 143
  artificial, 89
general chemistry
  chair, 38, 93, 200
  laboratories, 34
  subjects, 22
  textbooks, 29, 196
General Electric, 1, 80
Generalitat de Catalunya, 64
Geneva, 29, 31, 38, 157

274 Index

Geneva Protocol, 85
German Federal Republic, 180–1
Gesellschaft für Chemische Industrie (Basel), 41
Gil, Salvador, 126
Giral, Francisco, xii, xiii, xvii, 13, 20–1, 53–4
  and Artur Bladé, 200
  and James Gerald Crowther, 52
  and Lázaro Cárdenas, 227
  and Lidia Rodríguez Hahn, 199
  and the Soviets, 53
  at La Marañosa, 86
  Ciencia española en el Exilio, 93
  disappointment, 215, 229
  exile, 110, 194–5
  exile university, 226, 228
  natural products, 111
  return to Spain, 213
Giral, José, 13–14, 33, 193, 202
  and Obdulio Fernández, 103
  exile, 110
  industrial entrepreneurship, 73
  Minister of the Navy, 85
  Nobel Prize nominator, 72
  Spanish Republic, 52
  Spanish Republic in exile, 193
Glick, Thomas, 9–10
Gmelin Institute for Inorganic Chemistry, 180
Godó, Count of, 110
Gómez, Amparo, 8
González Barredo, José María, 128–9
González Núñez, Fernando, 97
González, Adolfo, 58, 206
Gramsci, Antonio, 226
Granada. *See* Jesús Yoldi. Adolfo Rancaño
  Pharmacy Faculty, 72
  Science Faculty, xvii, 62, 72
Granados, Francisco (anarchist), 170
Grimau, Julián, 170
gunpowder, 76, 89, 213
Guye, Phillipe, 29. *See* Moles
Guzmán, Julio, 34, 38, 57, 59, 92, 125
  electrochemistry, 62
gyroplane (autogiro), 113. *See* Juan de la Cierva

Haber, Fritz, 29, 66–7, 190
  and Hugo Stolzenberg, 84
  Nobel Prize, 83
Haber–Bosch (process), 77
Hahn, Otto, 160
Haldane, John B.S., 100

Heisenberg, Werner, 156
Herf, Jeffrey, 133
Herrán, Néstor, 8–9
Herrera Oria, Ángel, 96
historiography, 9
  of chemistry, 12
  of exile, 203
  of Francoism, xvi, 9
  of Spanish exile, xvi
history and memory, 228–31
Hoechst (chemical company), 183, 185, 220
Hönigschmid, Otto, 57
honorary doctorate, 71–2, 109, 130, 160, 178–9, 181
hormones, 129, 157, 179, 198, 206, 213, 220, 232
Howard, Frank A, 137. *See* autarky
Huelva
  industrial area, 72, 74, 185
  pyrites, 76, 185
hydrogenation, 61, 78, 89, 139, 141, 143, 147, 182, 198

I.G. Farben, 48, 77, 142, 190
Ibáñez Martín, José, 112–13, 116, 118, 161
Ibarz, José, 176
Ibys, Instituto de Biología y Sueroterapia, 183, 204–6
ideological agents
  chemists as, 13, 116, 226, 231
Imperial Chemical Industries (ICI), 99, 183, 200
Industria Nacional Química Farmacéutica, 195
industrial chemistry, xvi, 40–50, 79, 81–3, 121, 123, 149, 154, 156, 170, 219
  chair, 151, 174
  conference, 49, 146, 171
  courses, 34
  doctorate programme, 150
  encyclopaedia, 44
  experimental training, 34
  International Society for, 187
  journals, 219
  research, 121–2
  secrets, 225
  Silver Age of, 54, 73–82
  training, 138, 161
industrial shelters, 203–9. *See* Ibys, Zeltia
INFQ, Madrid, 7, 14, 16, 36, 54, 56–7, 59, 61–7, 71, 78–9, 95–7, 100–1, 103, 105–10, 117–8, 128, 135, 152, 198, 206, 217, 220–21, 224, 229
Ingold, Christopher, 29, 99, 157–8,

inorganic chemistry, 70, 78, 113, 180
  Adela Barnés, 196
  chair, 26, 36, 38, 109, 123, 213
  Emilio Jimeno, 46
  laboratories, 64
insecticides, 49, 154, 185, 220
Insituto Politécnico Nacional, Mexico, 199
Institución Libre de Enseñanza, 21, 96, 106
Institut Català de Cultura, 200
Institut d'Estudis Catalans, xviii, 23
Institut de France, 112
Institut de Química, 64, 87
Institut de Química Aplicada, 27, 43, 203
Institut the Fisiologia, 200
Instituto de Catálisis y Petroleoquímica, 183
Instituto Alonso Barba, 118, 129, 144, 150, 157, 170, 181
Instituto Cajal, 129
Instituto de Biología y Sueroterapia. See Ibys
Instituto de Edafología, 129, 144
Instituto de Estudios Pirenáicos, 127
Instituto de Industrias Pesqueras, 136
Instituto de Investigaciones Astronómicas, 161
Instituto de Investigaciones Químico Industriales, 121
Instituto de Investigaciones Técnicas, 170
Instituto de Investigaciones Textiles, 136
Instituto de la Grasa, 136, 144, 179
Instituto de Metalurgia y Mecánica, 80
Instituto de Química (Mexico), 224
Instituto de Química aplicada (Oviedo), 78
Instituto de Química Física Rocasolano, 118
Instituto de Química Orgánica, 117
Instituto de Tecnología Alimentaria, 136
Instituto del Carbón, 78, 136
Instituto del Combustible, 143, 147
Instituto del Hierro y del Acero, 122
Instituto Físico-químico, 43. *See* Novellas
Instituto Nacional de Edafología y Agrobiología, 127
Instituto Nacional de Física y Química, Madrid. *See* INFQ
Instituto Nacional de Higiene, 204
Instituto Nacional de Industria (INI), 122, 137
Instituto Nacional del Libro Español, 149
Instituto Politécnico Nacional (IPN), Mexico, 193, 194
Instituto Químico de Sarriá (IQS), 44, 132, 145, 177, 232
  Technical Services, 130
Instituto Químico-técnico. See Novellas
instrumental revolution, 11, 176, 216
intellectuals, 21, 120, 214
  and Republican elites, 192
  and the SPSL, 98

chemists as, 13, 16, 217, 226–8
Francoist, 112
liberal, 231
organic. *See* Antonio Gramsci
repression against, 194
silent resistance, 214
International Agency for Development, 177
International Conference, 48, 66, 138, 149, 156, 162, 224
  American Chemical Society, 159
  industrial chemistry, 146, 171
  IUPAC, 14, 69, 77, 84, 102, 109, 221–2
  of History of Science, 63
  on coal, 139
  on industrial chemistry, 48
  on oligo-elements, 128
  UNESCO, 161, 213
International Education Board, 55, 107
International Exhibition
  Barcelona 1929, 32–3, 46–9
  Philadelphia 1926, 32
International Monetary Fund, 165
International Olympic Committee, 145
International Union of Pure and Applied Chemistry, 69, 157. *See* IUPAC
isotopes, 68, 70, 99, 101
Istituto di chimica industriale, Milan, 78
Istituto Nazionale Medico Farmacologico Serono
Istituto di chimica, Rome, 140
Istituto Seroterapico Milanese
Istituto per la reconstruzione industriale, 137
IUPAC, 32, 67, 70, 222
  Amsterdam Conference, 222
  Stockholm Conference, 160
  Washington Conference, 222
Izquierda Republicana, 62, 93, 106, 108

Jeger, Oskar, 232–3
Jesuit chemists, 46, 102
Jimeno, Emilio, xvii, 14–15, 46, 81–2, 95–6, 110–11, 139, 159, 223
  and José Ibarz, 176
  anti-Catalanism, 96
  anti-liberalism, 120
  cleansing committees, 110
  conservative ideology, 118
  metallurgy, 46, 80
  National Bureau of Standards, 128
  rector, 139
  school, 80
  support to the IQS, 102
  totalitarian chemistry, 120, 122, 126, 129, 210

Index

Joliot-Curie
  Frédéric and Irene, 109, 212
  laboratory, 99
Joliot-Curie, Frédéric, 107
Judt, Tony, 165, 170
Junta para Ampliación de Estudios e Investigaciones Científicas (JAE), xviii, 7–8, 14, 16, 23–25, 28, 30–3, 39, 41, 46, 50, 53–6, 62–3, 65, 72–3, 76, 78, 81, 95–8, 103, 105–6, 112, 116–18, 120, 134, 138, 150, 156, 158, 200, 202, 214, 216, 218, 223–4, 232, 234

Kaiser Wilhelm Gesellschaft, 136
Kaiser Wilhelm Institute for Biochemistry, 179
Kamerlingh-Onnes, Heike, 24
Karrer, Paul, 71, 157, 233
Kragh, Helge, 5
Krauch, Carl, 190
Krause, Friedrich, 53
Krausism, 53–4, 81–2, 97, 214
Krige, John, 11
Krupp–Renn process, 143
Kuhn, Richard, 20, 70, 233
  and Francisco Giral, 20

Laboratori d'Estudis Superiors de Química, Barcelona, 35
Laboratori de Química Orgànica, Barcelona, 89, 90
laboratories, 2, 6, 12, 14, 29–30, 33, 35–6, 43–4, 50, 63, 78–9, 81, 107, 109, 111, 118, 121, 153, 200, 207, 221, 223, 233
  culture, xiv, 58
  customs, 95, 209
  equipment, 25, 97, 176, 232
  ETH, 29
  experiments, 79
  facilities, 153
  INFQ, 59
  pharmaceutical, 78
  reforms, 23, 28, 42, 54
  research, 123
  Residencia de Estudiantes, 34
  scale, 83, 92, 216
  Science Faculty, Madrid, 37
  steroids, 233
  training, 42
  university, 79
Laboratorio de Investigaciones Bioquímicas, Zaragoza, 38
Laboratorio de Investigaciones de Química Orgánica, Mérida (Venezuela), 201
Laboratorio de Investigaciones Físicas (LIF), Madrid, 34, 57
Laboratorio Químico del Ebro, 44
Laboratorios Andreu, 78
Laboratorios Esteve, 77
Laboratorios Hormona, 195
Laín Entralgo, Pedro, 191
Langmuir, Irving, 1, 165
Le Boucher, León, 94
Le Chatelier, Henry-Louis, 24, 71
League of Nations, 54, 85–6, 191
Leipzig, 12, 29, 38, 57, 62
León, Andrés, 58, 98, 103, 207
Leuven/Louvain, 157
  University, 44
Lewis, Gilbert Newton, xv, 32, 70
Lewis, Warren, 151
Ley de Ordenación Universitaria, 115
liberal chemistry, 29, 33, 68, 191, 203, 213
liberalism, 14, 116, 124, 189–91, 214–5, 227
  economic, 121
Liga protectora de la población civil contra de la guerra química, 85
Lipperheide, José, 145, 183
Llopis, Antonio, 76–7, 80
London, 29–30, 63, 199
  Imperial College, 49, 101
  University College, 97, 157
López Rodó, Laureano, 166
Lora-Tamayo, Emilio, xvii
Lora-Tamayo, Manuel, xi–xiii, xvii, 15, 95, 110, 115–17, 125, 134, 173, 222, 224
  Academia Pontificia de Ciencias, 127
  and Adolf Butenandt, 179
  and Christopher Ingold, 157
  and fascist Italy, 139
  and Francisco Buscarons, 214
  and Franco, 170
  and José Pascual, 115, 126, 170
  and Kurt Alder, 171, 220
  and León Le Boucher, 95
  and Opus Dei, 168
  applied chemistry, 125
  applied science, 183
  as chemist-historian, 222
  as organic chemist, 146
  as organic intellectual, 227
  autarky, 136
  books to the Caudillo, 156
  Catholic universities, 173
  diene synthesis, xiii, 217
  diplomacy, 155
  European journey. *See* José Pascual
  history and memory, 230

Index

industrial chemistry, 161
Instituto Alonso Barba, 144
IUPAC Conference (Amsterdam), 222
political neutrality, 157, 167, 172, 187
resignation, 174
school, 144, 158, 178
technocracy, 167–75, 228
university reform, 172
university repression, 210
visit to Germany, 179
lubricants, 143, 147, 182

macromolecules, 31, 157
Madinaveitia, Antonio, 20, 30, 33, 59, 66–7, 93, 110, 115
 and Adolfo González, 206
 and Ignacio Ribas, 98
 and Richard Willstätter, 57
 and the INFQ, 58, 78–9
 and the LIF, 34
 as Republican chemist, 53, 65, 82, 97, 103, 225
 chemical weapons, 86, 92
 exile, 194–5, 213
 industrial chemistry, 81
 Instituto de Química, 197, 199
 natural products, 111, 220
 Nobel Prize nominator, 72
 school at the INFQ, 58, 61
Madinaveitia, José, 206
magnetism, 36, 57
magnetochemistry, 34, 38
Malet, Antoni, 8
Mallorca, 102
Manchester, 99
Manhattan Project, 100. *See* Harold C. Urey
Marañón, Gregorio, 206
March, Juan, 142, 176–7, 232
Marcilla, Juan, 79
Marconi, Guglielmo, 139
marginalisation, xii, 15, 84, 109–10, 151, 206–7, 222–3
Marker, Rosell, 195
Marquina, Mariano, 34
Martí Franquès, Antoni de, 55, 225
Martín Panizo, Fernando, 141
Marxism, 52–3, 63, 82, 201, 214. *See* John B. S. Haldane
Mascareñas, Eugenio, 26, 28
Masonry, 95–6, 108, 118, 225
Masriera, Miguel, xii–xiii, 110, 154, 213–14
 and Antonio García Banús, 36
 and José Pascual, 89
 cosmochemistry, xiii, 104
 Insitut de Química, 87

internal exile, xii
liberal free trade, 49
liberal shelter, 210–12
marginalisation, 109–10
nuclear power, 176
science journalist, 160
technical chemistry, 49
Mathematical Institute (Oxford), 160
Matignon, Camille, 48, 67, 70
Max-Planck Institute for Photochemistry, 232, 235
Max-Planck Society, 158, 179–80
McAdams, William, 151
medical chemistry, 72, 105, 136
medicine
 and chemistry, 5, 11
 chemistry applied to, 38
 faculties, 22, 25, 39, 72, 201
 history of, 191
 laboratories, 200
 Nobel Prize, 20, 24, 171
memory, 16
 and history, 218. *See* History and memory
 exile, 18, 213
 university in exile. *See* Francisco Giral
metals, 20, 31, 89, 121, 136, 139, 143
Mexico, xii, xiii, xvii, 191, 196–7, 199, 200–1, 213
 Antonio Madinaveitia, 197
 editorial Atlante, 224
 exile, 100–1, 192–200
 Francisco Giral, 202, 228
 María Teresa Toral, 196
 Modesto Bargalló, 196
 natural products, xii
Meyer, Kurt, 157
Meyer, Richard, 180
Milan, 31, 47, 78, 140
militant history, 17
mineral acids, 76, 121
minerals, 20, 49, 77
mining, 46, 78, 136, 139, 143, 197, 202, 225
modernisation, 14, 16, 24, 28, 35, 39, 50, 52, 54–6, 73, 117, 182, 217, 226
 and Krausism, 53
 chemical industry, 186
 economic, 164–5, 168
 JAE project, 33
 laboratories, 36, 232
 material culture, 50
 non-democratic, 15
 paradoxes, 221–3
 rhetoric, 33, 187, 217
 technocratic, 172

modernisation (cont.)
  theories, 166
  university, 210
modernity, 14, 23, 25, 40, 50, 54, 221, 234
  against backwardness, 50
  alternative, 9
  and Catalanism, 203
  and he INFQ, 56
  and the FNICER, 78
  conservative, 127, 130
  economic, 169
  industrial chemistry, 40
  laboratories, 50
  reactionary, 133–4
  Republican, 16
  rhetoric, 221
molecular biology, xv, 11
Moles, Enrique, xvii, 20, 32, 39, 67, 110, 222
  and Ángel Vián, 101
  and Augusto Pérez Vitoria, 212
  and José Agell, 42
  and Lora-Tamayo, 222
  and the INFQ, 56, 65, 97
  and the RSEFQ, 39
  as chemist–historian, 55
  as Republican chemist, 54, 65
  atomic weights, 16, 34, 38, 118, 220
  chemical education, 49
  chemical weapons, 86, 92
  commemorative practices, xvii, 229
  exile, 99, 195
  Ibys, 206
  IUPAC Conference, 32
  laboratories, 36
  moral ambiguity, 216
  physical chemistry, 58–9, 61–2, 103, 225
  pure–applied chemistry, 42, 48–9, 78, 219
  repression, 109
  school, 36, 61, 63, 101, 150–1, 196
  war weapons, 217
Monterrey, 196
Monsanto (chemical company), 183, 185
moral ambiguity of chemistry, 16, 216–31
Morocco, 76, 84
Moscow, 52–3
Movimiento Nacional, 106, 125, 142
Mulhouse, 41
multiplets, 30, 57, 62, 106, 111. *See* Catalán
Munich, 12, 29–31, 57, 104, 168, 179
Muñoz del Castillo, José, 24
Muñoz, Eugenio, 199–200
Murcia, 144, 152
  University, 213

Mussolini, Benito, 3, 54, 90–1, 120, 135, 139–40, 164, 226
mustard gas (yperite), 84–5, 87, 90, 206

National Catholicism, 116, 126, 130, 135, 232
National Socialism, 4, 188, 190
nationalism, 55, 96, 125, 131, 136–7, 165, 228
  and anti-liberalism, 137
  and totalitarianism, 125
  autarchic, 136
  Primo de Rivera, 55
  Spanish, 96, 120, 131, 165, 228
  Catalan, 23, 105, 203–4, 225
Natta, Giulio, 140
natural theology, 134, 141
Navy
  Minister, 52, 85. *See* José Giral
Nazi
  areligious worldview, 133
  army, 142
  collaboration, 142–3, 156
  concentration camps, xv
  experts, 164
  mass extermination, 167
  occupation, 107
  party, 84, 179
  policies, 167, 179
  regime, xv, 156, 162, 187
  science, 3
  terror, 99
  war criminals, 190
Nazi Germany, 3, 54, 83, 91, 100, 108, 138, 147, 155, 162–3
Negrín, Juan, 100
neo-salvarsan, 77
neo-scholasticism, 115
Nernst, Walther, 12
networks, 13
  European, 25
  international, xiv, 28, 54, 71, 181, 221, 225–6
  professional, 4
  religious, 82, 130, 225
neutral expertise, 167–74. *See* Lora-Tamayo
neutrality, 10, 16, 33, 84–5, 99, 135, 157, 165, 167, 170, 173, 227
New York, 29, 34, 46, 159, 198
Nobel Prize winner, 1, 8, 20, 24, 30–1, 49, 100, 139, 157, 171, 179, 190, 199, 217, 232
Nogareda, Carlos, 145
nomenclature (chemical), 2, 42, 50, 71, 128, 160

Index

nopal gum, 198
North American Committee for Help to Democratic Spain, 100
Novellas, Francesc, 43
nuclear magnetic resonance, 11, 176
Nye, Mary Jo, 5, 51

Observatorio del Ebro, 44
Ochoa, Severo, 171, 229
oil crisis, 215
Opus Dei, xi, 15, 113, 126, 128–30, 133–4, 166, 168, 172–3, 191, 225. *See* Albareda
Oreskes, Naomi, 17
Orfeó Català de Mèxic, 200
organic chemistry, xi–xii, 20, 27, 30, 98, 105, 115–17, 122
   Barcelona, 50
   chair. *See* Giral. Pascual, Lora-Tamayo
   Colombia. *See* García Banús
   courses, 22
   industrial, 150. *See* José Prats
   INFQ, 79, 197
   Instituto Nacional de Física y Química (INFQ). *See* Madinaveitia
   IQS, 233
   laboratories, 34–5, 37, 59, 64
   Laboratorio de Química Orgánica. *See* García Banús
   Leuven/Louvain, 44
   Mexico, 198–9
   natural products, 132
   plant pigments, 24
   reaction mechanisms, 160
   school, 174, 176, 186, 225. *See* Lora-Tamayo, Calvet, García Banús
   section, 70, 170
   synthesis, 29
   textbooks, 46, 194
   Zurich, 31, 57, 232
Organisation for Economic Co-operation and Development, 165
Orozco, Fernando, 197, 199
Ostwald, Karl Wilhelm Wolfgang, 32
Ostwald, Wilhelm, 12, 29
   Physical Chemistry Institute (Leipzig), 46, 57, 62
   school, 80
Othmer, Donald, 152
Oviedo, 231
   chemical industry, 72
   Civil War, 92
   Emilio Jimeno and Franco, 118
   research institute, 64
   science faculty, 72

   University, 46, 78–9
Oxford, xviii, 29, 105
   F. D. Chattaway, 29
   Isaiah Berlin, 189
   José Ignacio Fernández, 160
   Robert Robinson, 58, 157

Palacio de la Química, Barcelona, 33, 47, 217
Palacios, Julio, 57
X rays, 61
Pallmann, Hans, 156
Palmer, Brian, 8
Palomares (atomic bombs), 174
Paneth, Friedrich, 101
Paris, 29–30, 38, 70, 98, 109, 160
   exile, 98–9, 107, 109, 195, 213
   Pasteur Institute, 12, 29–30
   Sorbonne, 24, 29, 178
   UNESCO, 213
Parravano, Nicola, 67, 71, 139
Pascual, José, xvii, 15, 30, 89, 95, 115, 134, 171, 210, 224–5
   and Enrique Moles, 222
   and Lora-Tamayo, 126, 170
   and the LQO, 90
   applied chemistry, 170, 220
   commemorative practices, xvii, 170
   European journey, 157–8
   industrial chemistry, 171
   international network, 221
   IUPAC, 160
   memory, 229
   school, 115, 176, 178, 186, 220, 225
   technocracy, 171, 174, 181, 187, 228
Pasteur Institute, 12, 29–30
Patronato Alfonso X el Sabio, 113
Patronato Juan de la Cierva, 113, 135–6, 178, 229
Patronato Instituto Químico de Sarriá, 145–6
Pauling, Linus, xv, 160, 196, 217
peace, 14, 65, 71, 102
   25 years of, 170, 185
   'atoms for peace', 176
pellagra, 100
Pellicer, José, 145
pensionados, 16, 28–33, 35, 39, 73, 76, 95, 155, 157, 226, 235. *See* JAE
Pérez Álvarez-Osorio, Rafael, 157
Pérez del Pulgar, José Agustín, 134
Pérez Vitoria, Augusto, 67, 97, 103
   chair, 213
   exile, 195
   UNESCO, 213
perfumes, 31, 47, 72, 76–7, 102
Perrin, Jean, 31

personal perfection, 128, 133
Pertierra, José Manuel, 78
Pestre, Dominique, 11
petrochemical(s)
  complex, 183, 185, 217
  industry, 5, 147, 182, 186
Peypoch, Ramon, 90, 204
pharmaceutical
  chemistry, 206
  entrepreneurs, 44, 146
  industry, 11, 98, 128, 150, 204, 207, 230
  laboratories, 44, 78, 196, 199
pharmaceuticals, 11, 47, 76, 89, 136, 149, 154, 182, 185, 194, 204, 206, 209
pharmacy, 5, 11, 42–3, 48, 52, 63, 200
  chemistry applied to, 43
  degree, 112, 200
  faculties, 22, 25, 30, 58, 64, 71–2, 74, 98, 101, 128–9, 229
Pharmazeutisches Institut (Dahlem), 200
Philippines, 22, 120
phosgene, 85, 87, 90, 92
photochemistry, 232–3
photography, 62, 109, 142
physical chemistry, 15, 24, 32, 38, 43, 46, 57, 62, 64, 70, 106, 113, 115–16, 127, 129, 143, 146, 150, 160, 180, 216, 225
  at the LIF, 34
  chair, 87, 128, 145, 150
  laboratories, 59
  Moles, 106
  Moles' school, 36, 38, 61, 150, 196
  Ostwald's school, 80
  Rocasolano Institute, 144
physiological chemistry, 22, 34
Pi Sunyer, August, 200, 206
Pictet, Amé, 31
Pietsch, Erich, 180–1
pigments, 24, 29, 44
Pi-Sunyer Bayo, César, 200
Pitaluga, Gustavo, 206
Pius XII, 127
Planell, Joaquín, 147, 171
plastics, xv, 11, 113, 136, 144, 149, 164, 182–6, 216
  Centro Español de Plásticos, Barcelona, 144
pluralism, 189–90, 210
Polémica de la Ciencia Española, 28
Politecnico di Milano, 31
popular chemistry, 43, 52, 122
Popular Front, 193
Porcioles, José María, 170
Portugal, 5, 9
Prats, José, 24, 41
Prelog, Vladimir, 157, 233

Presas, Albert, 8
presentism
  motivational, 17
  new, 17
Preston, Paul, 17
Prieto, Carlos, 196
Primo de Rivera's dictatorship, 2, 4, 7, 10, 13–14, 16, 25, 33, 44, 46, 50, 52, 55, 84, 92, 107, 190, 203, 217, 221
Prison Notebooks
  Antonio Gramsci, 227
Proctor, Robert, 3, 10
professional 'calling', 168
profesor agregado, 173
Public Instruction
  Minister, 57, 68
Public Works
  Minister, 14, 142, 231
Puertollano, 72, 147, 182–3, 185
pure–applied
  chemistry, 42, 48, 120, 186, 218–21
  dichotomy, 41
Putnam, Robert, 163
pyrites, 76, 143, 147

quantum chemistry, 11, 62, 160, 199

radiochemistry, 160
Raman spectrum, 70
Ramón y Cajal, Santiago, 20, 24, 229
Ramoneda, Josep, 18
Ramsay, William, 29, 48
  fellowship, 99
Rancaño, Adolfo, 62, 87
raw materials, 15, 73–7, 81, 89, 121, 125, 134–5, 138–9, 142–3, 147, 151, 186, 207, 211, 213
  analysis, 89, 162
  and God's providence, 131
  control of, 216
  foreign, 154
  geography, 138
  geopolitics, 217
  in autarchy, 220
  in exile, 111
  local, 120, 138, 202
  national, 121, 138, 144, 162
  pyrite ores, 77
  restrictions, 111
Rawls, John, 214
reaction mechanisms, 157, 160
reactionary modernism, 133. *See* Jeffrey Herf
Real Academia de Ciencias Exactas, Físicas y Naturales, Madrid, 27, 55

Index

Real Academia Nacional de Farmacia, Madrid, 230
Real Academia Sevillana de Ciencias, 230
Reial Acadèmia de Ciències i Arts de Barcelona, xviii, 170
Real Sociedad Española de Física y Química. *See* RSEFQ
refining, 142, 182
refugees, 201–2
  from Nazism, 99
  in Mexico, 199
  internal, 16, 192, 203–13
  political, 191
regenerationist movement, 23
Reid, John T., 176
religion and chemistry, 126–34
REPENSA refinery, 182
Reppe, Walter, 142
repression, xviii, 14, 17–18, 84, 93, 97, 99, 100–1, 103, 106, 108, 117, 123, 125, 138, 155, 166, 168, 187, 194, 196, 203, 207, 209–10, 222–3, 229
  academic, 93
  internal, 223
  police, 209–10
  political, 123, 168
  scientific, xviii
republic of chemistry, 29
Republican
  army, 85–6, 95
  cause, 17, 90
  chemistry, 54, 61, 73, 81, 95, 225
  chemists, 14, 193
  culture, 16, 20, 53, 105
  dream, 16, 54, 67, 81–2, 225
  elites, 192
  exile, 214
  government, 56, 65, 79, 97, 100–3, 107–8, 145, 193, 199
  government in exile, 13
  opposition to Primo de Rivera, 52
  order, 14
  parliament, 64
  parties, 225
  policies, 62, 78–9, 95
  political thought, 53
  positivism, 81
  regime, 7, 84
  science, 134, 225
  university, 228. *See* Francisco Giral
  values, 79, 103, 116, 193, 195, 200, 204, 213
republicanism, 14, 54, 65
research school
Residencia de Estudiantes, xviii, 34, 68

Ribas, Ignacio, 58, 66–7, 98, 110, 115, 214
  and Antonio Madinaveitia, 79, 98
  Zeltia, 207
Richards, Theodore, 32
Rideal, Eric K., 145
Rif War, 84
Riotinto, 142, 185
Rius de la Pola, Pilar, 199
Rius Miró, Antonio, 15, 95, 101, 103, 113, 161
  and Albareda, 150
  and Fernández-Ladreda, 159
  and González Barredo, 129
  and José Pascual, 134
  books to the Caudillo, 156
  chemical engineering, 151, 154
  electrochemistry, 160
  physical chemistry, 225
  school, 115
  technical chemistry, 150
rivalry
  among nations, 26
  Jesuits–Opus Dei, 130
  US–Soviet, 161
Robinson, Robert, 29, 58, 71, 105, 157, 207
Roca-Rosell, Antoni, 8
Rocasolano, Antonio de Gregorio, 14, 31, 49, 92, 110, 113, 118, 129, 227
  Acción Española, 96
  against the JAE, 95–6, 110
  and Albareda, 143
  and Bermejo, 118
  and Franco, 97
  death, 128
  Instituto, 159, 229
  school, 38–40, 92, 113, 150, 221
Roche, Jean, 179
Rockefeller (The), 57, 61, 96, 102, 106, 118. *See* INFQ
Rockefeller Foundation, 33, 56, 107, 198
  International Education Board (IEB), 56
Rodríguez Mourelo, José, 32
Rodríguez-Hahn, Lidia, 199
Rome, 128, 139–40
Romero, Ana, 8
Romo, Jesús, 198
Roosevelt, Theodore, 100
Roqué, Xavier, 8–9
Roquero, César, 196
Rosenkranz, George, 198
RSEFQ, 13, 23, 39, 48, 50, 103, 115, 150, 159–60, 222, 224
rubber, 66, 143, 182, 184
  natural, 198
  synthetic, 136–7, 142, 167

Ruiz Giménez, Joaquin, 209
Russell, Colin A.
  chemist–historians, 6
Russia, 5, 52
Rutherford, Ernst, 48
Ruzicka, Leopold, 157, 232–3

Sabatier, Paul, 29, 31, 46, 48–9
Sabiñánigo, 72, 91
Saeta (publisher), 129
Sainz Rodríguez, Pedro, 107, 112
Salamanca
  industrial chemistry, 151
  organic chemistry, 98
  Science Faculty, 72
  Universidad Pontificia, 173
  University, xiii, 20–2, 145, 209, 228
  university laboratories, 79
  war weapons, 92
Salazar dictatorship, 164
Samaranch, Juan Antonio, 145
Sánchez Ron, José Manuel, 8, 106
Sánchez-Bella, Florencio, 128
Sandoval, Alberto, 199
Sandoz (pharmaceutical company), 176, 183
Santander, 72, 86, 185
  Aula de Cultura Científica, 213
  Chemistry Week, 159
  Summer School, 66–7, 72
Santesmases, María Jesús, 8
Santiago (de Compostela)
  Pharmacy Faculty, 72
  University, 93, 105
  university laboratories, 79
Santos, Ángel, 128–9
Santos, Eduardo, 200
Sanz, Manuel, 132
Saraiva, Tiago, 9, 137
Schaffner, Kurt, 232
Scherrer, Paul, 57
Schlenk, Wilhelm, 194
Schmidlin, Julius, 41
School of Physiology (Barcelona), 200
science policies, xiii, 8, 78, 108, 112, 126, 158, 194, 229–30
science popularisation
  as shelter, 203
  Miguel Masriera, 110, 210, 212
  UNESCO Conference, Madrid, 161
science-based industry, 41, 49, 73, 80, 142, 154, 218
second Industrial Revolution, 10
Second World War, 52, 109, 156, 158–9, 162–4, 167, 189–90, 217, 220–1
  German defeat, 149

Nazi Germany, 83, 142, 147
totalitarianism. *See* Hannah Arendt
secular
  chemists, 45
  education, 53, 81, 95
  science, 54
secularism, 14, 96–7, 116, 130–1
selection committees (*oposiciones*), 95
self-sufficiency, 15, 140, 154, 163–7, 204.
  See autarky, autarchy
Servicio de Pólvoras y Explosivos, 86
Seville, 39, 55, 136, 144, 179, 230
  Academy of Science, 230
  Instituto de Química, 64
  Lora-Tamayo and Pascual, 115, 157
  Military hospital, 90
  repression, 117
  Science Faculty, 72
  University, 30, 93, 115, 136, 157
Shell (chemical company), 183–4
shelter, 126, 212–13
  academic, 98, 214
  chemical, 214
  gas, 85
  in exile, 202
  industrial, 192, 203–4, 214
  laboratory, 90
  liberal, 203, 210
Silver Age, 28, 42, 82, 116, 156, 191, 202, 214, 223, 232
  chemists, 23, 57, 95–6, 102–3, 153, 161, 194
  industrial, 219, 225
  of Spanish culture, 21
  of Spanish Science, 21
Sindicato Español Universitario (SEU), 115, 123
Sindicato Nacional de Industrias Químicas, 121, 123–4, 146, 206
Skłodowska-Curie, Marie, 65
socialism, 3, 53, 192, 196, 206, 221
  and communism, 86
  Antonio Madinaveitia, 61, 81, 227
Société de chimie industrielle, 44, 48
Society for the Protection of Science and Learning. *See* SPSL
Society of Jesus, 15, 44, 79, 102, 126, 130, 133–4, 178, 225, 227
soil chemistry, xi, 112, 127–8, 143, 156, 179, 220, 225, 227
Sommerfeld, Arnold, 57, 62
Sorbonne, 12, 24, 29, 109, 178
Sorolla, Joaquín, 98
Soviet
  Academy of Sciences, 160

science, 52–3, 168
US rivalry, 161
Soviet Union, 108, 199
spectroscopy, 11, 15, 30, 34, 57, 59, 103, 106, 109–10, 160, 176, 220
   chair, 106
   INFQ, 62, 100
   quantum chemistry applied to, 62
Speer, Albert, 164
SPSL, xviii, 98–9, 105
stabilisation plan, 165, 182
Staudinger, Hermann, 29, 31, 36, 58, 66, 160, 206, 233
steel, 87, 122, 136, 143, 149, 182
steroids, xviii, 68, 129, 157, 179, 195, 198–9, 220, 232–4
Stockholm, 105, 160
Stolzenberg, Hugo, 84
student protests, 174, 209
Suanzes, Juan Antonio, 122, 137, 181–2
summer school, 224
   Béjar, 145
   Santander, 72
Sureda, José, 31
surface chemistry, 176
Sweden, 5
Syntex, 195, 198
synthetic
   dyestuffs, 24
   fibres, 147, 164
   fuel, 77, 147
   gasoline, 143
   molecules, 216
   organic chemistry, 29
   perfumes, 47
   products, 44, 143, 182, 217
   rubber, 136–7, 142, 167
   sulphonamides, 77

Tarragona, 183
   macrochemical industry, 185
   polo químico, 186
Taylorism, 163
teargas, 90
technical chemistry, 49, 64, 78, 105, 113, 120, 138, 149–51, 153, 225
   chair, 212
technocracy, xv, 1, 10, 14–15, 165–6, 169, 174, 187–8, 214, 216, 228
   and corporations, 183
   and José Pascual, 174, 187
   and López Rodó, 166
   and Lora-Tamayo, 166–7, 181, 187
terpenes, 16, 200, 232

tetraethyl lead, 87
theology, 45
Tiselius, Arne, 160
Todd, Alexander, 157, 160
Todt, Fritz, 164
Toral, María Teresa, 100–1, 196, 200
totalitarian chemistry, 15, 122–3, 126
totalitarianism, xii, xiii, xvi, 1, 2, 10, 14, 83, 98, 115, 125, 134, 138, 156, 161, 164, 166, 191, 204, 210, 223–4, 227
   and Albareda, 126, 129
   and Bermejo, 118
   and chemical engineering, 151
   and Emilio Jimeno, 118, 121
   and Gumersindo García, 123
   and Ibáñez Martín, 116
   and Opus Dei, 126–7, 129
   and the CSIC, 112
   autarchy, 186
   liberal opposition to, 189
   resistance to, 16
Toulouse, 29, 46, 49, 130
Trade and Industry
   Minister, 68, 142
Tribunal de Orden Público, 170
Trowbridge, Augustus, 56

Uhthoff, José, 27, 203
UNAM, 196–200
UNESCO, 161, 213
Union Carbide (chemical company), 80, 183
Unión de Profesores Universitarios Españoles en el Extranjero, 193
unit operations, 138, 151, 153–4, 218. *See* chemical engineering
United States, 5, 12, 36, 46, 68, 99–100, 128, 137, 149, 151, 162, 164, 168, 172, 175, 197, 199
Universidad Nacional Autónoma de México. *See* UNAM
Universidad Central de Venezuela, 201
Universidad de Navarra, xviii, 173
Universidad International de Verano, Santander, 66
university, xiv, 33, 35, 43, 46, 49, 64, 87, 98, 105, 107, 115–16, 118, 123, 129, 136, 138, 150, 153, 170–6, 179, 209–10, 212, 214, 220, 224, 226, 228–30, 233–4
   and industry, 219
   buildings, 36
   Catalan system, 96
   chairs, 8, 16, 61, 65, 93, 106, 130, 213
   chemists, 152

university (cont.)
  crisis, 210, 214
  culture, 48
  departments, 29, 39
  faculties, 217
  institutes, 65
  laboratories, 79
  professors, xiv, 154
  rectors, 14, 64, 93, 115, 180
  reforms, 14, 22, 61, 172
  research, 49, 65
  research institutes, 64, 81
  sites, 153
  Spanish system, 43, 48
  students, 172
  system, 53, 64
  undergraduates, 30
  upheavals, 187
Urbain, Georges, 24, 29–30
  and León Le Boucher, 94
Urey, Harold C., 100
Urgoiti, Nicolás María, 204
US Department of State, 152
USSR, 36, 52
Utrecht, 157

Valencia, 55, 79, 98, 107, 136, 144
  customs laboratory, 95. See Le Boucher
  Republican government, 98, 103
  Science Faculty, 72
  University, 45, 160
Valley of the Fallen, 165
Valls Taberner, José, 145
Vián, Ángel, 14, 101, 153, 174
  and Enrique Moles, 101
  chemical engineering, 151
  industrial chemistry, 151, 174
Vichy government, 107
vitamins, 61, 68, 129, 157, 206–7, 209
Vitoria, Eduardo, 24, 44, 79, 81–2, 129, 132, 227
  and God's providence, 131
  and Paul Sabatier, 46
  and the Civil War, 102
  and the Society of Jesus, 126
  Catholicism, 210
  Jesuit chemistry, 45
  natural theology, 131–3, 141
  school. See IQS
  textbooks, 45–6
Volterra, Vito, 139
von Bayer, Adolf, 29
von Euler, Hans, 67, 105
von Papen, Franz, 179

Vonnegut, Kurt, 1, 165

Walden, Paul, 71
Walker, Mark, 3, 17
  fellow travellers, 187
Walker, William, 151
Wall Street Crash, 68, 137, 163
Warfare Service, 206
War. See First World War, Second World War, Civil War, Cold War.
Washington, DC, 29, 147
  Bureau of Standards, 40, 46, 128
  IUPAC Conference, 32, 222
Watson, James, xv
Weber, Max, 127, 168
Weiss, Pierre, 57
Werner, Alfred, 233. See coordination chemistry
Wieland, Heinrich, 29–30, 105
  and Calvet, 104
  and García Banús, 35
  and Sureda, 31
Willstätter, Richard, 24, 29, 57, 66–7, 105
  and Antonio Madinaveitia, 57, 61
  and José Prats, 41
  and the INFQ, 56
  and the Republican government, 65
  and the Spanish liberal chemists, 29, 57
Wilson, Christopher L., 99
Wise, Norton, 137
Wohlsteller, Albert, 168
wolfram, 142
Woodward, Robert F., 177

X-rays, 11, 61, 176

Yoldi, Jesús, 14, 93–4, 110. See Granada

Zaragoza
  Industrial Engineering School, 150
  Instituto de Química, 64
  Science Faculty, 72
  University, 32, 38, 92, 96, 113, 118, 128, 209
  university laboratories, 79
Zeeman, Pieter, 24
Zeltia, 207
Zhdanov, Yuri, 160
Zurich, 29, 31, 35, 41, 58, 206, 232
  ETH, 12, 29, 32, 35, 41, 50, 57, 156–7, 232–3